INTEGRATED REGIONAL ASSESSMENT OF GLOBAL CLIMATE CHANGE

The aim of integrated regional assessment (IRA) is to promote a better understanding of – and more informed decisions on – how regions contribute to and respond to global environmental change. Understanding the regional implications of global environmental change is important, because it is at regional levels that global environmental change mitigation must be practiced and that human impacts will be felt. This book provides both a detailed treatment of the methodological challenges of IRA and a set of international examples illustrating the practice and results of assessments at the regional scale.

The first nine chapters of the book address IRA as a concept and process and set the stage from a methodological perspective. They address questions of scale, uncertainty, and quantitative versus qualitative approaches, as well as some particular conceptual frameworks for assessment. The next five chapters illustrate a range of IRA activities that have combined qualitative and quantitative approaches in innovative ways. The final five chapters look at IRA as a process from an implementation perspective.

This volume is the culmination of the START/CIRA/IHDP initiative responding to the needs of the global research programs International Geosphere–Biosphere Programme (IGBP), the World Climate Research Programme (WCRP), and the International Human Dimensions Programme on Global Environmental Change (IHDP). The book is ideal for researchers and policy makers in environmental science and policy.

C. GREGORY KNIGHT is a Professor of Geography and Professor of International Environmental Affairs at The Pennsylvania State University. He joined the faculty of Penn State in 1971, where he has served as a department head, as well as Vice Provost and professor in the School of International Affairs. His work on resource management includes development in Africa, energy management in the United States, and global climate change impacts on water resources. He was founding director of the National Science Foundation-funded Center for Integrated Regional Assessment, and is an honorary member and consultant to the Scientific Coordination Center for Global Change in the Bulgarian Academy of Sciences.

JILL JÄGER has worked as a consultant on energy, environment, and climate for numerous national and international organizations and has an extensive publication record. She was Deputy Director of the International Institute for Applied Systems Analysis (1994–1998) and Executive Director of the International Human Dimensions Programme on Global Environmental Change (1999–2002). She was a member of the Core Group of the EU-funded MATISSE (Methods and Tools for Integrated Sustainability Assessment) project (2005–2008).

INTEGRATED REGIONAL ASSESSMENT OF GLOBAL CLIMATE CHANGE

C. GREGORY KNIGHT

JILL JÄGER

CAMBRIDGE
UNIVERSITY PRESS

CAMBRIDGE
UNIVERSITY PRESS

University Printing House, Cambridge CB2 8BS, United Kingdom

One Liberty Plaza, 20th Floor, New York, NY 10006, USA

477 Williamstown Road, Port Melbourne, VIC 3207, Australia

4843/24, 2nd Floor, Ansari Road, Daryaganj, Delhi - 110002, India

79 Anson Road, #06-04/06, Singapore 079906

Cambridge University Press is part of the University of Cambridge.

It furthers the University's mission by disseminating knowledge in the pursuit of education, learning and research at the highest international levels of excellence.

www.cambridge.org
Information on this title: www.cambridge.org/9781108447089

© International START Secretariat 2009

First published 2009
First paperback edition 2017

A catalogue record for this publication is available from the British Library

Library of Congress Cataloging in Publication data
Knight, C. Gregory.
Integrated regional assessment of global climate change / C. Gregory Knight, Jill Jäger.
p.   cm.
Includes index.
ISBN 978-0-521-51810-9
1. Climatic changes.   2. Global environmental change.   I. Jäger, Jill.   II. Title.
QC903.K57   2009
304.2′5–dc22   2009022114

ISBN 978-0-521-51810-9 Hardback
ISBN 978-1-108-44708-9 Paperback

Contents

The authors

SARA BERESFORD served as Project Coordinator for the Assessments of Impacts and Adaptation to Climate Change project at START until 2004. Her interests lie in promoting environmental awareness and change in human behavior through effective communication of scientific information to the public. She currently serves as the Managing Director of the EcoFocus Film Festival in Athens, GA, USA.

LIVIA BIZIKOVA is a Project officer with the Measurement and Assessment Team at the Institute for Sustainable Development (IISD). From 2005 till 2008, she worked as a postdoctoral fellow on linkages between climate change and sustainable development with the Adaptation and Impact Research Group, Environment Canada and the University of British Colombia.

IAN BURTON, Scientist Emeritus, Meteorological Service of Canada and Professor Emeritus, University of Toronto, works on natural and man-made hazards and disasters, risk assessment, environment and development, climate change adaptation, and the interface between science and policy. He was an IPCC Lead Author and is a Fellow of the Royal Society of Canada.

SUBASH DHAR is a Scientist at the UNEP Risoe Centre, Denmark. He worked on MARKAL and AIM families of energy–environment models. His current research interests include gas markets in India, climate change, and sustainable development. He has worked on many international projects and published in international journals.

THEA DICKINSON manages projects for the Clean Air Partnership (a science and policy NGO in Toronto, Canada) and is a partner in Burton Dickinson Consulting Ltd. She has over eight years experience in environmental and health research. She has authored several publications including *The Compendium of Adaptation Models for Climate Change*.

ANN FISHER, Professor Emerita of Environmental Economics at the Pennsylvania State University, led the Mid-Atlantic Regional Assessment (MARA) and the Consortium for Atlantic Regional Assessment (CARA). She examined how society might be affected by and respond to global climate change.

AMIT GARG is Associate Professor in the Public Systems Group at the Indian Institute of Management, Ahmedabad, India. His current research interests include carbon finance, energy policy, climate change vulnerability, adaptation and sustainable development, greenhouse-gas emission projections, and public management. He has co-authored four books, over 20 international papers and contributed to three IPCC reports.

JILL JÄGER was a Senior Researcher at the Sustainable Europe Research Institute in Vienna at the time of writing. She was Executive Director of the International Human Dimensions Programme on Global Environmental Change (IHDP) from April 1999 until October 2002. Her research focuses on integrated sustainability assessment and linking knowledge to action in the area of global environmental change.

MIKIKO KAINUMA is a chief of the Climate Policy Assessment Research Section at the National Institute for Environmental Studies, Japan. She has been a leader of the project for developing the Asia–Pacific Integrated Model (AIM) since 2003. Her research focuses on economic analysis of national/international policy for reducing greenhouse gases. She is a lead author of the IPCC Fourth Assessment Report.

BERND KASEMIR is a founding director of Sustainserv, a consulting firm focused on sustainable management and sustainability reporting. At the time of the work discussed here, he was a research fellow at the Swiss Federal Institute of Aquatic Science and Technology (Eawag) and at Harvard's Kennedy School of Government.

C. GREGORY KNIGHT is Professor of Geography and Professor of International Affairs at the Pennsylvania State University, USA. He was founding director of the Center for Integrated Regional Assessment and is an honorary member of the Scientific Coordination Center for Global Change in the Bulgarian Academy of Sciences. An IPCC reviewer, his work on global climate change in southeastern Europe focuses on water.

NEIL LEARY is Director of the Center for Environmental and Sustainability Education at Dickinson College, Pennsylvania. Previously he was a Senior Scientist with the global change SysTem for Analysis, Research and Training (START) in Washington, USA. He directed the Working Group II Technical Support Unit of the

Intergovernmental Panel on Climate Change and co-edited the IPCC 2001 report on impacts, adaptation, and vulnerability.

LOUIS LEBEL is the founding Director of the Unit for Social and Environmental Research (USER). The unit carries out interdisciplinary, action-oriented research on environmental change in Thailand and, with collaborators, elsewhere in Southeast Asia. Since 1997 he has been the Science Coordinator for the START Global Environmental Change Programme in Southeast Asia.

RIK LEEMANS is Professor of Environmental Systems Analysis at Wageningen University. He directs the WIMEK graduate school, chairs the Earth System Science Partnership and contributed to the IPCC and Millennium Ecosystem Assessment. Research interests include integrated assessments of biodiversity, land-use/land-cover change, biogeochemical cycles, ecosystem services, and sustainable development.

DIANA LIVERMAN is Director of Oxford University's Environmental Change Institute and holds a university Chair in Environmental Science in the School of Geography. Her research focuses on the human dimensions of global environmental change including climate change policy and impacts, the social causes and consequences of land-use change, and environmental management in the Americas.

ELIZABETH L. MALONE is a senior research scientist at the Joint Global Change Research Institute. She is interested in policy-relevant sociological studies that integrate disparate worldviews, data sources, and scientific approaches. Research areas include comparative methods for analyzing country, sector, and local vulnerabilities to climate change and associated approaches to social–environmental scenarios.

YUZURU MATSUOKA is a Professor of Engineering at Kyoto University, Japan. He is one of the founding members of the AIM project, which started in 1991. His research focuses on integrated analysis of climate change problems using computer simulation models. He is a lead author of the IPCC Third Assessment Report.

TSUNEYUKI MORITA was a director of the Social and Environmental Systems Research Division at the National Institute for Environmental Studies, Japan. He was a coordinating lead author of the IPCC Third Assessment Report in Working Group III. He was one of the founding members of the AIM project and led the project from 1991 until his death in 2003.

DARLA K. MUNROE is Assistant Professor of Geography and Adjunct Assistant Professor of Agricultural, Environmental, and Development Economics at the Ohio

State University, USA. She is an economic geographer with research interests relating to spatially-explicit land-use models, land markets and regional development, and land-use change at the rural–urban interface.

CLAUDIA R. NIERENBERG is Senior Advisor, Climate Program Office, National Oceanic and Atmospheric Administration. She has been responsible for the design and implementation of research programs to expand our understanding of the integrated societal effects of climate variability and change, and options for improving the use of scientific information in practical settings.

COLIN POLSKY is Associate Professor of Geography, and Research Associate Professor at the George Perkins Marsh Institute, Clark University. He is a human–environment geographer specializing in vulnerability assessments using a mixed-methods analytical approach, with specific focus on social causes of sub-urbanization in the USA, and associated social and ecological outcomes and responses.

ROGER S. PULWARTY, Director of the National Integrated Drought Information System at the National Oceanic and Atmospheric Administration focuses on climate in the Americas, social and environmental vulnerability, water resources, developing information services to support planning and adaptation to climate variability, and change in the USA and the Caribbean. He is a lead author on the IPCC Working Group 2.

P. R. SHUKLA is Professor in the Public Systems Group at the Indian Institute of Management, Ahmedabad, India. He is a lead author of several international reports on energy and environment, including IPCC reports. His publications focus on modeling and policies for developing countries. He is a consultant and adviser to international organizations, companies, and the Government of India.

CAITLIN SIMPSON is Program Manager of the Regional Integrated Sciences and Assessments at the Climate Program Office of the National Oceanic and Atmospheric Administration. She fosters integrated research among sciences to analyze how climate affects resource management and how climate science, forecasts, and impacts information could improve management and policy decisions.

KIYOSHI TAKAHASHI is a Senior Researcher of the Climate Risk Assessment Research Section at the National Institute for Environmental Studies, Japan. He joined the AIM project in 1995. His research focuses on assessment of climate change impact on agriculture and water resources at the global scale. He is a lead author of the IPCC Fourth Assessment Report.

MARJOLEIN B. A. VAN ASSELT holds the Risk Governance chair in the Faculty of Arts and Social Sciences, Maastricht University, the Netherlands. She is a member of the Dutch Scientific Council for Government Policy (WRR) and of the Young Academy, part of the Royal Netherlands Academy of Arts and Sciences (KNAW). Her research focuses on uncertainty, risk, and foresight.

RICHARD WARRICK, formerly Deputy Director of the International Global Change Institute at the University of Waikato (NZ), is Professor of Climate Change Adaptation at the University of the Sunshine Coast, Australia. His interests lie in climate change impacts and adaptation, and integrated models and tools for assessment and decision making. He was a lead author in all four of the IPCC assessment reports.

BRENT YARNAL is Professor of Geography at the Pennsylvania State University, USA where he was recently Director of the Center for Integrated Regional Assessment. His research focuses on the local dimensions of global change, bridging the physical and social sciences, and integrating climate change, natural hazards, land-use change, and the use of environmental information in decision making.

YONGYUAN YIN, environmental scientist with the Adaptation and Impacts Research Division, Environment, Canada, studies climate change in China and Canada to link adaptation policies with societal sustainability, climate change vulnerability assessment and adaptation policy evaluation, environmental sustainability, land use and water planning, sustainable forest management, and government policy.

Foreword

Global climate change is occurring at an unprecedented rate and affecting all facets of the Earth's ecosystems as well as our societies. Human activities lie at the very heart of global change and alter the Earth's environment from a local to global scale and from biogeochemical to hydrological and ecological processes. Human behavior is, for example, responsible for a variety of large-scale changes, which range from systematic developments such as climate change to cumulative impacts such as the decrease in biological diversity. Quite suitably, the current era of Earth's history has been named the Anthropocene. Consequently, there is an increasing realization that these human–environment interactions give rise to complex and dynamic socio-ecological systems in which both anthropogenic and biophysical drivers play central roles in environmental changes. Climate change is the most prominent one. It affects socio-ecological systems and significantly impacts on human development, well-being, and vulnerability. Inevitably, the human dimensions of climate change are increasingly becoming issues of governance, adaptation strategies, and scientific agendas. It is impossible then to address climate change without placing humans and society at the center of the debate.

As a biogeophysical phenomenon, climate change occurs over a range of spatial scales. The regional implications of global climatic change are vast and it is at this level that assessment of environmental change, and society's response and adaptation to change are crucial. With the vast number and variety of actors at a regional level in a social system it is important to include integrated systems in the assessments of change. Appreciation of this large-scale impact of global climate change and the types of impact on society requires new perspectives such as that of "integrated regional assessments." This book addresses both the dearth of knowledge of the regional dimensions of global climate change and describes the combined approach of a regional perspective with integrative assessments. Integration is a key term highlighting the interdisciplinary nature that involves cross sectoral stakeholder collaboration from the realms of science, policy, and society. The choice of scale in this approach is significant as it is one attribute that can guide the examination of global environmental problems. The regional implications of climate change are significant as it is at this level that impacts are felt by humans and that mitigation and adaptation must be practiced. For instance, it is here that the ecosystem services paradigm becomes important. The related processes of environmental change affect all ecosystems whether directly or indirectly. Consequently, changes of ecosystem services will directly affect the people within this region that rely on these services for their well-being and security.

This book sets out what is known about integrated regional assessment, specifically focusing on climate change, and addresses a number of important but difficult questions. What are the concepts and processes of integrated regional assessment? Why is a regional level of spatial specificity needed in environmental assessments? What is the global context of integrated regional assessment? How can integrated regional assessments be applied to scientific agendas and deployed at local and national levels? How can integrated regional assessments be used in collaboration with national assessment processes and scientific agendas? The spatial context of integrated regional assessment is emphasized in a range of case studies that have combined qualitative and quantitative approaches in innovative ways. By raising and attempting to address these questions in this volume, the editors and authors endeavor to highlight an approach of environmental change research which has previously received little focus.

Andreas Rechkemmer
Former Executive Director, International Human Dimensions
Programme on Global Environmental Change
Bonn

1

Integrated regional assessment

C. GREGORY KNIGHT AND JILL JÄGER

1.1 The context of integrated regional assessment

Recognition that human activities have impacts on the environment at regional and global scale has a long history, including persistent questions about human effects on the environment (and the influences of the environment on mankind) from the time of Greek philosophers (Glacken 1967). Seminal publications that document concern about human impacts on the environment include the work of George Perkins Marsh, *Man and Nature* (1864), the landmark conference in 1955, *Man's Role in Changing the Face of the Earth* (Thomas 1956), and the latter's 1990 successor, *The Earth as Transformed by Human Action* (Turner *et al.* 1990a).

While humans interact with many aspects of the environment, the focus of this book is largely on the interactions with the climate system, since this issue has received considerable recent attention in integrated regional assessments.

The specter of climate change due to greenhouse gases became a public policy as well as a scientific issue in the 1970s, but human understanding of the possible effects of greenhouse gases has a much longer history (Fleming 1998; Clark *et al.* 2001a; Weart 2003). The absorption and re-emission of long-wave (heat) radiation by atmospheric gases, which has become popularly known as the "greenhouse effect," was described by Jean Baptiste Joseph Fourier in 1824; in 1862, John Tyndall identified carbon dioxide and water vapor as such gases; and in 1896, Svante August Arrhenius suggested that carbon dioxide from the burning of coal could result in atmospheric warming. In the 1930s and 1940s, Guy Stewart Callendar (1949) studied CO_2 and climate, and in the 1950s Gilbert N. Plass (1956) continued research related to greenhouse-gas warming through decades in which human-induced climate change had yet to become a popular issue. In the 1970s, the *Study of Critical Environmental Problems* (1970) and the *Study of Man's Impact on Climate* (1971) were prepared for the 1972 United Nations Conference on the Human Environment in Stockholm. From Russia, Budyko (1974) noted the threat of rapid climate change. Evidence of increased atmospheric CO_2 concentrations, such as those of the Mauna Loa observatory (Keeling *et al.* 1976) and elsewhere, inspired researchers to alert the public to the possible impacts of global warming. Attention soon turned to the assessment of climate change impacts (Kates *et al.* 1985; Jacobson and Price 1991; Kuhn *et al.* 1992).

The Intergovernmental Panel on Climate Change (IPCC), created by the World Meteorological Organization and the United Nations Environment Programme in 1988, shared the 2007 Nobel Peace Prize, drawing ever more attention to its periodic summary and synthesis of global climate change knowledge (now moving toward a fifth cycle). Since the second IPCC assessment, regional dimensions have received increasing attention in its assessments.

Over the past two decades climate change research has been pursued within a large number of national and international programs. In parallel to the development of official bodies, numerous university-based and independent research institutes redefined their mission or were created to examine global climate change; their work is widely cited in this chapter and book. Box 1.1 lists a selection of regional climate assessments that have been carried out.

The 1994 United Nations Framework Convention on Climate Change (UNFCCC) now has 192 national signatories, although some nations failed to ratify the Kyoto Protocol (1997, coming into force from 2005), which established greenhouse-gas emission targets. Periodic reporting on emissions through greenhouse-gas inventories (industrialized nations), "national communications" (UNFCCC 2003) and "national adaptation programmes of action" (UNFCCC 2001) in the least developed countries are required (see Box 1.1). All of these documents contribute, in part, to national assessments of climate change.

Climate change is just one of the changes interacting with human development. The recognition of a wide range of global changes led to the evolution of the concept of an "Earth system science" in the 1980s. The important human role in global *systemic* change such as ozone depletion and greenhouse-gas emissions and in *cumulative* change such as soil erosion and biodiversity loss was increasingly recognized (Turner *et al.* 1990b). Research groups, whose work was at scales from local to global, examined the interactions between human agency and environmental change, including feedbacks between these components, in both conceptual/qualitative and quantitative aspects. Climate change became an increasingly important motivation for such studies, as evidenced by a number of international and national initiatives.

The International Geosphere–Biosphere Programme (IGBP) was launched in 1986 to address scientific issues of global change under the aegis of the International Council for Science (ICSU). In 1991, Diversitas was founded to address issues of biodiversity with sponsorship by the United Nations Educational, Scientific

---

Box 1.1
Selected regional assessments of climate change

International assessments
United States Country Studies Program (1993–1994; Smith and Lazo 2001)

Africa and the Middle East (16 countries)
Asia and Pacific (14 countries)
Central and Eastern Europe and Former Soviet Union (nine countries)
Latin America (eight countries plus seven countries in the Central American
Regional Study)

United Nations Framework Convention on Climate Change (www.unfccc.int)

National Communications (Annex 1, Organization for Cooperation and
Development Industrialized Countries plus Countries in Transition)
40 initial, 39 second, 38 third, 38 fourth country submissions includes
European Community Multi-Nation Report

National Communications (non-annex 1, developing countries)
134 initial, three second, one third country submissions
National Adaptation Programmes of Action (48 UN-defined least-developed
countries)
30 country submissions

United Nations Environment Programme, Country Studies on Climate Change
Impacts and Adaptations Assessment (O'Brien 2000)

Antigua and Barbuda
Cameroon
Estonia
Pakistan

Intergovernmental Panel on Climate Change (www.ipcc.ch)

IPCC Regional Impacts Studies (Watson *et al.* 1998)
Africa
The Arctic and Antarctic
Australasia
Europe
Latin America
Middle East and Arid Asia
North America
Small Island States
Temperate Asia
Tropical Asia

IPCC Third and Fourth Assessment Reports (McCarthy *et al.* 2001; Parry *et al.* 2007)

> Africa
> Asia
> Australia and New Zealand
> Europe
> Latin America
> North America
> Polar Regions (Arctic and Antarctic)
> Small Island States

Continental, national, and regional assessments

Africa

> Climate change and Africa (Low 2005)

AIACC (Assessments of Impacts and Adaptation to Climate Change Regional Studies (limited sectors; Leary *et al.* 2007)

> Africa (11 studies)
> Asia (five studies)
> Latin America (five studies)
> Small Islands (three studies)

Arctic

> Arctic Climate Impacts Assessment (ACIA 2005)

Asia

> Asia Pacific Network (Kainuma *et al.*, Chapter 11, this volume)
> Southeast Asia Regional Network (Lebel, Chapter 16, this volume)

Australia

> Climate Change in the Cairns and Great Barrier Reef Region (Crimp *et al.* 2004)
> Climate change impacts for Australia (Preston and Jones 2006)

Baltic Sea Region

> BALTEX assessment of Climate Change (2008)

Brazil

> WAVES-water availability, vulnerability of ecosystems and society in Northeastern Brazil (Ferreira *et al.* 2000; Gaiser *et al.* 2003)

Canada

> Mackenzie Basin impact study (Cohen 1995, 1996, 1997; Cohen *et al.* 1997)

Canada country study (Mayer and Avis 1998)
Regional assessments (six)
Great Lakes–St. Lawrence Basin Project (Mortsch and Mills 1996; Mortsch and
  Quinn 1998)
Reducing Canada's Vulnerability to Climate Change Program (Natural Resources
  Canada 2007)
  Local and regional studies (nine with limited sectors)
  From impacts to adaptation: Canada in a changing climate (Lemmen *et al.* 2008)
Regional assessments (six)

China

  National assessment report on climate change (Ding *et al.* 2006; Chen *et al.* 2005)

Europe

  Vulnerability and Adaptation to Climate Change in Europe (EEA 2005)
  Adapting to Climate Change in Europe (European Commission 2007)

France

  Impacts Potentiels du Changement Climatique en France au XXIe Siècle (France
    2006)

India

  Climate Change and India: Vulnerability Assessment and Adaptation (Shukla *et al.*
    2003)

New Zealand

  Vandaclim (Warrick *et al.* 1999; Warrick, Chapter 14, this volume)

Switzerland

  Climate in Human Hands (Swiss Agency for the Environment 2002)

United Kingdom

  East Anglia (Lorenzoni 2000a, 2000b)
  East Midlands (East Midlands Sustainable Development Round Table 2000)
  Regional Climate Change Impact and Response Studies in East Anglia and North
    West England – REGIS (Holman *et al.* 2002)
  UK Climate Impacts Programme (UKCIP 2008)

United States

  United States Global Change Research Program (National Assessment Synthesis
    Team 2000)

National Assessment of Climate Change (NACC)
   Mega Regions (nine)
   State and Multi-state Regions (18)
Climate Change Impacts for the Conterminous USA (Rosenberg and Edmonds
   2005a)
   Other US Regional Assessments
   California (California Climate Change Center 2006)
   Consortium for Atlantic Regional Assessment (Dempsey and Fisher 2005;
   Fisher *et al.* 2006)
   MINK (Missouri, Iowa, Nebraska, Kansas; Rosenberg 1993a)
   NOAA RISA Program (Pulwarty, Chapter 18, this volume).
   Regional Integrated Sciences and Assessments (eight state or multi-state
   regions, additional region to be added in 2008)
   Pew Center on Climate Change (Ebi 2007)
   Regional Impacts of Climate Change (four regions, selected sectors)
   Susquehanna River Basin Integrated Assessment (Yarnal 1998)
   Union of Concerned Scientists
   Northeast Climate Impacts Assessment (Union of Concerned Scientists 2006a)
   Confronting Climate Change in the Northeast (Frumhoff *et al.* 2007), plus nine
   state-level brochures issued in 2007
   Our Changing Climate, Assessing the Risks to California (Union of Concerned
   Scientists 2006b)
   Climate Change in Pennsylvania (Union of Concerned Scientists 2008)

**Some world and regional integrated regional assessment models with regional
resolution**

   AIM – 21 geopolitical regions (Kainuma *et al.*, Chapter 11, this volume).
   DICE – 13 regions (Nordhaus and Boyer 2000)
   ESCAPES – four regions (Rotmans *et al.* 1994)
   ICAM 3 – 12 regions (Dowlatabadi 2002)
   IMAGE 2 – 17 regions (Leemans, Chapter 15, this volume)
   MERGE – nine regions (Manne *et al.* 1995)
   TARGETS – six regions (Rotmans and de Vries 1997)

and Cultural Organization (UNESCO), the Scientific Committee on Problems of the Environment (SCOPE), and the International Union of Biological Science (IUBS), joined by ICSU as a sponsor in 1996. The International Human Dimensions Programme on Global Environmental Change (IHDP) was established in 1996 under the sponsorship of ICSU and the International Social Science Council. The World Climate Research Programme (WCRP) was established in 1980 by ICSU and the World Meteorological Organization, joined from 1993 by UNESCO's Intergovernmental Oceanographic Commission. Global change research was thus coordinated internationally by the four programs (IGBP, Diversitas, IHDP, and WCRP) that began to work closely together.

IGBP hosted the initial meeting in 1990 that led to START, the Global Change SysTem for Analysis Research in Training, which focuses on regional research and capacity-building activities in the transitional and developing countries (Eddy *et al.* 1991). START's capacity-building activities were carried out under the auspices of IGBP, WCRP, and IHDP, with the START secretariat having opened in 2002 and regional committees and networks being established in 2002–2005.

The joint activities of the international global change research programs led to the organization of the 2001 Global Change Open Science Conference in Amsterdam. This conference demonstrated the emergence of a global community of researchers and stakeholders who recognized that "global change" is much more than climate change and that there is a need to consider the Earth as a system, with all of its interlinked changes. As a result of the Amsterdam conference IGBP, IHDP, WCRP, and Diversitas formed the Earth System Science Partnership (ESSP) for integrated study of the Earth system from both natural and human perspectives. In addition to four joint projects on crosscutting themes of carbon, water, food, and health, ESSP plans for a series of "integrated regional studies," the first being the Monsoon Asia Integrated [Regional] Study (MAIRS; Jäger, Chapter 10, this volume).

This brief history of various programs at international and selected regional and national levels suggests the rapid growth of attention to global change, including climate change, and its impacts. During the same period in which these institutions and programs developed, integrated regional assessments of global climate change have multiplied in number, while being motivated and executed via a variety of routes (see Box 1.1).

1.2 Integrated regional assessment

"Integrated regional assessment": strangely enough, a search on the world wide web generates relatively few citations for this phrase, when in fact hundreds of integrated regional assessments (or close relatives) have been done and are being done worldwide.[1] These assessments may be of a kind, but titles often mask their commonalty. Country studies, national communications, climate assessments, integrated assessments, and climate impact studies – these are among the many phrases that label integrated regional assessments. What does integrated regional assessment (IRA) mean?

---

[1] At the time of publication, there is no Wikipedia article on "integrated regional assessment," nor on "regional assessment" or "integrated assessment."

Assessment in this context means the application of scientific knowledge and human experience to important decisions, typically with a policy dimension. Usually this means mustering existing knowledge rather than undertaking new science, although an assessment may well identify knowledge gaps and call for new research. Assessment is a process, although the resulting activities (such as reports, world wide web sites, and presentations) are also referred to as assessments. In the examples described in this volume, the decisions to be made are both personal and public, involving the identification of impacts of, and vulnerabilities to, climate change, seeking ways to adapt to the changes that cannot be avoided, while contributing to mitigation of the causes of the changes.

Integrated has two connotations. Contribution from a diversity of relevant sciences and human knowledge is one of them: crossing boundaries of disciplines and experience, incorporating multiple perspectives to understand a real-world problem. The second connotation is acknowledgement that phenomena outside the laboratory do not occur in isolation, thus integration across domains of impact and influence is imperative. These are not really different, since integration across phenomena requires, *ipso facto*, collaboration among experts, those from scientific disciplines and those steeped in local knowledge and experience.

Integrated assessment is, then, ". . . the practice of combining different strands of knowledge to accurately represent and analyze real world problems of interest to decision-makers" (Dowlatabadi *et al.* 2000). Rosenberg and Edmunds (2005b, p. 1) defined integrated assessment as ". . . an analytical approach that knits together knowledge derived from a variety of disciplinary sources to gain insights from the analysis of interactions." In particular, integrated assessment of global climate change brings multiple, complementary perspectives to understanding present and potential future impacts cascading from the climate system to terrestrial and aquatic ecosystems, and via those impacts (or directly, in some cases) to society.

Regional means that the decisions we want to inform have a spatial locus, a place or places inhabited by individuals, families, communities, societies, and nations, or places vital to the health of planet Earth (e.g., oceans or polar regions). Places provide the necessities of human life – shelter, sustenance, health, and safety – as well as values of human experience – belief, beauty, fulfillment, and pleasure. Human-induced global climate change comes from local places (and regions) and is experienced in those and other places. Places do not exist in isolation. They are linked by regional and global climatic, oceanic, atmospheric, ecological, economic and social processes. Thus, regional assessment of climate change applies science and experience to understanding climate change in a particular geographic space (Easterling 1997; Knight 2001). Regional assessments often focus on specific phenomena, such as water, ecosystems, forests, agriculture, or human health.

Integrated regional assessment, thus, is a process that uses a diversity of interrelated information and expertise to apply knowledge for stakeholder and policymaker decisions concerning regional dimensions of global change. Integrated regional assessment extends all three component terms to cross both disciplinary boundaries for analysis and impact boundaries by making explicit the linked and cascading impacts of environmental and socio-economic change in regions. Integrated regional assessment also implies close involvement of stakeholders and policy makers in the assessment process, not just as passive recipients of what scientists think is important.

One of the fundamental approaches to studying a technical vocabulary is to ask, "What kinds of X are there?" and "What is X a kind of?" These are useful questions here. In answer to the "What kinds of" question, there are many kinds of integrated regional assessments, as we illustrate in the next section. On the other hand, integrated regional assessment is a kind of public policy tool to address important questions about global change at the interface of science and society. As such, it is part of a family of related assessment procedures and policy-development and decision-making processes (see Box 1.2). One closely related process is environmental impact assessment (EIA). Similarities include:

- both often develop future scenarios;
- both try to identify potential impacts, both positive and adverse;
- both try to suggest ways to avoid, mitigate, or adapt to negative consequences;
- both rely on expert judgment;
- both are interdisciplinary;
- both are assessments; and
- both have public and stakeholder involvement.

Differences between EIA and IRA include:

- EIA is generally limited to the local/regional scale; IRA bridges to a global scale, on both causal and impact dimensions;
- in EIA there are decision choices about causal events; in IRA such choices may not exist, at least not locally;
- in IRA, "business as usual" is not the same as the EIA "do nothing" alternative;
- EIA does not address inevitable change;
- IRA may be more truly integrated than EIA;
- IRA has no legislated process requirements, EIA does; and
- EIA focuses on inventories, IRA on (sometimes linked) models and chains of effects.

In view of these similarities and differences, it is helpful if integrated regional assessment practitioners are aware of the environmental impact assessment literature.

Box 1.2

Related assessment types and terminologies

Climate impact assessment
Demographic impact assessment
Development impact assessment
Ecological impact assessment
Economic and fiscal impact assessment
Environmental auditing
Environmental impact assessment
Health impact assessment
Impact assessment
Integrated ecological assessment
Integrated impact assessment
Participatory integrated assessment
Project evaluation
Public consultation
Public participation
Risk assessment
Social impact assessment
Strategic environmental assessment
Strategic impact assessment
Sustainability impact assessment
Technology assessment

Source: International Association for Impact Analysis (GRDC 2004) and authors.

Finally, we need to emphasize that integrated regional assessments are *not* predictions of the future. It is important to make a clear distinction between the terms prediction, forecast, projection, and scenario (McCracken 2001). A *prediction* is a statement about the future based on current conditions and reasonably well-specified methods of anticipating the future. A *forecast* connotes prediction, the credibility of which depends on the person doing the predicting; it also implies a time frame in which the forecasts can be tested and refined. In contrast to both prediction and forecast, a *projection* is based on one of several or many initial conditions and specification of how those conditions may change in the future. In most climate change assessments, there are a number of initial assumptions and future trajectories of driving forces and mitigation actions, all used as input to a variety of different climate models. Such projections create climate *scenarios*, neither predictions nor forecasts, as bases for assessment of what *could* happen in future. General circulation models of the atmosphere are retrodictions when they account for the climate record; for the future they generate projections which can be summarized as scenarios; they are decidedly not predictions. Derivation of impacts from these scenarios is similarly of scenario, not prediction, status, whether the derivations come from loosely linked qualitative analyses or structured mathematical models. Thus the output of an integrated regional assessment process is neither a prediction nor a forecast. Rather it is a plausible set of scenarios that could occur under specified conditions.[2]

1.3 The integrated regional assessment process

There are several ways to look at integrated regional assessments, including their origins, methods used in assessment, the scale of the geographical area under consideration, and the degree of integration.

1.3.1 Assessment origins

Integrated regional assessments can originate in many ways. *Programmatic assessments* originate in legal requirements, such as the United States National Assessment on Climate Change (NACC) as mandated by the Global Change Research Act in 1990, or international agreements, such as reporting required under the United Nations Framework Convention on Climate Change. Such program requirements may have common elements (such as specification of climate scenarios and sectors for analysis), but allow freedom for the responsible parties to include other elements of regional importance. As such, one might refer to these IRAs and IRA-like documents as having a common design, with reporting in a consistent format that is amenable to aggregation to other levels. For example, the 18 largely multi-state regions in the NACC were subsequently aggregated into nine mega-regions and a national synthesis for the United States.

IRAs may also result from a process of *review and consolidation*. An example is the literature-based continental assessments included in the IPCC Working Group II on impacts and adaptation, which in recent reports included six contiguous continental regions plus categories for the polar regions and small islands (McCarthy *et al.* 2001; Parry *et al.* 2007). For most authors of the literature summarized in such reviews, there was no intent to have produced a truly integrated assessment, but rather to assess some impact category in a particular spatial venue. Review writers were able, however, to synthesize this literature into coherent statements about significant issues and impacts in the respective areas.

A variant on the review and consolidation process is that of *review and collaboration*, wherein researchers find common ground that synergistically leads to a

---

[2] Although their title may overstate the case, Pilkey and Pilkey-Jarvis's *Useless Arithmetic* (2007) reminds us of the limitations of model-based "predictions."

shared work that amounts to an IRA. Examples include the Task Force on Climate Change in Texas (North *et al.* 1995), the *Climate Change Impacts of the Conterminous USA* special journal issue (Rosenberg and Edmonds 2005a), the climate change–drought analog research project on Bulgaria (Knight *et al.* 2004), and climate change in Africa (Low 2005). Syntheses also emerge from thematic conferences, such as climate change in northern North America (Peterson and Johnson 1995) and climate change and the Amazon (Malhi *et al.* 2008).

Yet another route by which IRAs evolve is through *organized research agendas* such as the RISA network of investigations (Pulwarty, Chapter 18, this volume) and agency-sponsored assessments such as the US Environmental Protection Agency's Global Change Research Program, some of which were elements of NACC (Great Lakes, Gulf Coast, and Mid-Atlantic; see Fisher and Kasemir, Chapter 8, this volume), and others such as the Consortium for Atlantic Regional Assessment (CARA; Fisher *et al.* 2006) and Climate's Long-term Impacts on Metro Boston (CLIMB; Kirshen *et al.* 2004, 2006) which were extensions of those elements.

Networked activities also drive the formation of IRAs; examples include the Asia Pacific Network (APN; Kainuma *et al.*, Chapter 11, this volume) and the Southeast Asian Regional Committee for START (SARCS; Lebel, Chapter 16, this volume), both linked at the global level to institutions in the Earth Systems Science Partnership (ESSP), themselves operating under the aegis of ICSU, the International Council for Science.

IRAs also emerge from *initiatives of individual scientists and groups* in response to research imperatives and funding opportunities. Examples include the CLIMPACTS project at the University of Waikato (Warrick, Chapter 14, this volume), the MINK study (Rosenberg 1993a), the Susquehanna River Basin Integrated Assessment (Yarnal 1998) and the university–government–industry–stakeholder-driven Mackenzie Basin Impact Study (Cohen 1996).

Non-governmental organizations are playing an increasing role in climate change impact assessment with sectoral, regional, and integrated approaches. For example, the Pew Center on Climate Change has focused on selected sectors in four regions in the United States, and the Union of Concerned Scientists has produced integrated assessments in California and the Northeast (see Box 1.1). Both organizations emphasize outreach to the public and policy makers with foci on adaptation and mitigation.

1.3.2 Assessment methods

IRAs can similarly be viewed from the perspective of the organizational and scientific methods utilized. On one organizational dimension, IRAs have been either *science-driven, science-driven with communication to policy makers and stakeholders*, or *largely policy maker and stakeholder driven with their strong influence on the research agenda*, the latter becoming increasingly the standard. Another organizational dimension is the degree of *active rather than advisory participation* in the assessment process by individuals from educational and research institutions, government, industry, policy communities, and stakeholders. Since many participants have both professional interest and a personal stake as community members in the assessment outcomes, roles are often obscured rather than singular and clear-cut. In addition, assessment often becomes a *process* beyond a *product*, wherein the assessment community joins in a movement to address issues of adaptation, mitigation, education, and outreach (see Box 1.3).

Box 1.3
Assessing the assessment process

Research by scholars and experience of practitioners have increasingly demonstrated that institutional aspects of assessment are critical to better understanding and structuring the connection between science and policy in the environment arena (Farrell and Jäger 2005; Mitchell *et al.* 2006a).

It is now clear that some assessments have more influence than others due to the *process* by which they are developed rather than just their *products*. Assessments are usefully viewed as *social communication processes* through which scientists, decision-makers, advocates, and the media interact to define relevant questions, mobilize certain kinds of experts and expertise, and interpret findings in particular ways (see, for example, Jasanoff and Martello 2004).

An assessment can be influential across a spectrum of policy relevant factors such as:

- the environment, e.g. European assessments have led to decreased forest damage due to acid deposition (Tuinstra *et al.* 1999; Eckley 2002);
- the strategies and behavior of key actors, e.g. the role of the Ozone Trends Panel in shifting corporate approaches to the regulation of CFCs (Parson 2003);
- the pool of management options (Clark *et al.* 2001b), e.g. the broad survey of the IPCC response strategies report in 1991;
- issue framing and agenda setting, e.g. the role of the 1986 Villach climate assessment in elevating the climate issue on the policy agenda (Torrance 2006);
- the building of scientific communities, the creation and maintenance of issue networks, and professional advancement (Haas 1992; Sabatier and Jenkins-Smith 1999), e.g. the creation of the "Villach Group" an active science and policy network around the issue of climate change (Haas and McCabe 2001).

That is, assessments exert any influence they may have in a variety of ways. But, why do some assessments have more influence than others?

The impact of an assessment depends at least partly on when it is conducted. Those conducted at an early stage of issue development are unlikely to lead to immediate and direct policy changes (Mitchell *et al.* 2006b, p. 310). Most assessments have influence with some audiences but not with others. The Global Environment Assessment project (see Jasanoff and Martello 2004; Farrell and Jäger 2005; Mitchell *et al.* 2006b) showed that three characteristics of assessments are important in distinguishing more from less influential ones.

salience: the perceived relevance or value of the assessment to particular groups who might employ it to promote any of the policy changes noted above;

credibility: the perceived authoritativeness or believability of the technical dimensions of the assessment process to particular communities;

legitimacy: the perceived fairness of the assessment process to particular communities.

The tactics adopted to promote one of these attributes can undermine another, but opportunities to promote different attributions simultaneously do exist. For instance, increasing participation in an assessment process in order to increase salience and legitimacy can also increase credibility by providing access to local knowledge and to data that would not otherwise be available.

A final finding regarding assessment processes is the value of building the capacity of various actors to be involved in producing assessments and to understand the findings of assessments. Assessment processes have gained influence with wider audiences by establishing a long-term goal and process to enhance the capacity of a range of scientists to participate substantively in assessments (Mitchell *et al.* 2006b, p. 328). Capacity building among potential users is equally important.

A further organizational dimension is the degree to which the assessment is *prescriptive, loosely prescriptive, selective*, or *innovative*. Prescriptive assessments are often part of programmatic assessment activities, with common structure, data sources, modeling and analytic approaches, and reporting standards. Loosely prescriptive assessments follow a general approach that builds upon assessment experience in other contexts and provide roughly comparable results with less uniformity. The US NACC is an example of the former; the UNFCCC reports are typical of the latter. Largely self-standing selective assessments may have greater scope for selecting unique approaches for organization, scientific method, and reporting; they may still be constrained by the state of the art and the experience of participants. Still other assessments may offer innovative directions that help to move the field of IRA forward as a concept and methodology. Particularly interesting in the latter sense are developments of modeling approaches at the regional level.

From the scientific viewpoint, integrated regional assessments and assessment components may range from *quantitative* (statistics or model based) to *qualitative* (use of experts, focus groups, and the like); they are frequently mixed in analytical mode (Malone, Chapter 3, this volume). Integrated regional assessment models are, by their nature, quantitative (Leemans, Chapter 15, this volume). In general, integrated regional assessments draw upon: (1) the results of quantitative scientific analyses, sometimes applying relevant methods with region-specific data; (2) substantiating quantitative conclusions, themselves qualified by notations of uncertainty in the underlying data, models, and simplifications of reality; and (3) translation of scientific insights into qualitative statements about the directions, magnitude, and importance of impacts.

An additional approach to regional assessments is the use of climate analogs and place analogies. Assessments can be informed by reference to places that have a similar climate now to climates anticipated under climate change scenarios (Union of Concerned Scientists 2006a); alternatively, past climate or climate events (like drought) could be imposed on the same place (Rosenberg 1993b; Knight *et al.* 2004).

Assessments also vary by the direction from which they proceed, from *bottom up*, *top down*, or a *mixture* of the two. Bottom-up assessment begins with high resolution (in time, space, or sectors) for a smaller region (AAG GCLP Team 2003) looking upward (to wider geographic and global processes), including contributions of the region to higher levels. Top-down assessment resolves from the global or (multi-nation) regional level to the national and local; examples are the integrated assessment models with regional or national resolution (Leemans, Chapter 15, this volume). Many integrated regional assessments are a mixture, drawing from wider global and regional processes, resolving down to local levels via techniques such as climate downscaling, geographic information systems, or case studies.

Integrated assessment models (IAMs), mentioned above, are formally quantitative, dynamic representations of climate change and impacts (Leemans, Chapter 15, this volume; Dowlatabadi 1995, 2002; Parson and Fisher-Vanden 1997), varying by the degree to which climate scenarios are exogenous or endogenous to the model, by sectoral and spatial specificity, by time step, by the nature of feedbacks, and mathematical formulation (e.g. process versus optimization strategies). Yohe (2000) rightly pointed out that the insights from integrated assessment models depend on policy alternatives, adaptation choices, system adaptation to both environmental change and policy, the ubiquitous existence of uncertainty, and optimization choices, particularly for IAMs that focus on economically efficient goal achievement.

1.3.3 Scale of the geographical area

Integrated regional assessments involve consideration of all scales from local to global. We can only touch on the rich ways in which investigators, stakeholders, and policy makers have conceived of regions of importance (see Box 1.1). Some examples include the following.

(1) *Local civil divisions/municipalities.* Examples include Canberra and Western Port, Victoria in the Australian Department of Climate Change Program "Settlements and Infrastructure" assessment projects (Australia 2008). The City of Aspen [Colorado] Canary Initiative (Katzenberger 2006) provides another example, as does the London [England] Climate Change Partnership (2002).

(2) *States* (provinces, prefectures). Nine northeastern state reports by the Union of Concerned Scientists are examples, as are other state-level activities in Box 1.1. The UK Climate Impacts Programme addresses 12 regions in that nation (UKCIP 2008).

(3) *Multi-province/state/prefecture regions.* The WAVES project in northeastern Brazil (Ferreira *et al.* 2000; Gaiser *et al.* 2003) is one example, as are the NACC regional studies in the United States.

(4) *National assessments.* Examples include: Australia's *Climate Change Impacts for Australia* (Preston and Jones 2006); France's *Impacts Potentiels du Changement Climatique en France au XXIe Siècle* (France 2006; *Potential Impacts of Climate Change for France in the 21st Century*); and Canada's *The Canada Country Study: Climate Impacts and Adaptation* (Mayer and Avis 1998).

(5) *Special places and heritage sites.* Important national sites were addressed in *Climate Change and the Historic Environment in the UK* (Cassar 2005). National parks are a focus in the United States (National Park Service 2006).

(6) *Vernacular regions.* For example, the North West of England is one focused study (Shackley *et al.* 1998). Puget Sound has received attention from the Climate Impacts Group (2007), and the Alps have been a locus of attention in Europe (Cebon *et al.* 1998). REGIS-Regional Climate Change Impact and Response Studies in East Anglia and North West England (Holman *et al.* 2002) is another example.

(7) *Natural regions.* For example, the Arctic is addressed in the Arctic Climate Impact Assessment (ACIA 2005), as are polar regions in recent IPCC Working Group II reports.

(8) *River basins.* The Climate Impacts Group at the University of Washington, one of the RISAs discussed in this volume (Pulwarty, Chapter 18) focused on the Columbia River as did the work of Cohen and colleagues (2000). Aerts and Droogers (2004) include short reports on assessments in a variety of basins. The Susquehanna River Basin Integrated Assessment (Yarnal 1998), the Great Lakes–St. Lawrence Basin Project (Mortsch and Mills 1996; Mortsch and Quinn 1998), and studies for Chesapeake Bay (Abler *et al.* 2002) are other examples.

One of the major challenges in work at any scale is the dynamic of interactions across scales, an issue addressed in this volume (Polsky and Munroe, Chapter 4).

1.3.4 Degree of integration

Integrated regional assessments are also differentiated by their degree of integration, from use of a fully integrated mathematical model, to linked models, linked analyses, and loosely integrated analyses, followed by topical analyses without formal links and, finally, hybrid approaches in which some formal linkage of components is accompanied by less formal connections between assessment components.

Fully integrated models are illustrated in this volume for the national (Warrick, Chapter 14), regional (Kainuma *et al.*, Chapter 11), and global levels (Leemans, Chapter 15). Here there is formal mathematical elaboration of the causal chain moving from climate to other phenomena, including in some cases an elaborate attention to feedbacks.

Linked models draw upon established precedents in various disciplines, where the output of one model is the input of another model, although the models are not formally linked and feedback mechanisms may be absent. Examples include some of the US NACC studies.

Linked analyses include some components that are less formally modeled and often qualitative in nature. Many of the integrated regional assessments that result from review and consolidation or review and collaboration are of this nature, including the reports of IPCC Working Group II.

Loosely integrated analyses include some assessments from review and consolidation or review and collaboration and are common in assessments resulting from research agendas and networked activities. Integration may be provided by use of common climate scenarios, but chains of analysis among phenomena may not be well developed.

Topical analyses include components that touch various domains of concern that are only incidentally linked and have little methodological commonalty. Examples include the early IPCC regional assessments (Watson *et al.* 1998) as well as many of the UNFCCC National Communications and most assessments from review and collaboration initiatives.

Finally, *hybrid approaches* combine several of the mentioned degrees of integration with somewhat disparate rigor among components of the assessment. The UNFCCC "National Communications" and "Adaptation Programmes" are examples. Experiments with new approaches and methods provide exciting new directions.

1.3.5 Manuals and guidelines

The technical literature on performing integrated regional assessments continues to evolve. It abounds with guidelines and manuals for undertaking assessments at various scales. Three documents are widely cited: the *International Handbook on Vulnerability and Adaptation Assessments* (Benioff *et al.* 1996); the *UNEP Handbook on Methods for Climate Change Impact Assessment and Adaptation Strategies* (Feenstra *et al.* 1998); and UNFCCC's *Compendium of Decision Tools to Evaluate Strategies for Adaptation to Climate Change* (Smith *et al.* 1999).

The general IPCC *Technical Guidelines for Assessing Climate Change Impacts and Adaptations* (Carter *et al.* 1994) suggest several steps: defining the problem; selecting the method; testing the method/sensitivity; selecting climatic scenarios; assessing biophysical/socio-economic impacts; assessing autonomous adjustments; and evaluating adaptation strategies. Focusing on the regional level, IPCC guidelines for its special report on regional impacts of climate change (Watson *et al.* 1998) provide a broad outline for regional chapters (see Box 1.4).

Box 1.4
Approach used in the IPCC regional assessment

Executive summary
Regional characteristics

- Biogeography (countries, ecosystems, socio-economic activities covered).
- Trends (key socio-economic and resource-use information based on data from existing international sources, compiled by the Technical Support Unit in cooperation with World Resources Institute).
- Major climatic zones.
- Observed trends for temperature and precipitation (based on IPCC, 1996, WG I, Chapter 3, extended and updated to cover a broader number of contiguous regions).
- Summary of available information on projections of future climate (based on Houghton *et al.* 1996, WG I, Chapter 6) and including updated material specific to the region used in regional impact assessments.

Sensitivity, adaptability, and vulnerability

Coverage of topics in this section will vary by region, depending on the most important sectors for each region; however, chapters organize the information into the following categories:

- ecosystems (including biodiversity);
- hydrology/water supply;
- food and fiber for human consumption (agriculture, forestry, and fisheries);
- coastal systems;
- human settlements and urbanization;
- human health; and
- other topics particularly relevant to each region (e.g., energy, transport, and tourism).

Integrated assessment of potential impacts

- Assessments of illustrative case examples related to ecosystems, water supply/basin management, and socio-economic activities.
- Integrated model results, if available.
- Lessons from past fluctuations/variability.

Source: Watson *et al.* (1998).

At other scales, and for other purposes, guideline examples include:

(1) UNFCCC guidelines for national adaptation programs (UNFCCC 2001);
(2) UNFCCC guidelines for national communications (UNFCCC 2003);
(3) United Nations, Economic and Social Commission for Asia and the Pacific's *Impact Assessment and Regional Response Strategies for Climate Change* (1999);
(4) the Government of Northern Ireland Policy Toolkit's *Practical Guide to Impact Assessment* (Northern Ireland, 1995);
(5) the US National Assessment of the Potential Consequences of Climate Variability and Change's *Socio-Economic Scenarios: Guidance Document* (National Assessment Synthesis Team 1998a) and *Scenarios & Data* (National Assessment Synthesis Team 1998b); and
(6) the CIRA framework for integrated regional assessment (Knight *et al.*, Chapter 9, this volume).

Guidance is also provided in literature reviews (Parsons and Fisher-Vanden 1995; Weyant *et al.* 1996) as well as within methodological discussions of specific regional assessments; examples include the MINK study (Rosenberg 1993b) and UK regional, multi-sectoral, and integrated assessments (Holman *et al.* 2005a, 2005b), among many others.

In addition, two specialized journals focus on the assessment process, although rarely on regional assessment of climate change: *Environmental Modeling and Assessment*, which was founded in 1996, and *Integrated Assessment*, begun in 2000. The European Forum on Integrated Environmental Assessment operated from 1997–2002 (Tol and Vellinga 1998); it has been succeeded by The Integrated Assessment Society (TIAS 2004–2009). Perhaps the best way for a new integrated regional assessment process to begin is for participants to become aware of activities similar in scope and scale, and to use information from them in formulating a strategy for the task at hand.

1.4 Questions addressed in this volume

This book is about integrated regional assessment, the nature of this process and its products, examining conceptual, organizational, and management issues. We

explore the nature and accomplishments of IRA through discussion of both fundamental underlying challenges and some specific IRA activities. With regard to the
former, we are well aware that the kinds of differentiating characteristics discussed
above generate many questions about strategy, methods, communications, and use
of IRA processes and products. Selection of specific IRAs for discussion clearly
cannot address all permutations of the hundreds of activities that are occurring
worldwide as scientists and communities try to anticipate, modify, or forestall the
impacts of climate change.

The respective chapters address the following specific issues.

(1) What is IRA and what are its features and variations?
(2) What are the intellectual and practical roots of integrated regional assessment and
     how is IRA similar to, yet different from, its disciplinary forebears and parallel
     methodologies?
(3) What are the tradeoffs and synergies between qualitative and quantitative approaches
     to IRA?
(4) Given that any region is embedded in larger-scale processes (up to global ones) and
     itself subsumes smaller, sub-regional scales, in what ways does scale matter?
(5) Given the magnitude and complexity of domains of knowledge and experience brought
     to IRA, how do we understand, limit, and communicate uncertainty?
(6) If plausible scenarios or storylines result from an IRA, how do we understand their
     implications for vulnerability and resilience in human–environment systems?
(7) How do we move our understanding of the risks and vulnerability from climate change
     to address issues of adaptation and beyond adaptation, to sustainable futures?
(8) If the goal of IRA is to provide policy makers and the public with perspectives on
     choices that will address climate change, how do we engage them in the IRA process,
     not just give them a product?
(9) Given the plethora of schema for IRA, both verbal and graphic, is it possible to suggest
     an exemplar that encapsulates the chain of thinking in an IRA, whether elaborated by
     quantitative or qualitative methods (or both)?
(10) What is the global context of IRA in terms of international and regional scientific
     agendas, and how can IRA both gain from and help to further those agendas?
(11) How can an integrated regional assessment model be used within a wider national
     and multi-national assessment process?
(12) Could an IRA assist in making decisions about managing regional energy resources
     when those decisions will have a strong impact on global climate change trajectories?
(13) Is there a way to develop an IRA that incorporates a rigorously defined approach to
     achieving specific goals for a region?
(14) How can integrated assessment models be developed and deployed for the national
     and local scale?
(15) What is the importance of regional detail as integrated assessment models become
     increasingly important for translating climate trajectories into biophysical and human

impacts, particularly at the global scale, and when those models help to understand important feedbacks between economic growth, technological change, and driving forces of climate dynamics?

(16) What kind of formal, structured entity is required to have a successful multi-national IRA collaboration, or are informal networks feasible?

(17) When a formal organizational structure is indicated, how might it evolve and work to address regional climate change issues?

(18) How can regional IRA activity be facilitated by a government-funded program that encourages local initiatives to address salient issues yet maintains a commitment to common goals of analysis and to fostering the development of science and strategies for doing an IRA?

(19) What have we learned in assessing the IRA as a process and a product, and what are the most significant future challenges?

In addressing these questions, this volume is divided roughly into three parts. The first nine chapters address integrated regional assessment as a concept and process, corresponding to the first nine questions and setting the stage from a methodological perspective. The next six chapters set the global context of integrated regional assessment and illustrate a range of activities that have combined qualitative and quantitative approaches in innovative ways. The final four chapters look at IRA as a process from an implementation perspective and summarize what we have learned and where future frontiers might take IRA. Clearly, all chapters contribute in some measure across these soft boundaries.

References

Abler, D. *et al.* 2002. Climate change, agriculture, and water quality in the Chesapeake Bay region. *Climatic Change*, **55**, 339–359.

ACIA 2005. *Arctic Climate Impact Assessment.* Cambridge: Cambridge University Press.

Aerts, J. and P. Droogers 2004. *Climate Change in Contrasting River Basins: Adaptation Strategies for Water, Food and Environment.* Cambridge, MA: CABI.

Arrhenius, S. 1896. On the influence of carbonic acid in the air upon the temperature of the ground. *Philosophical Magazine and Journal of Science*, **41**, 237–276.

Association of American Geographers GCLP Team 2003. *Global Change and Local Places: Estimating, Understanding and Reducing Greenhouse Gases.* Cambridge: Cambridge University Press.

Australia 2008. *Australia's Settlements and Infrastructure – Impacts of Climate Change.* Canberra: Australian Government, Department of Climate Change. http://www.greenhouse.gov.au/impacts/settlements.html

BALTEX Assessment of Climate Change Author Team 2008. *Assessment of Climate Change for the Baltic Sea Basin.* Berlin: Springer-Verlag.

Benioff, R. and J. Warren 1995. *Steps in Preparing Climate Change Action Plans: a Handbook.* Washington, DC: US Country Studies Program.

Benioff, R. *et al.* 1996. *International Handbook on Vulnerability and Adaptation Assessments.* Boston, MA: Kluwer.

Budyko, M. I. 1974. *Izmeneniya klimata (Climate Change)*. Leningrad: Gidrometeoizdat.

California Climate Change Center 2006. *Our Changing Climate, Assessing Risks to California*. Sacramento, CA: California Energy Commission.

Callendar, G. S. 1949. Can carbon dioxide influence climate? *Weather*, **4**, 310–314.

Carter, T. R. *et al.* 1994. *Technical Guidelines for Assessing Climate Change Impacts and Adaptations*. University College London, UK and the Centre for Global Environmental Research, National Institute for Environmental Studies, Japan.

Cassar, M. 2005. *Climate Change and the Historic Environment*. London: Centre for Sustainable Heritage, University College London with support from English Heritage and the UK Climate Impacts Programme.

Cebon, P. 1998. *Views from the Alps: Regional Perspectives on Climate Change*. Cambridge, MA: MIT Press.

Chen, Yiyu *et al.* 2005. *Measures to Adapt and Mitigate the Effects of Climate and Environment Changes, Assessment of Climate and Environment Changes in China (II)*. Beijing: Science Press.

Clark, W. C. *et al.* 2001a. Acid rain, ozone depletion, and climate change: an historical overview. In *The Social Learning Group, Learning to Manage Global Environmental Risks*. Cambridge, MA: MIT Press, pp. 22–55.

Clark, W. C., J. Jäger, and J. v. Eijndhoven 2001b. Managing global environmental change: an introduction to the volume. In *The Social Learning Group, Learning to Manage Global Environmental Risks*. Cambridge, MA: MIT Press.

Climate Impacts Group 2007. *Preparing for Climate Change: a Guidebook for Local, Regional, and State Governments*. Climate Impacts Group, University of Washington and King County in association with ICLEI – Local Governments for Sustainability. http://www.ecy.wa.gov/climatechange/docs/CCguidebook_localgovs_2007.pdf

Cohen, S. J. 1995. An interdisciplinary assessment of climate change on northern ecosystems: the Mackenzie Basin Impact Study. In D. L. Peterson and D. R. Johnson (eds.), *Human Ecology and Climate Change: People and Resources in the Far North*. Washington, DC: Taylor and Francis, pp. 301–306.

Cohen, S. J. 1996. Integrated regional assessment of global climatic change: lessons from the Mackenzie Basin Impact Study (MBIS). *Global and Planetary Change*, **11**(4), 179–185.

Cohen, S. J. (ed.) 1997. *Mackenzie Basin Impact Study: Final Report*. Downsview, Ontario: Canadian Climate Center, Atmospheric Environment Service, Environment Canada.

Cohen, S. J. *et al.* 1997. Mackenzie Basin Impact Study, Final Report, Summary of Results. http://www.taiga.net/mbis/summary.html

Cohen, S. J. *et al.* 2000. Climate change and resource management in the Columbia River Basin. *Water International*, **25**(2), 253–272.

Crimp, S. *et al.* 2004. *Climate Change in the Cairns and Great Barrier Reef Region: Scope and Focus for an Integrated Assessment*. Canberra: Australian Greenhouse Office.

Dempsey, R. and A. Fisher 2005. Consortium for Atlantic Regional Assessment: information tools for community adaptation to changes in climate or land use. *Risk Analysis*, **25**(6), 1495–1509.

Ding, Yihui *et al.* 2006. National assessment report of climate change (I): climate change in China and its future trend. *Advances in Climate Change Research* 2006, **2**(1), 1–8.

Dowlatabadi, H. 1995. Integrated assessment models of climate change: an incomplete overview. *Energy Policy*, **23**(4/5), 289–296.

Dowlatabadi, H. 2002. Scale and scope in integrated assessment: lessons from ten years with integrated climate assessment model (ICAM). *Integrated Assessment*, **3**(2–3), 122–134.

Dowlatabadi, H., J. Rotmans, and P. Martins 2000. Integrated assessment: a new international journal. *Integrated Assessment*, **1**(1).

East Midlands Sustainable Development Round Table 2000. *The Potential Impacts of Climate Change in the East Midlands*. Entec UK Limited: Warwickshire, England. http://www.ukcip.org.uk/pdfs/east_midlands_tech.pdf

Easterling, W. E. 1997. Why regional studies are needed in the development of full-scale integrated assessment modelling of global change processes. *Global Environmental Change*, **7**(4), 337–356.

Ebi, K. L. *et al.* 2007. *Regional Impacts of Climate Change, Four Case Studies in the United States*. Arlington, VA: Pew Center on Climate Change.

Eckley, N. 2002. Dependable dynamism: lessons for designing scientific assessment processes in consensus negotiations. *Global Environmental Change*, **12**, 15–23.

Eddy, J. A. *et al.* 1991. *Global Change SysTem for Analysis, Research and Training (START): Report of a meeting at Bellagio, December 3–7, 1990*. International Geosphere–Biosphere Programme Report No. 15. Boulder, CO: University Center for Atmospheric Research (UCAR) Office for Interdisciplinary Earth Studies.

European Commission 2007. *Green Paper from the Commission to the Council, the European Parliament, the European Economic and Social Committee and the Committee of the Regions – Adapting to climate change in Europe – options for EU action.* {SEC(2007) 849}

European Environmental Agency 2005. *Vulnerability and Adaptation to Climate Change in Europe*. Copenhagen: EEA.

Farrell, A. E. and J. Jäger (eds.) 2005. *Assessments of Regional and Global Environmental Risks*. Washington, DC, Resources for the Future.

Feenstra, J. F., I. Burton, J. B. Smith, and R. S. J. Tol (eds.) 1998. *UNEP Handbook on Methods for Climate Change Impact Assessment and Adaptation Strategies*. Amsterdam: Vrije Universiteit.

Ferreira, L. G. R., T. Gaiser, and K. Stahr 2000. *The WAVES Program: Conception, Results and Perspectives, German–Brazilian Workshop on Neotropical Ecosystems – Achievements and Prospects of Cooperative Research*. Hamburg: Overviews on Bilateral Cooperation in Environmental Research and Development.

Fisher, A. *et al.* 2006. *Center for Atlantic Regional Asssessment*. University Park, PA. http://www.cara.psu.edu/

Fleming, J. R. 1998. *Historical Perspectives on Climate Change*. Oxford: Oxford University Press.

Fourier, J.-B.-J. 1824. Remarques generales sur le temperatures du globe terrestre et des espaces planetaires. *Annales de Chimie et de Physique*, Ser. 2, **27**, 136–167.

France 2006. *Impacts Potentiels du Changement Climatique en France au XXIe Siècle*. Paris: Premier Ministre et le Ministère de l'Aménagement du Territoire et de l'Environment.

Frumhoff, P. C. *et al.* 2007. *Confronting Climate Change in the U.S. Northeast*. Cambridge, MA: Union of Concerned Scientists.

Gaiser, T. *et al.* (eds.) 2003. *Global Change and Regional Impacts*. Berlin: Springer-Verlag.

GDRC 2004. *Different Types of Impact Assessments*. Global Development Research Center. http://www.gdrc.org/uem/eia/impactassess.html

Glacken, C. 1967. *Traces on the Rhodian Shore: Nature and Culture in Western Thought from Ancient Times to the End of the Eighteenth Century*. Berkeley, CA: University of California Press.

Haas, P. M. 1992. Introduction: epistemic communities and international policy coordination. *International Organization*, **46**(1), 1–35.

Haas, P. M. and D. McCabe 2001. Amplifiers or dampeners: international institutions and social learning in the management of global environmental risks. In *The Social Learning Group, Learning to Manage Global Environmental Risks*. Cambridge, MA: MIT Press.

Holman, I. P. *et al.* 2002. *REGIS – Regional Climate Change Impact Response Studies in East Anglia and North West England*. London: Department for Environment, Food & Rural Affairs.

Holman, I. P. *et al.* 2005a. A regional, multi-sectoral and integrated assessment of the impacts of climate and socio-economic change in the UK: Part I. Methodology. *Climatic Change*, **71**, 9–41.

Holman, I. P. *et al.* 2005b. A regional, multi-sectoral and integrated assessment of the impacts of climate and socio-economic change in the UK: Part II: Results. *Climatic Change*, **71**, 43–73.

Houghton, J. T. *et al.* (eds.) 1996. *Contribution of Working Group I to the Second Assessment of the Intergovernmental Panel on Climate Change*. Cambridge: Cambridge University Press.

Jacobson, H. K. and M. F. Price 1991. *A Framework for Research on the Human Dimensions of Global Environmental Change*. Paris: ISSC Standing Committee on the Human Dimensions of Global Change, ISSC with the cooperation of UNESCO.

Jasanoff, S. and M. L. Martello (eds.) 2004. *Earthly Politics: Local and Global in Environmental Governance*. Cambridge, MA: MIT Press.

Kates, R. W., J. H. Ausubel, and M. Berberian 1985. *Climate Impact Assessment*. Scientific Committee on Problems of the Environment (SCOPE), vol. 27. Chichester: John Wiley.

Katzenberger, J. 2006. *Climate Change and Aspen: an Assessment of Impacts and Potential Responses*. Aspen, CO: Aspen Global Change Institute. http://aspenglobalwarming.com/

Keeling, C. D. *et al.* 1976. Atmospheric carbon dioxide variations at Mauna Loa Observatory, Hawaii. *Tellus*, **28**(6), 538–551.

Kirshen, P. *et al.* 2004. *Climate's Long-term Impacts on Metro Boston (CLIMB), Final Report*. Boston, MA: Tufts University. http://www.tufts.edu/tie/pdf/CLIMBFV1–8_10pdf.pdf

Kirshen, P., M. Ruth, and W. Anderson 2006. Climate's long-term impacts on urban infrastructures and services: the case study of Boston. In R. Matthias, K. Donaghy, and P. Kirshen (eds.), *Regional Climate Change And Variability: Impacts And Responses*. Northampton, MA: Edward Elgar, pp. 190–252.

Knight, C. G. 2001. Regional assessment. In A. S. Goudie (ed.), *Encyclopedia of Global Change*, vol. 2. New York: Oxford University Press, pp. 304–308.

Knight, C. G., I. Raev, and M. P. Staneva (eds.) 2004. *Drought in Bulgaria, a Contemporary Analog for Climate Change*. Aldershot: Ashgate Press.

Kuhn, W. *et al.* 1992. *Pathways of Understanding: the Interactions of Humanity and Global Environmental Change*. Consortium for International Earth Science Information Network. University Center, MI.

Leary, N., J. Kulkarni, and C. Seipt 2007. *Assessment of Impacts Adaptation to Climate Change*. Washington, DC: International START Secretariat.

Lemmen, D. S. *et al.* (eds.) 2008. *From Impacts to Adaptation: Canada in a Changing Climate 2007*. Ottawa: Government of Canada.

London Climate Change Partnership 2002. *London's Warming: the Impacts of Climate Change on London.* http://www.london.gov.uk/approot/gla/environment/londons_ warming_tech_rpt_all.pdf

Lorenzoni, I. *et al.* 2000a. A co-evolutionary approach to climate impact assessment: Part I. Integrating socio-economic and climate change scenarios. *Global Environmental Change*, **10**, 57–68.

Lorenzoni, I. *et al.* 2000b. A co-evolutionary approach to climate impact assessment: Part II. A scenario-based case study in East Anglia. *Global Environmental Change*, **10**(2), 145–155.

Low, P. S. (ed.) 2005. *Climate Change and Africa.* Cambridge: Cambridge University Press.

Malhi, Y. *et al.* 2008. Climate change, deforestation, and the fate of the Amazon. *Science*, **319**, 169–172.

Manne, A., R. Mendelsohn, and R. G. Richels 1995. MERGE: a model for evaluating regional and global effects of GHG reduction policies. *Energy Policy*, **23**, 17.

Marsh, G. P. 1864. *Man and Nature or, Physical Geography as Modified by Human Action.* New York: Scribner.

Mayer, N. and W. Avis (eds.) 1998. *The Canada Country Study: Climate Impacts and Adaptation, National Cross-Cutting Issues Volume.* Toronto: Environment Canada.

McCarthy, J. J. *et al.* 2001. *Contribution of Working Group II to the Third Assessment Report of the Intergovernmental Panel on Climate Change (IPCC).* Cambridge: Cambridge University Press.

McCracken, M. 2001. Prediction versus projection – forecast versus possibility. *WeatherZine*, **26**. http://sciencepolicy.colorado.edu/zine/archives/1–29/26/ guest.html

Mitchell, R. B. *et al.* (eds.) 2006a. *Global Environmental Assessments: Information and Influence.* Cambridge, MA: MIT Press.

Mitchell, R. B., W. C. Clark, and D. W. Cash 2006b. Information and influence. In R. B. Mitchell *et al.* (eds.), *Global Environmental Assessments: Information and Influence.* Cambridge, MA: MIT Press.

Mortsch, L. D. and B. N. Mills (eds.) 1996. *Great Lakes–St. Lawrence Basin Project Progress Report #1: Adapting to the Impacts of Climate Change and Variability.* Burlington: Environment Canada.

Mortsch, L. and F. Quinn 1998. Great Lakes–St. Lawrence Basin Project: what have we learned. In L. D. Mortsch *et al.* (eds.), *Adapting to Climate Change and Variability in the Great Lakes–St. Lawrence Basin, Proceedings of a Binational Symposium.* Burlington: Environmental Adaptation Research Group, Environment Canada.

National Assessment Synthesis Team 1998a. *US National Assessment of the Potential Consequences of Climate Variability and Change, Socio-Economic Scenarios: Guidance Document.* Washington, DC: US Global Change Research Program.

National Assessment Synthesis Team 1998b. *US National Assessment of the Potential Consequences of Climate Variability and Change, Scenarios & Data.* Washington, DC: US Global Change Research Program.

National Assessment Synthesis Team 2000. *The Potential Consequences of Climate Variability and Change.* Washington, DC: U.S. Global Change Research Program.

National Park Service 2006. *Climate Change in National Parks.* Brochure.

Natural Resources Canada 2007. *Reducing Canada's Vulnerability to Climate Change.* Ottawa: Natural Resources Canada.

Nordhaus, W. D. and J. Boyer 2000. *Warming the World: Economic Models of Global Warming.* Cambridge, MA: MIT Press.

North, G. R., J. Schmandt, and J. Clarkson 1995. *The Impact of Global Warming on Texas: a Report of the Task Force on Climate Change in Texas.* Austin, TX: University of Texas Press.

North West Regional Assembly 2003. *Spatial Implications of Climate Change for the North West.* Manchester: Centre for Urban & Regional Ecology, School of Planning & Landscape, University of Manchester, and Tyndall Center North.

Northern Ireland 1995. *The Policy Toolkit.* Belfast: Office of the First Minister and Deputy Minister. http://www.ofmdfmni.gov.uk/policy-toolkit

O'Brien, K. (ed.) 2000. *Developing Strategies for Climate Change: the UNEP Country Studies on Climate Change Impacts and Adaptations Assessment.* Oslo: Center for International Climate and Environmental Research.

Parry, M. *et al.* 2007. *Climate Change 2007 – Impacts, Adaptation and Vulnerability, Contribution of Working Group II to the Fourth Assessment Report of the IPCC.* Cambridge: Cambridge University Press.

Parson, E. A. 2003. *Protecting the Ozone Layer: Science and Strategy.* Oxford: Oxford University Press.

Parson, E. A. and K. Fisher-Vanden 1995. *Searching for Integrated Assessment: a Preliminary Investigation of Methods, Models, and Projects in the Integrated Assessment of Global Climatic Change.* University Center, MI: Consortium for International Earth Science Information Network (CIESIN).

Parson, E. A. and K. Fisher-Vanden 1997. Integrated assessment models of global climate change. *Annual Review of Energy and the Environment,* **22**, 589–628.

Parthier, B. and D. Simon (eds.) 2000. *Climate Impact Research: Why, How and When: Joint International Symposium.* Berlin: Akademie-Verlag.

Peterson, D. L. and D. R. Johnson (eds.) 1995. *Human Ecology and Climate Change: People and Resources in the Far North.* Washington, DC: Taylor and Francis.

Pilkey, O. H. and L. Pilkey-Jarvis 2007. *Useless Arithmetic, Why Environmental Scientists Can't Predict the Future.* New York: Columbia University Press.

Plass, G. N. 1956. The carbon dioxide theory of climatic change. *Tellus,* **8**, 140–154.

Preston, B. L. and R. N. Jones 2006. *Climate Change Impacts on Australia and the Benefits of Early Action to Reduce Global Greenhouse Gas Emissions.* Aspendale: CSIRO.

Rosenberg, N. J. 1993a. *Towards an Integrated Impact Assessment of Climate Change: the MINK Study.* Dordrecht: Kluwer Academic Publishers.

Rosenberg, N. J. 1993b. A methodology called "MINK" for study of climate change impacts and responses on the regional scale. *Climatic Change,* **24**(1–2), 1–6.

Rosenberg, N. J. and J. A. Edmonds (eds.) 2005a. *Climate Change Impacts for the Conterminous USA.* Dordrecht: Springer [reprint of *Climatic Change,* **69**(1)].

Rosenberg, N. J. and J. A. Edmonds 2005b. Climate change impacts for the conterminous USA: an integrated assessment: from MINK to the "Lower 48." *Climatic Change,* **69**(1), 1–6.

Rotmans, J. and B. de Vries 1997. *Perspectives on Global Change: the Targets Approach.* Cambridge: Cambridge University Press.

Rotmans, J., M. Hulme, and T. E. Downing 1994. Climate change implications for Europe: an application of the ESCAPE Model. *Global Environmental Change,* **4**(2), 97–124.

Sabatier, P. A. and H. C. Jenkins-Smith 1999. The advocacy coalition framework: an assessment. In P. A. Sabatier (ed.), *Theories of the Policy Process.* Boulder, CO: Westview Press.

Shackley, S. *et al.* 1998. *Changing by Degrees: the Impacts of Climate Change in the North West of England. PT1: Summary.* http://www.snw.org.uk/pdf_files/summary%20report.pdf

Shukla, P. R. *et al.* (eds.) 2003. *Climate Change and India: Vulnerability Assessment and Adaptation*. Hyderabad: Universities Press (India) Pvt. Ltd.

Smith, J. *et al.* 1999. *Compendium of Decision Tools to Evaluate Strategies for Adaptation to Climate Change*. Bonn: UNFCCC Secretariat.

Smith, J. B. and J. K. Lazo 2001. A summary of climate change impacts from the U.S. Country Studies Program. *Climatic Change*, **50**, 1–29.

Study of Critical Environmental Problems 1970. *Man's Impact on the Global Environment: Assessment and Recommendations for Action*. Cambridge, MA: MIT Press.

Study of Man's Impact on Climate 1971. *Inadvertent Climate Modification*. Cambridge, MA: MIT Press.

Swiss Agency for the Environment, Forests and Landscape 2002. *Climate in Human Hands: New Findings and Perspectives*. Bern: SAEFL. http://www.bafu.admin.ch/php/modules/shop/files/pdf/phpcC0in1.pdf

Thomas, W. L. (ed.) 1956. *Man's Role in Changing the Face of the Earth*. Chicago, IL: University of Chicago Press.

TIAS 2004–2009. The Integrated Assessment Society. http://www.tias.uni-osnabrueck.de

Tol, R. S. J. and P. Vellinga 1998. The European forum on integrated environmental assessment. *Environmental Modeling and Assessment*, **3**, 181–191.

Torrance, W. E. F. 2006. Science or salience: building an agenda for climate change. In R. B. Mitchell *et al.* (eds.), *Global Environmental Assessments: Information and Influence*. Cambridge, MA: MIT Press.

Tuinstra, W., L. Hordijk, and M. Amann 1999. Using computer models in international negotiations: the case of acidification in Europe. *Environment*, **41**(9), 33–42.

Turner, B. L., II *et al.* (eds.) 1990a. *The Earth as Transformed by Human Action: Global and Regional Changes in the Biosphere over the Past 300 Years*. Cambridge: Cambridge University Press.

Turner, B. L., II *et al.* 1990b. Two types of global environmental change: definitional and spatial-scale issues in their human dimensions. *Global Environmental Change*, **1**(1), 14–22.

Tyndall, J. 1862. On radiation through the Earth's atmosphere. *Philosophical Magazine*, **4**(25), 200–206.

UK Climate Impacts Programme 2008. http://www.ukcip.org.uk

UNFCCC 2001. *Guidelines for the Preparation of National Adaptation Programmes of Action*. Decision 28/CP.7, FCCC/CP/2001/13/Add.4

UNFCCC 2003. *Reporting on Climate Change: User Manual for the Guidelines on National Communications from Non-Annex I Parties*. http://unfccc.int/files/essential_background/application/pdf/userman_nc.pdf

Union of Concerned Scientists 2006a. *Climate Change in the U.S. Northeast*. Cambridge, MA: Union of Concerned Scientists.

Union of Concerned Scientists 2006b. *Our Changing Climate, Assessing the Risks to California*. Cambridge, MA: Union of Concerned Scientists.

Union of Concerned Scientists 2008. *Climate Change in Pennsylvania*. Cambridge, MA: Union of Concerned Scientists.

United Nations, Economic and Social Commission for Asia and the Pacific 1999. *Impact Assessment and Regional Response Strategies for Climate Change*. United Nations ST/ESCAP/2030.

Warrick, R. A. *et al.* 1999. VandaClim: a training tool for climate change vulnerability and adaptation assessment. In C. Kaluwin and J. E. Hay (eds.), *Climate Change and Sea Level Rise in the South Pacific Region. Proceedings of the Third SPREP Meeting*,

New Caledonia, August, 1997. Apia, Samoa: South Pacific Regional Environment Programme, pp. 147–156.

Watson, R. T., M. C. Zinyowera, and R. H. Moss 1998. *The Regional Impacts of Climate Change, an Assessment of Vulnerability.* Cambridge: Cambridge University Press.

Weart, S. 2003. *The Discovery of Global Warming.* Harvard, MA: Harvard University Press.

Weyant, J. *et al.* 1996. Integrated assessment of climate change: an overview and comparison of approaches and results. In J. P. Bruce, H. Lee, and E. F. Haites (eds.), *Climate Change 1995: Economic and Social Dimensions of Climate Change.* Cambridge: Cambridge University Press.

Yarnal, B. 1998. Integrated regional assessment and climate change impacts in river basins. *Climate Research,* **11**, 65–74.

Yohe, G. W. 2000. Integrated assessment of climate change. In B. Parthier and D. Simon (eds.), *Climate Impact Research: Why, How and When? Joint International Symposium,* Berlin, October 28–29, 1997. Berlin: Akademie Verlag, pp. 135–163.

2

Integrated regional assessment: overview and framework

BRENT YARNAL

2.1 Introduction

The original emphasis of global environmental change research was understanding global-scale aspects of environmental change. For more than a decade, however, there has been a growing interest in the regional[1] implications of global environmental change. For example, the United States Global Change Research Program (USGCRP, 1998) challenged American researchers to:

- Develop regional-scale estimates of the timing and magnitude of global environmental change.
- Advance regional analyses of the environmental and socio-economic consequences of global environmental change in relation to other stresses.
- Conduct integrated assessments of the environmental and societal implications of global environmental change.

The motivation for regionally specific global environmental change studies comes from two related quarters (Yarnal 1998). First, society in general and decision makers in particular are demanding more socially relevant science (e.g., Rubin *et al.* 1991–92, Yarnal 1996). Second, early climate change impacts studies (e.g., Titus *et al.* 1991) suggested that impacts could be severe in some regions, thus raising the concerns of decision makers representing regional constituencies. Although subsequent research on the physical manifestations of climate change (Houghton *et al.* 1996) suggested that the early scenarios probably exaggerated these impacts, the scientific community now understands that climate change and other global environmental changes are not just global in scope, but have significant regional dimensions. The publications of Watson and colleagues (Watson *et al.* 1998), the

---

[1] Although the geography literature argues that there are significant differences among these terms, the global environmental change literature often uses the terms region, locale, area, and place interchangeably. For simplicity, I will refer to these terms collectively as *region*.

United States National Assessment of the Potential Consequences of Climate Variability and Change (National Assessment Synthesis Team 2000), and the regional chapters in the Third Assessment Report of the Intergovernmental Panel on Climate Change (IPCC; McCarthy *et al.* 2001), which formally assessed the regional impacts of climate change, underscored this significance. *Climate Change 2007* continued the tradition of regional assessments (IPCC 2007).

Integrated Regional Assessment (IRA) is a significant research area appraising the regional details of global environmental change. One goal of this chapter is to introduce and explain this approach to global environmental change research. To reach this goal, it is necessary first to define the older, more inclusive model of integrated assessment of global environmental change and then to show how Integrated Regional Assessment is the regional form of that model.[2]

A second goal of the paper is to compare IRA with the themes of other areas of inquiry and to determine whether this seemingly new field is rediscovering or building upon a scholarly landscape explored many decades earlier. Accordingly, the chapter takes a brief look at the some of the fundamental ideas of geography, regional science, environmental impact assessment, and spatial analysis. The chapter then identifies themes shared by these fields and IRA, but also explores new themes important to IRA. The chapter concludes that the idea of treating regions as holistic, integrated entities is not new, but that the combination of IRA's central themes of change, scale interactions, stakeholder interaction, vulnerability, and uncertainty makes IRA a fundamentally different field.

2.2 Introduction to integrated regional assessment

2.2.1 Integrated assessment

The integrated assessment of global environmental change – or simply Integrated Assessment (IA) – was practiced in the 1970s and 1980s,[3] but its conceptualization was fuzzy and it did not become an important research focus until the 1990s (Rotmans and Van Asselt 1996, Weyant *et al.* 1996, Rothman and Robinson 1997). With this more prominent position, it became necessary to define IA more clearly. For example, Jepma and Munasinghe (1998: 153–154) defined IA as "a conceptual framework for addressing complex problems by combining a broad

---

[2] The discussion is a significant modification of Yarnal (1998).

[3] Integrated assessment of not necessarily global environmental issues, such as acid deposition, existed long before the global environmental change community assumed the term "integrated assessment" in the 1990s (Toth and Hizsnyik 1998). The Global Environmental Assessment Project at Harvard University (http://www.ksg.harvard.edu/gea/) reviewed these assessments in an effort to shape an understanding of the relationships among science, assessment, policy, and management in the social responses to global environmental change.

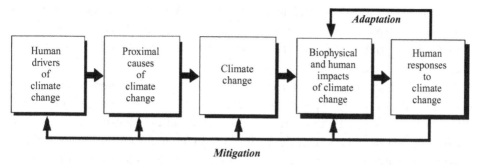

Figure 2.1. Simplified framework of an integrated assessment of climate change (from Yarnal 1998; used by permission of *Climate Research*).

and transdisciplinary range of knowledge in a systematic manner." As part of their definition of integrated assessment, Rotmans and Van Asselt (1996) proposed a framework based on a cyclical process involving communication and consensus building among scientific, policy, and societal stakeholders. From these and other definitions that have been put forward (e.g., Parson 1996a), integrated assessment is a process characterized by its comprehensive scope and the integration of the many disciplines from which it draws; it differs from most disciplinary research in that it aims to inform policy and decision making, rather than to advance knowledge for its intrinsic value (Weyant *et al.* 1996).

The goals of IA follow directly from these ideas: to provide a systematic conceptual framework in which to structure present knowledge about global environmental change, to assess potential responses to global environmental change, and to determine the importance of global environmental change relative to other matters of human concern (Weyant *et al.* 1996, Carter *et al.* 2007). A very simple integrated assessment framework is presented in Figure 2.1.

An important mode of integrated assessment activity is the formal computer model called an Integrated Assessment Model (IAM). There are several advantages to modeling the components of the global environmental change problem, including the ability to analyze dynamic behavior of complex systems, show interrelationships among various issues, and treat uncertainties explicitly (Rotmans and Van Asselt 1996). Furthermore, IAMs promote communication among participants from diverse disciplines by imposing common language, data standards, and research protocols. Nonetheless, IAMs have weaknesses. They involve more precise representation of model components than current knowledge warrants (Weyant *et al.* 1996). In addition, mathematical expressions cannot capture the complexities, subtleties, and ambiguities presented by a process comprised of complex human–environment interactions (Rotmans and Van Asselt 1996). For a summary of IAMs of one important global environmental change – climate change – see Weyant and colleagues (1996, especially their table 10.1); for an evolutionary typology of IAMs of climate change, see Schneider (1997); and for an updated history of IAMs of climate change (including what they call a "new wave" in IAMs after 2001), see Schneider and Lane (2005).

Integrated assessments do not need to be based on formal mathematical models, however. Non-modeling approaches are especially relevant to research involving the right-hand side of Figure 2.1, where societal and decision-making concerns are found. Non-model assessments range from individual authors with interdisciplinary and policy competence (e.g., Smil 1993), to collaborative interdisciplinary research teams sharing empirical knowledge (e.g., Cohen 1995), to expert panels contributing disciplinary expertise (Parson 1997). In addition, approaches that mix model and empirical methodologies have proved to be effective (e.g., Rosenberg 1993).[4]

Regardless of the approach, practitioners of integrated assessment have identified many areas of concern (e.g., Morgan and Dowlatabadi 1996, Parson 1996a, Risbey *et al.* 1996, Rotmans and Van Asselt 1996; Toth 2003). Parson (1996b) distilled these worries to three fundamental dilemmas confronting integrated assessment. First, there is a tension between the traditions and standards of disciplines and the interdisciplinary synthesis and needs of IA. Second, IAs require long-term projections of society and the environment, but the social sciences and biophysical sciences have limited ability to predict the future. Third, given the uncertainty and bias that cloud the scientific and political enterprises, there is no clear way to bridge the gap between scientific expertise and policy making.

One dilemma not raised by Parson (1996b), but emphasized by others (e.g., Risbey *et al.* 1996) and specific to models, is that most IAMs – and therefore many integrated assessments – suffer from problems of spatial resolution. Specifically, IAMs tend to address large geographical areas and aggregated socio-economic factors. Neither current global models of environmental change nor IAMs have the spatial resolution necessary to deal with regionally specific impacts of global environmental change or mitigation measures. The climate change community has been confronting the problem of translating global data to regional scales for more than two decades (Kattenberg *et al.* 1996). Integrated assessment modelers, however, have put increased effort into understanding the regional implications of global environmental change (Leemans, Chapter 15, this volume).

There were a few geographically focused studies of the impacts of global environmental change in the late 1980s and early 1990s. Examples included the studies of agricultural regions susceptible to climate variation and change conducted by

---

[4] Many early projects would be called integrated assessments today, such as the work of Rosenberg (1993), but were not because they preceded the adoption of the term by the global environmental change community.

Table 2.1. *The major tasks for integrated regional assessments*
(from Pulwarty 2000).

(1) Characterizing the current state of knowledge of climate variability on all timescales, social and environmental impacts, and levels of criticality within a region or sectors within a region.

(2) Assessing vulnerability to climate on the seasonal and decadal to centennial scales including how social transformations such as demographic changes influence social and environmental vulnerability to climate risks including abrupt changes.

(3) Developing pilot projects and prototypes for demonstration of opportunities, acceptability, and use of climate information and to enhance collaboration among researchers, decision makers, and the public.

(4) Improving decision-support dialogues and developing awareness with respect to climate impacts on regional system outputs, capacity for action, and management, e.g., water, agriculture, fisheries, energy, human health, and private sector investments for enhanced economic productivity.

(5) Iteratively refining mechanisms of interaction and learning among the research and programmatic communities, through clearer problem definitions, understanding of public goals and expectations, and capacity building to realize the benefits afforded by developments in climate research, products, and service.

Parry *et al.* (1988) and Rosenberg (1993) and the analyses of potential influence of climate change on river basins reported by Riebsame *et al.* (1995) and Strzepek and Kaczmarek (1996). These and other studies alerted the global environmental change community and decision makers in climatically sensitive regions to the importance of assessing each region's vulnerability and adaptability to global environmental change. From this alert, the idea of integrated regional assessment emerged.

2.2.1.1 Integrated regional assessment

Integrated regional assessment of global environmental change (or simply integrated regional assessment) is in most ways the regional equivalent of the larger-scale integrated assessments. Integrated regional assessment is an interdisciplinary, iterative process that involves scientific researchers, policy makers, and societal stakeholders. Its aim is to promote a better understanding of – and more-informed decisions on – how regions contribute to and respond to global environmental change. According to Pulwarty (2000), the main tasks of integrated regional assessment include understanding impacts, characterizing vulnerability, and spotting opportunities through close work with regional stakeholders and decision makers (Table 2.1). This section addresses some of the general issues concerning integrated regional assessment.

Integrated assessment modelers have yet to develop the capacity to translate from the global to the regional scales with precision.[5] Yet, to capture the unique characteristics of a region and to see how those traits result in varying interactions between the region and the larger world, integrated regional assessments need to be as specific as possible. Most scientists working on IRAs, however, either have been unable to quantify and parameterize their data because of the inherent difficulty in generalizing place-based data or have been unwilling to do so because they believe it compromises the uniqueness of the places they are studying. Consequently, unlike integrated assessment (which relies heavily on IAMs), integrated regional assessment has not usually involved formal mathematical modeling at this point in its development (some notable efforts in this direction are described in Chapters 11–15 of this volume). The principal approaches in integrated regional assessment have entailed collaborative interdisciplinary research teams sharing empirical, disciplinary knowledge of regions or have used mixed-method and empirical techniques. An example of a mixed-method approach was the MINK project (Rosenberg 1993), which combined regional economic modeling with empirical analyses and analog climate scenarios to estimate the impacts of climate change on the agricultural heartland of the United States.

No matter what the approach, and unlike larger-scale integrated assessments, regional teams have been more willing to limit the overall scope of the assessment. To help achieve the necessary geographic detail and integration and to make the analysis more manageable, they have limited their integrated regional assessments by sector (e.g., energy production) or by issue (e.g., vulnerability to drought). In addition, the assessments often have not been end to end, but have addressed various components of the global environmental change problem (e.g., specific boxes and arrows in Figure 2.1).

Although project design can, and often must, constrain the range of any one regional assessment project,[6] it is possible to accomplish full integration by conducting multiple, parallel projects, with each project focusing on one aspect of the region's global environmental change problems. In this approach, some detailed assessments have focused on regional issues contained in the right-hand side of Figure 2.1 (i.e., consequences of global environmental change),[7] while others have

---

[5] Applications of agent-based models have advanced the integration of IAMs and regional stakeholder interactions (Pahl-Wostl 2005).

[6] Integrated regional assessments are often restricted because of lack of funds or boundaries on the scope of research (e.g., when political considerations put mitigation of greenhouse-gas emissions "off limits"). More often, however, limitations result from the magnitude of work involved in comprehensively assessing all elements of the global environmental change problem in a region.

[7] The Fourth Assessment Report of the IPCC focused on climate change impact, adaptation, and vulnerability (CCIAV) assessment (Carter *et al.* 2007). In addition, Füssel and Klein (2006) covered the evolution of climate change vulnerability assessment and developed a framework specifically aimed at that component of integrated regional assessment.

concentrated on issues germane to the figure's left-hand side (i.e., causes of global environmental change). Still others give attention to the middle box (e.g., downscaling the effects of global environmental change to regions). In due course, such multiple studies can be synthesized into a single end-to-end appraisal of the region's contributions and responses to global environmental change.[8] This long-range strategy promotes detailed disciplinary and interdisciplinary knowledge not possible in larger-scale integrated assessments, thus contributing not only to regional integration, but also to the intellectual underpinnings of the greater enterprise.

Integrated regional assessments concentrating on the right-hand side of Figure 2.1 spotlight the impacts of global environmental change. Integrated assessments of global environmental change impacts are an outgrowth of the seminal work of Kates (1985) on climate impact assessment. Box 2.1 briefly presents one example of an integrated regional assessment for illustrative purposes, the Mackenzie Basin Impact Study. This volume presents many other examples in more detail.

In summary, integrated regional assessment treats regions as holistic, integrated entities and thereby provides scientists and stakeholders with a framework that helps them comprehend the totality of regional variations in the causes and consequences of global environmental change. Understanding the regional implications of global environmental change is important because it is at the regional level that global environmental change mitigation must be practiced and that human impacts will be felt. However, the paradigm is immature and needs further elaboration. Part of that elaboration includes a deeper understanding of the intellectual roots of integrated regional assessment. The following section briefly explores that ancestry.

2.3 The roots of integrated regional assessment

2.3.1 Geography

Integrated regional assessment is the outcome of diverse intellectual underpinnings, but none are more important than the discipline of geography. Geography has two grand themes, both of which contribute to integrated regional assessment. The first theme concerns space, place, and region. The second explores the interactions between humans and their environment (e.g., Turner 2002).

Dominant perspectives within geography's spatial theme have changed over time. The first important perspective was the German School of the early twentieth century, which focused on regions and landscape (e.g., Hettner 1928, Schlüter 1928, Spethmann 1928; see Elkins 1989). In essence, the German geographers thought

---

[8] Iterative approaches have also been developed in contrast to end-to-end approaches. See Chapter 18.

Box 2.1

Example Integrated Regional Assessment: The Mackenzie Basin Impact Study

The six-year Mackenzie Basin Impact Study (MBIS), initiated by Environment Canada in 1990, is an early example of an integrated regional assessment (Cohen 1995, Cohen 1997a). Not only were impacts, vulnerability, and adaptations part of the assessment, but stakeholders were integral to the process and included representatives from aboriginal groups, industry, colleges and institutes, and municipal, territorial, provincial, and federal governments (Cohen 1997b). A working committee composed of representatives from aboriginal organizations, the private sector, and governments steered the interdisciplinary effort. Especially important was the systematic approach used by MBIS. This approach incorporated qualitative information into a quantitative analysis and explicitly built a bridge between science and policy (Yin & Cohen 1994).

Many individual impact assessments (e.g., Lonergan *et al.* 1993) were developed to answer the project's overarching, integrative questions (Yin & Cohen 1994):

- What are the implications of climate change for achieving regional resource development objectives? Should governments within the Mackenzie Basin alter their current resource-use policies or plans in anticipation of climate change?
- Does climate change increase land-use conflicts among different regional economic and social sectors? If potential conflicts are identified, how serious might they be and how could compromises be reached?
- What are the possible tradeoffs for various public responses to climate change? For example, should regional parks and forests be managed to anticipate change or to preserve existing conditions?

Specific regional impacts scenarios were produced for: basin runoff quantities and timing; lake levels; ice formation, break up, and location; permafrost thaw and accompanying landslides; peatland migration; forest growth, fire frequency, and forestry potential; caribou; wheat production; tourism; oil and gas production costs; transportation systems and other infrastructure; and economic development patterns.

MBIS concluded that climate change presents significant threats and opportunities for each region. The study stressed that climate change is not just about climate, energy, and greenhouse gases – it is also about the effects of climate change on a region's ecosystems, resource systems, and human systems. The study also concluded that the consequences of climate change would be unique to each region, which in turn means that a wide range of regional stakeholders should be part of national and international response strategies to address the impacts of climate change.

that the main goals of the discipline were the description and understanding of regions. By mid-century, the intellectual center of geography migrated to America, where the Midwestern Tradition, led by the ideas of Hartshorne (1939 and 1959; see Entrikin and Brunn 1989), and the Berkeley School, dominated by Sauer (e.g., 1941), competed. The Midwestern Tradition elaborated on the spatial theme, while the Berkeley School paid more attention to human–environment studies. Although both camps had a profound effect on the course of geography, the Midwestern Tradition proved to have the greater influence. Of importance to integrated regional assessment, the Midwestern Tradition emphasized the synthesis of diverse knowledge in understanding how regions operate. Perhaps in reaction to the view of geography as regional holism, a new perspective emerged in the mid-1950s (e.g., Abler *et al.* 1971) that stressed a quantitative, reductionist, and positivist approach to spatial studies and that sounded the apparent death knell for the earlier regional perspective of the Midwestern Tradition (Smith 1989; Richardson 1992). Countering this "quantitative revolution," a qualitative, social-theoretical perspective arose – led first by neomarxist ideas (e.g., Harvey 1969) and followed by postmodern geographies (e.g., Soja 1989) – and was firmly in place by the mid-1980s. Led by these alternative viewpoints, which stress the need for contextualizing theory, contemporary geography is returning to a more holistic regional approach that draws together the quantitative and qualitative perspectives of the last few decades, but is reminiscent of the Midwestern Tradition (Abler *et al.* 1992).

Differences among the spatial perspectives have been the focus of bitter internecine wars that at times have exhausted geography intellectually and slowed the discipline's progress. Two points emerge from these battles. First, no matter what the perspective, geography concentrates on the region as the primary unit of inquiry. Second, parallel tensions permeate the regional studies of each perspective. One is the tension between ideographic (the specific, the unique) and nomothetic (the general, the universal). Geographers have spent much time developing methodologies to differentiate regions from one another, while at the same time devising theories that explain what all regions have in common. Another important tension exists between pattern and process. Much geographic study focuses on identifying landscape patterns with no regard for the processes that led to those patterns. At the other extreme, much other geography concentrates on theorizing the biophysical, socio-economic, and historical processes that create patterns, but often ignores patterns in the real world. A third tension pits narrow, focused studies against broad, all-inclusive research. For instance, not all geographers agree on the value of case studies for understanding the grand design of regions. Although these tensions show that geographers continue to debate fundamental questions on how to study and understand regions, they nonetheless demonstrate that geographers spend much of their time thinking about regions.

The other pervasive theme in geography is the interactions of humans with their environment (e.g., Knight 1992). Although their primary focus was on space, place, and region, the German School and the ensuing Midwestern Tradition paid careful attention to the physical environment as the backdrop for human activities. Some adherents embraced the ideas of environmental determinism, which builds on Darwinian notions of natural selection and states that much of human nature and behavior results from the environmental context (Huntington 1925). Most geographers, however, rejected environmental determinism and distanced themselves from it partly because they realized that "people are not merely passive actors on the natural stage" (Hart 1982: 7), but mostly because of its racist overtones. The extreme reaction to environmental determinism probably set human–environment studies back decades. The Berkeley School, however, continued with the human–environment theme by drawing upon Marsh (1864) and elaborating on his idea that human agency was changing the face of Earth.[9] This school spent less time worrying about regions and more time addressing the historical, cultural, and ecological contexts of human activity and its imprint. Gilbert White and his students founded and maintained the Natural Hazards School (Kates and Burton 1986a, 1986b), one element of the human–environment wing of the Midwestern Tradition that survived environmental determinism. This strand had particular influence on the present approach to climate impact assessment (see Kates 1985). Together, the Berkeley and Natural Hazards Schools of geography have contributed much to conceptualizing the human dimensions of global environmental change (HDGEC) paradigm (National Research Council 1992; Rayner and Malone 1998), which in turn is fundamental to integrated regional assessment.

HDGEC research is the study of the societal causes and consequences of changes in the global environment, as well as the analysis of individual and institutional responses to these changes (National Research Council 1999). HDGEC studies bring the social sciences into global environmental change research. In this case, "social sciences" indicates not only traditional disciplines (e.g., political science, sociology, anthropology, psychology, history, economics, and human geography), but also integrative interdisciplinary fields (e.g., environmental psychology, natural hazards, risk analysis and management, and human ecology). HDGEC researchers work closely with biophysical scientists, especially those from interdisciplinary fields (e.g., climatology, ecology, and physical geography). Geographers have

---

[9]   The idea that humans are determining the Earth's fate is somewhat opposite to the basic tenants of environmental determinism, in which the environment determines humanity's nature and fate. Thus, the Berkeley School escaped the taint of environmental determinism.

played a central role in the intellectual development of HDGEC research because of the discipline's human–environment and regional traditions, as well as its social science and biophysical science contributions (Liverman *et al.* 2003).

2.3.1.1 Other contributors to integrated regional assessment

Besides geography, many disciplines have components that have added to the scholarly foundations of integrated regional assessment – environmental sciences and social sciences among them. Instead of discussing scattered disciplinary contributions, however, it may be more useful to address three interdisciplinary efforts – regional science, environmental impact assessment, and spatial analysis – that have made more focused contributions to the ideas behind integrated regional assessment.

Regional science and regional studies. Many of the questions that will confront the social sciences in the future will require spatial analyses of integrated economic, social, political, and environmental phenomena. Regional science is an interdisciplinary field that links many social science disciplines, including economics, geography, urban and regional planning, sociology, finance, and political science. It studies the impact of spatial location on the socio-economic activities of individuals, firms, industries, and governments. The field tends to focus on dynamic interactions among regional units – cities and their suburbs, small regions within nations, and nations within world regions.

Regional science analyses treat complex, multi-dimensional problems, but almost all investigations have an economic focus. Thus, economists and economic geographers are dominant in regional science. Regional science has a strong tradition of creating and advancing the economic theories and methods for spatial analysis, emphasizing quantitative methods and mathematical models.

In contrast, regional studies (also known as area studies) is an umbrella term for a diverse approach to regional issues. Some programs focus exclusively on regional development and take rigorous analytical approaches. Similar to regional science, these programs are typically rooted in economics. In contrast, other regional studies programs support eclectic, largely non-quantitative methodologies for understanding the socio-economic and cultural context of a specific region and include humanistic, as well as social science techniques, with roots in cultural geography and anthropology, history, and political science. Regional studies promote holistic, interdisciplinary research; environment is a growing concern of regional studies experts.

Regional science and regional studies contribute to integrated regional assessment by means of their focus on regions and through their promotion of integrated, interdisciplinary research. Although environment is just one of many themes addressed by scientists from these fields, they add to the broad understanding of

linkages between the natural and human worlds. Both fields are especially concerned with the impacts of globalization on regions.

Environmental impact assessment. Environmental impact assessment (EIA) is
a process of predicting and evaluating the impacts of projects on the environment
(Canter 1996). EIA is interdisciplinary, systematic, and predictive. Planners and
decision makers use the results of EIA to prevent the environmental degradation
that normally comes from a development project.

When EIA originated in the early 1970s with the passage of environmental
impact legislation, it only considered impacts on the biophysical environment
caused by specific projects (Porter and Fittipaldi 1998). This focus did not last
long. The institutionalization of EIA, with its public disclosure and consultation
processes, improved opportunities for public participation and led to the inclusion
of social, risk, and health factors within a single integrated framework. EIA increasingly has been built into policy, planning, and regulatory frameworks and, in turn,
used to establish monitoring, auditing, and dispute resolution procedures for government entities. EIA has even become fundamental to international environmental
protection and sustainability conventions.

The evolution of EIA into a more holistic, general process led it away from
its original project-specific basis. So that EIA could focus on narrow concerns,
the concept of Strategic Environmental Assessment (SEA) was created to cover
the broader picture (Dalal-Clayton and Sadler 1999). According to Sadler and
Verheem (1996), SEA is a systematic process for evaluating the environmental
consequences of proposed policies or programs – not of specific projects. While EIA
is geographically localized, highly detailed, and project focused, SEA has a wide
geographical perspective and low level of detail, providing a vision and framework
for areas or sectors (Dalal-Clayton and Sadler 1999). The aim of SEA is not only to
ensure that environmental considerations are included at the earliest possible stage
of decision making, but also to make certain that they are on par with economic
and social considerations. The European Union has formally embraced SEA to
help integrate environmental concerns into larger decisions taken by the European
Union and its member states (ICON Consultants 2001; European Environmental
Bureau 2003).

EIA, and more recently SEA, contribute to integrated regional assessment
through their focus on interdisciplinary, holistic, and systematic approaches to
establishing the environmental implications of human activity. Further, the EIA
concentrations on process, stakeholder input, and policy have added immensely
to the integrated regional assessment paradigm. The broadening of EIA into SEA
blurs the distinction from integrated regional assessment.

Spatial analysis. Spatial analysis refers to a suite of methods and tools used
to identify and evaluate patterns and processes over Earth's surface. Traditional

tools used by geographers have included maps, aerial photographs, and statistical analysis. Regional scientists have tended to use econometrics and input–output analysis. Improved digital technologies wrought a profound transformation in these tools during the last decade or two so that today geographic information systems (GIS), remote sensing, and geocomputation are the prevailing spatial analysis tools. Regional scientists are starting to use computable general equilibrium (CGE) models and other advanced economic analytical methods. In the past, geographers and regional scientists were the main practitioners of spatial analysis. Now, a wide variety of other scientists also engage in spatial analysis, including agricultural scientists, archeologists, architects, biologists and ecologists, civil engineers, Earth and atmospheric scientists, environmental scientists and assessors, forest resource managers, landscape architects, planners, natural resource managers, risk assessors, and more.

The evolution of the sciences towards a more holistic and interdisciplinary paradigm is leading to the development of spatial analytical environments that mix tools from various fields. These problem-solving environments integrate modules for visualization, geocomputational and statistical analysis, risk assessment, environmental impact analysis, cost–benefit analysis, and other purposes into unified software packages. The ultimate aim of these systems is to provide support to scientists and decision makers, particularly in complex contexts where biophysical and socio-economic worlds intersect. Integrated regional assessment is one such context.

Recapitulation. Regional science, EIA and SEA, and spatial analysis have all contributed to integrated regional assessment. Together with geography, they provide the fundamental ideas behind integrated regional assessment. Yet, integrated regional assessment is more than just the sum of these ideas; it filters them through several themes to create a unique perspective on global environmental change.

2.4 Themes of integrated regional assessment

Because it seeks to foster a better understanding of the regional implications of global environmental change, integrated regional assessment clearly shares the themes of region and human–environment interaction with geography and regional science. The explicit aim of integrated regional assessment to synthesize the biophysical and social sciences through the holistic study of regions echoes the German School and Midwestern Traditions of geography. In addition, emerging debates among practitioners of integrated regional assessment over the unique versus the universal, pattern versus process, and intensive versus extensive studies mirror familiar, protracted controversies in geography. But it is the other themes of particular importance to integrated regional assessment – especially change, scale interactions, stakeholder interaction, and risk, vulnerability, adaptation, mitigation, and uncertainty – that make this new field stand out.

Geography, regional science, regional studies, and integrated regional assessment focus on synthesis and holism. The distinction between the older fields and integrated regional assessment comes from the way in which they try to achieve these intents. Geography, regional science, and regional studies have commonly involved individual researchers attempting to understand all aspects of a region's physical and human environments and their interactions (e.g., Hart 1982). Yet, even within the discipline of geography, commentators have long protested that, "there seems no place to go for the capable competent geographer who is not a genius" (Platt 1957: 190). In contrast, integrated regional assessment has put the emphasis on integration through collaboration among scientists from many disciplines (e.g., Schneider 1997). There are many advantages to this approach. The total knowledge of many scholars will be greater than the knowledge of any one person (although the individual may more easily grasp the interactions). Perhaps equally important is that interdisciplinary collaboration introduces different scientific viewpoints; no matter how broadly trained or smart, one scholar cannot grasp the depth of economics, political institutions, or biology that an economist, political scientist, or biologist can. Related to this idea, many perspectives countermand the personal and scientific biases of the individual.

Integrated regional assessment concentrates on the regional implications of global environmental change. Geography, regional science, and regional studies aim to understand the dynamic, contemporary region through the synergisms of historical human and biophysical processes. Similarly, specialists in integrated regional assessment study historical processes, but they stress the concept of change – of agents and structures that led to change in the past, that are causing change today, and that will lead to change in the future. The weight is on global environmental change, although notable global-scale human changes, such as economic globalization and cultural diversity loss, also fall under the domain of integrated regional assessment. The other fields may note the importance of large-scale change, but it is rarely the focus of inquiry.

Related to its concentration on global-scale change, scale interaction is a major constituent of integrated regional assessment (see Polsky and Munroe, Chapter 4, this volume). To understand how regions contribute and respond to global-scale biophysical and human changes means that grappling with interactions among scales is integral to integrated regional assessment. Despite the attention given to scale interaction by the field, most note that it is a difficult subject requiring better conceptualization and understanding. Geography and regional science also have been concerned with scale interaction, including global to regional linkages, for a long time. Nevertheless, a perusal of the literature shows that geography and regional science have produced no better insights on scale interaction than scientists from other fields have and that most geographers or regional scientists currently thinking about such problems are involved in integrated regional assessment (e.g., Easterling 1998).

Geography, EIA, SEA, and integrated regional assessment emphasize interaction with stakeholders. In the case of geography,

Good regional geography must be grounded on keen sensitivity to the needs, wishes, and values of the people who live in the region . . . The good scholar must learn to look at the real world . . . He must learn to talk to ordinary people, not just to other scholars, and above all he must learn to listen, and to listen carefully, if he wishes to understand, and to help others understand, this fascinating world.

(Hart 1982: 29)

Geographers and specialists in integrated regional assessment want to know how a given region works. The big difference between the two is that the traditional geographer is only concerned with drawing information from the region's inhabitants. In integrated regional assessment, the scientist wants the exchange of knowledge to be interactive so that not only do the locals provide information and help guide the assessment, but that they also gain information that they want and need to confront global environmental change in ways appropriate to their socioeconomic and biophysical context. The accent in integrated regional assessment, therefore, is on engagement and interaction, not just extraction.

Unlike geography, EIA and SEA both *mandate* stakeholder interaction. In EIA, the point is to hear the concerns and needs of developers and of those people that might be affected by the development project, but again the flow of information is from the stakeholder to the assessor. The broader vision of SEA is much more like integrated regional assessment, in which information flows both ways (Fisher and Kasemir, Chapter 8, this volume). The difference is that SEA is concerned with development in the largest sense, whereas integrated regional assessment is focused on global environmental change.

Vulnerability (e.g., Kasperson and Kasperson 2001; Leary and Beresford, Chapter 6, this volume), adaptation (e.g., Smit *et al.* 2000; Dickinson *et al.*, Chapter 7, this volume), mitigation (e.g., Metz *et al.* 2001), and related concepts are significant issues for scientists involved in integrated regional assessment. These ideas also captivate geographers of the Natural Hazards School who, along with social scientists concentrating on disasters (e.g., Quarantelli 1998; Mileti 1999), have spent considerable effort on defining these terms (e.g., Kates 1978; Mitchell 1990). As noted earlier, climate impact analysis – an important element of many integrated regional assessments – evolved from the Natural Hazards School. Consequently, hazards geographers, who follow the human–environment theme of geography more closely than the spatial theme, tend to be much closer to scientists working in integrated regional assessment than they are to most regional geographers. Regional geography, regional science, and regional studies have spent little effort on the concepts of vulnerability, etc., but are now starting to address these ideas.

Uncertainty (see van Asselt, Chapter 5, this volume) is important to integrated regional assessment because of the field's foci on decision making and change. For example, statisticians and philosophers (e.g., Dubins and Savage 1976 and Ramsey 1990, respectively) suggest that it is possible to anticipate an individual's likely decision based on the personal value of the decision and on beliefs about the likelihood of the uncertain situation to occur. Others think that the world in which we live is multi-variate, complex, and interrelated; an event could occur next week, next year, or next decade that could affect the interrelationships of variables. The more complex the global environment appears to a decision maker, the less able he/she is to anticipate the probability of success of a particular strategy (e.g., Morrison and Mecca, 1989). Integrated regional assessment, therefore, uses such thinking to determine how uncertainty affects the willingness of local decision makers to address issues of global environmental change and regional vulnerability, adaptation, and mitigation. Only those geographers and regional scientists concerned with the concept of uncertainty in risk assessment (e.g., Kasperson *et al.* 1988), which is closely allied with integrated regional assessment, focus on this topic. Uncertainty is fundamental to specialists in EIA and SEA.

2.5 Conclusions

The first purpose of this chapter was to explain integrated regional assessment. Integrated regional assessment is a collaborative, multi-disciplinary process that addresses the regional implications of global environmental change. Integrated regional assessment tries to determine how regions contribute to, and are affected by, global environmental change, and thereby to help regional stakeholders make informed decisions on how to deal with the problems. Ideally, this type of process is comprehensive and involves scientists, decision makers, and societal stakeholders in an interactive, iterative process. When integrated regional assessment works, it presents an ideal way of bringing together science, society, and policy to face the challenges of global environmental change.

The second purpose of this chapter was to determine whether integrated regional assessment is simply a recycling of previously developed thoughts. Indeed, integrated regional assessment uses the themes of holistic, integrated study of regions and of human–environment interactions promulgated by geography, regional science, and others. Nevertheless, the themes emphasized by integrated regional assessment – that is, change, scale interactions, stakeholder interaction, vulnerability, and uncertainty – are not unique, but taken together make it a significantly different field. Just as geography can be identified as a discipline defined not by distinct subject matter, but by the themes of space and human–environment interaction, integrated regional assessment can be differentiated from other fields by its emphases.

References

Abler, R., J. S. Adams, and P. Gould 1971. *Spatial Organization*. Englewood Cliffs, NJ: Prentice-Hall.

Abler, R. F., M. G. Marcus, and J. M. Olson 1992. Afterword. In *Geography's Inner Worlds: Pervasive Themes in Contemporary American Geography*, ed. R. F. Abler, M. G. Marcus, and J. M. Olson. New Brunswick, NJ: Rutgers University Press, pp. 391–401.

Canter, L. W. 1996. *Environmental Impact Assessment*, 2nd edn. New York: McGraw-Hill.

Carter, T. R., R. N. Jones, X. Lu *et al.* 2007. New assessment methods and the characterisation of future conditions. In M. L. Parry, O. F. Cantiani, J. P. Palutikof *et al.* (eds.), *Climate Change 2007: Impacts, Adaptation and Vulnerability. Contribution of Working Group II to the Fourth Assessment Report of the Intergovernmental Panel on Climate Change*. Cambridge: Cambridge University Press, pp. 133–171.

Cohen, S. J. 1995. An interdisciplinary assessment of climate change on northern ecosystems: the Mackenzie Basin impact study. In *Human Ecology and Climate Change: People and Resources in the Far North*, ed. D. L. Peterson and D. R. Johnson. Washington, DC: Taylor and Francis.

Cohen, S. J. (ed.) 1997a. *Mackenzie Basin Impact Study: Final Report*. Downsview: Canadian Climate Center, Atmospheric Environment Service, Environment Canada.

Cohen, S. J. 1997b. Scientist–stakeholder collaboration in integrated assessment of climate change: lessons from a case study of Northwest Canada. *Environmental Modeling and Assessment*, **2**, 281–293.

Dalal-Clayton, B. and B. Sadler 1999. *Strategic Environmental Assessment: a Rapidly Evolving Approach. Environmental Planning Issues* No. 18. London: International Institute for Environment and Development. http://www.nssd.net/References/KeyDocs/IIED02.pdf.

Dubins, L. E. and L. J. Savage 1976. *Inequalities for Stochastic Processes: How to Gamble if you Must*. New York: Dover Publications.

Easterling, W. E. 1998. Why regional studies are needed in the development of full-scale integrated assessment modeling of global change processes. *Global Environmental Change*, **7**, 337–356.

Elkins, T. H. 1989. Human and regional geography in the German-speaking lands in the first forty years of the twentieth century. In *Reflections on Richard Hartshorne's The Nature of Geography*, ed. J. N. Entrikin and S. D. Brunn. Washington, DC: Association of American Geographers.

Entrikin, J. N. and S. D. Brunn (eds.) 1989. *Reflections on Richard Hartshorne's The Nature of Geography*. Washington, DC: Association of American Geographers.

European Environmental Bureau 2003. Strategic Environmental Assessment. http://www.eeb.org/activities/env_impact_assessment/main.htm.

Fussel, H.-M. and R. J. T. Klein 2006. Climate change vulnerability assessments: an evolution of conceptual thinking. *Climatic Change*, **75**, 301–329.

Hart, J. F. 1982. The highest form of the geographer's art. *Annals of the Association of American Geographers*, **72**, 1–29.

Hartshorne, R. 1939. The nature of geography: a critical survey of current thought in the light of the past. *Annals of the Association of American Geographers*, **29**, 171–645.

Hartshorne, R. 1959. *Perspective on the Nature of Geography, Association of American Geographers Monograph Series 1*. Chicago, IL: Rand McNally and Co.

Harvey, D. 1969. *Explanation in Geography*. London: Edward Arnold.

Hettner, A. 1928. Neue Wege in der Landerkunde (New directions in regional geography). *Zeitschrift fur Geopolitik*, **5**, 273–275.

Houghton, J. T., L. G. Meiro Filho, B. A. Callander, N. Harris, A. Kattenberg, and K. Maskell (eds.) 1996. *Climate Change 1995 – The Science of Climate Change*. Cambridge: Cambridge University Press.

Huntington, E. 1925. *The Character of Races as Influenced by Physical Environment, Natural Selection and Historical Development*. New York: C. Scribner's Sons.

ICON Consultants 2001. *SEA and Integration of the Environment into Strategic Decision-Making*. European Commission Contract No. B4–3040/99/136634/MAR/B4. http://europa.eu.int/comm/environment/eia/sea-studies-and-reports/sea_integration_xsum.pdf

Intergovernmental Panel on Climate Change 2007. *Climate Change 2007: Impacts, Adaptation and Vulnerability. Contribution of Working Group II to the Fourth Assessment Report of the Intergovernmental Panel on Climate Change*, ed. M. L. Parry, O. F. Canziani, J. P. Palutikof *et al.* Cambridge: Cambridge University Press.

Jepma, C. J. and M. Munasinghe 1998. *Climate Change Policy: Facts, Issues and Analyses*. Cambridge: Cambridge University Press.

Kasperson, J. X. and R. E. Kasperson 2001. *International Workshop on Vulnerability and Global Environmental Change*, 17–19 May 2001. *A Workshop Summary*. Report 2001–01. Stokholm: Stockholm Environment Institute (SEI). http://www.sei.se/dload/2002/Vulnerability%20report2.PDF

Kasperson, R. E., O. Renn, P. Slovic, *et al.* 1988. The social amplification of risk: a conceptual framework. *Risk Analysis*, **8**, 177–204.

Kates, R. W. 1978. *Risk Assessment of Environmental Hazard*, vol. 8, *SCOPE*. Chichester: John Wiley and Sons.

Kates, R. W. 1985. The interaction of climate and society. In *Climate Impact Assessment*, ed. R. W. Kates, J. H. Ausubel, and M. Berberian. Chichester: John Wiley and Sons.

Kates, R. W. and I. Burton (eds.) 1986a. *Geography, Resources, and Environment, vol. II. Themes from the Work of Gilbert F. White*. Chicago, IL: The University of Chicago Press.

Kates, R. W. and I. Burton (eds.) 1986b. *Selected Writings of Gilbert F. White*. Chicago, IL: The University of Chicago Press.

Kattenberg, A., F. Giorgi, H. Grassl, *et al.* 1996. Climate models – projections of future climate. In *Climate Change 1995: the Science of Climate Change*, ed. J. T. Houghton, L. G. Meira, Filho, B. A. Callander, N. Harris, A. Kattenberg, and K. Maskell. Cambridge: Cambridge University Press.

Knight, C. G. 1992. Geography's worlds. In R. F. Abler, M. G. Marcus, and J. H. Olson (eds.), *Geography's Inner Worlds: Pervasive Themes in Contemporary American Geography*. New Brunswick, NJ: Rutgers University Press, pp. 9–26.

Liverman, D., B. Yarnal, and B. L. Turner, II 2003. Human dimensions of global change. In *Geography in America at the Turn of the 21st Century*, ed. G. Gaille and C. Willmott. Oxford: Oxford University Press.

Lonergan, S., R. Difrancesco, and M.-K.Woo 1993. Climate change and transportation in northern Canada: an integrated impact assessment. *Climatic Change*, **24**, 331–351.

Marsh, G. P. 1864. *Man and Nature: Physical Geography as Modified by Human Action.* New York: Scribner.

McCarthy, J. J., O. F. Canziani, N. A. Leary, D. J. Dokken, and K. S. White (eds.) 2001. *Climate Change 2001: Impacts, Adaptation & Vulnerability. Contribution of Working Group II to the Third Assessment Report of the Intergovernmental Panel on Climate Change (IPCC).* Cambridge: Cambridge University Press.

Metz, B., O. Davidson, R. Swart, and J. Pan (eds.) 2001. *Climate Change 2001: Mitigation. Contribution of Working Group III to the Third Assessment Report of the Intergovernmental Panel on Climate Change (IPCC).* Cambridge: Cambridge University Press.

Mileti, D. 1999. *Disasters by Design: a Reassessment of Natural Hazards in the United States.* Washington, DC: The Joseph Henry Press.

Mitchell, J. K. 1990. Human dimensions of environmental hazards: complexity, disparity, and the search for guidance. In *Nothing to Fear: Risks and Hazards in American Society*, ed. A. Kirby. Tucson, AZ: The University of Arizona Press.

Morgan, M. G. and H. Dowlatabadi 1996. Learning from integrated assessment of climate change. *Climatic Change*, **34**, 337–368.

Morrison, J. L. and T. V. Mecca 1989. Managing uncertainty: environmental analysis/forecasting in academic planning. In *Higher Education: Handbook of Theory and Research*, ed. J. C. Smart. New York: Agathon Press.

National Assessment Synthesis Team 2000. *Climate Change Impacts on the United States: the Potential Consequences of Climate Variability and Change.* Cambridge: Cambridge University Press.

National Research Council 1992. *Global Environmental Change: Understanding the Human Dimensions*, ed. P. Stern, O. Young, and D. Druckman. Washington, DC: National Academy Press.

National Research Council 1999. *Human Dimensions of Global Environmental Change: Research Pathways for the Next Decade.* Washington, DC: National Academy Press.

Pahl-Wostl, C. 2005. Actor based analysis and modeling approaches. *Integrated Assessment* [Online] 5. http://journals.sfu.ca/int_assess/index.php/iaj/article/view/167/110

Parry, M. L., T. R. Carter, and N. T., Konijn (eds.) 1988. *The Impact of Climatic Variations on Agriculture. Vol. 1, Assessment in Cool Temperate and Cold Regions; Vol. 2. Assessments in Semi-Arid Regions.* Dordrecht: Kluwer.

Parson, E. A. 1996a. Integrated assessment and environmental policy making: in the pursuit of usefulness. *Energy Policy*, **23**, 463–475.

Parson, E. A. 1996b. Three dilemmas in the integrated assessment of climate change. An editorial comment. *Climatic Change*, **34**, 315–326.

Parson, E. A. 1997. Informing global environmental policy-making: a plea for new methods of assessment and synthesis. *Environmental Modeling and Assessment*, **2**, 267–279.

Platt, R. S. 1957. A review of regional geography. *Annals of the Association of American Geographers*, **47**, 187–190.

Porter, A. L. and J. J. Fittipaldi (eds.) 1998. *Environmental Methods Review: Retooling Impact Assessment for the New Century.* Fargo, ND: The Press Club.

Pulwarty, R. 2000. NOAA-Office of Global Programs: the Regional Assessments Program. *The ENSO Signal*, Issue 12, January 2000. http://www.ogp.noaa.gov/library/ensosig/ensosig12.htm

Quarantelli, E. L. 1998. *What Is a Disaster?* London: Routledge.

Ramsey, F. P. 1990. *Philosophical Papers*. D. H. Mellor (ed.). New York: Cambridge University Press.

Rayner, S. and E. Malone (eds.) 1998. *Human Choice and Climate Change*. Washington, DC: Batelle Press.

Richardson, B. C. 1992. Places and regions. In R. F. Abler, M. G. Marcus, and J. M. Olson (eds.), *Geography's Inner Worlds: Pervasive Themes in Contemporary American Geography*. New Brunswick, NJ: Rutgers University Press, pp. 27–49.

Riebsame, W. E., K. M. Strzepek, Jr., J. L. Wescoate, *et al.* 1995. Complex river basins. In K. M. Strzepek and J. B. Smith (eds.), *As Climate Changes: International Impacts and Implications*. Cambridge: Cambridge University Press, pp. 57–91.

Risbey, J., M. Kandlikar, and A. Patwardhan 1996. Assessing integrated assessments. *Climatic Change*, **34**, 369–395.

Rosenberg, N. J. (ed.) 1993. *Towards an Integrated Impact Assessment of Climate Change: the MINK Study*. Dordrecht: Kluwer.

Rothman, D. S. and J. B. Robinson 1997. Growing pains: a conceptual framework for considering integrated assessments. *Environmental Monitoring and Assessment*, **46**, 23–43.

Rotmans, J. and M. van Asselt 1996. Integrated assessment: a growing child on its way to maturity. An editorial essay. *Climatic Change*, **34**, 327–336.

Rubin, E. S., L. B. Lave, and M. G. Morgan 1991–92. Keeping climate research relevant. *Issues in Science and Technology*, Winter, 47–55.

Sadler, B. and R. Verheem 1996. Strategic environmental assessment: status, challenges and future directions. Amsterdam: Ministry of Housing, Spatial Planning and the Environment.

Sauer, C. 1941. Foreword to historical geography. *Annals of the Association of American Geographers*, **31**, 1–24.

Schluter, O. 1928. Die analytische Geographie der Kulturlandschaft erlautert am Beispiel der Brucken (The analytical geography of the cultural landscape, as illustrated by the example of bridges). *Zeitschrift der Gesellschaft fur Erdkunde zu Berlin; Sonderband zur Hundertjahrfreir der Gesellschaft Berlin* 388–411.

Schneider, S. H. 1997. Integrated assessment modeling of global climate change: transparent rational tool for policy making or opaque screen hiding value-laden assumptions? *Environmental Modeling and Assessment*, **2**, 229–249.

Schneider, S. and J. Lane 2005. Integrated assessment modeling of global climate change: much has been learned – still a long and bumpy road ahead. *Integrated Assessment* [Online] 5. http://journals.sfu.ca/int_assess/index.php/iaj/article/view/169

Smil, V. 1993. *Global Ecology: Environmental Change and Social Flexibility*. New York: Routledge.

Smit, B., I. Burton, R. Klein, and J. Wandel 2000. An anatomy of adaptation to climate change and variability. *Climatic Change*, **45**, 233–251.

Smith, N. 1989. Geography and museum: private history and conservative idealism in *The Nature of Geography*. In J. N. Entrikin and S. D. Brunn (eds.), *Reflections on Richard Hartshorne's The Nature of Geography*. Washington, DC: Association of American Geographers, pp. 91–120.

Soja, E. W. 1989. *Postmodern Geographies: the Reassertion of Space in Critical Social Theory*. London: Verso.

Spethmann, H. 1928. *Dynamische Landerkunde (Dynamic Regional Geography)*. Breslau: Hirt.

Strzepek K. M. and Z. Kaczmarek 1996. Editorial: the United States Country Studies Program on water resources vulnerability and adaptation to climate change: scope and significance. *Water Resources Development*, **12**, 109–110.

Titus, J. G., R. A. Park, S. P. Leatherman, *et al.* 1991. Greenhouse effect and sea-level rise: potential loss of land and the cost of holding back the sea. *Coastal Management*, **19**, 171–204.

Toth, F. 2003. State of the art and future challenges for integrated environmental assessment. *Integrated Assessment* [Online] 4. http://journals.sfu.ca/int_assess/index.php/iaj/article/view/144/99

Toth, F. L. and E. Hizsnyik 1998. Integrated environmental assessment methods: evolution and applications. *Environmental Modeling and Assessment*, **3**, 193–207.

Turner, B. L., II 2002. Contested identities: human–environment geography and disciplinary implications in a restructuring academy. *Annals of the Association of American Geographers*, **92**, 52–74.

USGCRP 1998. *Our Changing Planet: the FY 1998 U.S. Global Change Research Program*. Washington, DC: Committee on Environment and Natural Resources, National Science and Technology Council.

Watson, R. T., M. C. Zinyowera, and R. H. Moss (eds.) 1998. *The Regional Impacts of Climate Change: an Assessment of Vulnerability*. Cambridge: Cambridge University Press.

Weyant, J., O. Davidson, H. Dowlatabadi, *et al.* 1996. Integrated assessment of climate change: an overview and comparison of approaches and results. In J. P. Bruce, H. Lee, and E. F. Haites (eds.), *Climate Change 1995: Economic and Social Dimensions of Climate Change*. Cambridge: Cambridge University Press.

Yarnal, B. 1996. The policy relevance of global environmental change research. *Global and Planetary Change*, **11**, 167–175.

Yarnal, B. 1998. Integrated regional assessment and climate change impacts in river basins. *Climate Research*, **11**, 65–74.

Yin, Y. and S. J. Cohen 1994. Identifying regional goals and policy concerns associated with global climate change. *Global Environmental Change*, **4**, 245–260.

3

Integrated regional assessment: qualitative and quantitative issues

ELIZABETH L. MALONE

3.1 Introduction

Integrated regional assessment is becoming increasingly important as more attention is focused on the impacts of climate and other global changes. Impacts are different and are experienced differently than global mean change suggests. A gradual increase of average temperature documents the reality of climate change, but impacts are likely to be more discontinuous and dramatic: extended droughts, severe storms, catastrophic heat waves, and so on. Changes in different regions may mean warmer, cooler, rainier, drier, or more uncertain weather. Patterns (including timing) of precipitation may change, changing established agricultural and other ways of living. Because people want to know with relative specificity what will happen in a given region, the quality and quantity of information available and the workings of models used for projections come under a great deal of scrutiny. When the economic impacts of the Kyoto Protocol were assessed for the United States, for example, the immediate challenges to modelers were to specify their input data and assumptions that drove the models' projections. In the same way, if adverse impacts of global change are projected, regional policy makers want to know what data and assumptions drive those results.

Thus, qualitative and quantitative issues are particularly significant in integrated regional assessment. This chapter first attempts to examine the terms "qualitative" and "quantitative" separately and in relation to one another. No research is purely one or the other, of course, so the degree of interdependence or overlap is particularly interesting. Strategies for integrating the two general approaches often produce uneasy compromises. However, integrated regional assessment provides opportunities for strong collaborations in addressing specific problems in specific places; this topic is taken up in the last two sections of the chapter.

3.2 Integration in theory and practice

The first questions to be asked are basic to the entire discussion, but rarely addressed explicitly. What is qualitative? What is quantitative? A standard answer to these questions goes as follows. Quantitative analysis is conducted by the use of accepted scientific methodologies to generate statistically significant numerical values within measured ranges of uncertainty. If quantitative methods are not available, or if the numerical values generated are not statistically valid, then the analysis is based on qualitative methods.

However, the standard answer begs the following questions. What determines "accepted scientific methodologies"? What guides a researcher to use one or more of these methods, and to choose what variables to measure and how to measure them? And, why should quantitative methods be universally preferred to qualitative ones?

A practical dividing line between quantitative and qualitative research and results is whether or not they are primarily based on numbers and statistics. If data tables, bar/pie charts, and results of regression analyses are the major features of a report, for example, it is likely to be thought of as a quantitative study – even if the numbers only represent a Likert scale[1] from "strongly agree" to "strongly disagree." If another report is mostly text, with 2×2 matrices of variables or process flow charts as the only graphics, it is likely to be considered a qualitative study – even if the flowchart represents highly precise chemical reactions.

A specific disciplinary example illustrates the complementary nature of the two. In analytical chemistry the difference between a qualitative and a quantitative analysis is the difference between naming the elements, radicals, or compounds composing the sample versus determining the precise percentage composition of the sample in terms of elements, radicals, or compounds. Note that the scientist cannot perform quantitative analysis without first having done qualitative analysis. Even if the researcher identifies the components after analyzing the composition of the sample, he or she needs to know what general class of compound is being sought in order to determine what type of analysis to perform.

In the classic science paradigm (Merton 1973), from (qualitative) theory a researcher derives (perhaps quantitatively) testable hypotheses and designs an experiment or study to test them. Data collection too must be guided by theory, since only a qualitative judgment determines that these data (and not some others) are the best test of a hypothesis. Theory specifies the statistical methods to use.

---

[1] A widely used technique for measuring attitudes, originated by the American psychologist R. A. Likert (1903–81). Likert's scale consists of a series of statements that a respondent uses to answer a question. The scale is normally five points: strongly agree, agree, neither agree nor disagree, disagree, strongly disagree. The data gained using the Likert scale are easily amenable to statistical analysis. See Likert (1932).

Substantively, the use of numbers may tell us little. Measurements may be precise or imprecise, percentages may be approximate, numbers may be missing or entered incorrectly. The assumptions of two seemingly parallel datasets may be different, so that aggregation seriously distorts both. We come quickly full circle to data *quality*, quality referring to both how and how well the measure or number reflects reality and to what it means, not how precise it is. Numbers are powerful, but they derive much of their power from the qualitative story they tell.

Quantities are often a numerical representation of a qualitative judgment on the part of the analyst rather than an empirical quantity. For example, estimates of land cover and land use depend heavily on (1) the definitions used by the estimator(s), (2) the tools used to survey the land, and (3) the viewpoint of the researcher (Meyer *et al.* 1998). Definitions of land-use categories, for example, often vary from place to place and study to study. Whether a survey of land uses is based on researcher observation or remote sensing (or another method) will affect how much land is assigned to which category. And a researcher interested in documenting the "degradation" of a region might very well overestimate the amount of degraded land (Thompson *et al.* 1986).

The distinction between quantitative and qualitative measurement is often rather fuzzy. Ordinal data on "low," "medium," or "high" values can easily be expressed as numerical categories and then manipulated as if they were higher-order data. Moreover, data may be misleadingly presented to transmit qualitative messages, as the classic *How to Lie with Statistics* (Huff 1993 (1954)) amply demonstrates, or ineptly interpreted, as Joel Best's books (2001, 2004) show.

However, without empirical data, and especially quantitative data, theories are – just theories. They may be intuitively compelling but false. They may be true for one person but not for another. Gathering data and performing statistical analyses provide support for hypotheses or disprove them. Quantification – measuring, weighing, counting – gives us the basis for evaluating the magnitude of impacts and pointers toward mitigation and adaptation.

Even uncertainty has both quantitative and qualitative aspects. Its most common usage implies that we are not sure, our knowledge not assured. In statistics, however, uncertainty has a specific meaning as the amount or percentage by which an observed or calculated value may differ from the true value. An uncertainty analysis may use sophisticated statistical sampling techniques (e.g., Monte Carlo sampling) to identify the source and amount of uncertainty in a given mathematical analysis. Some analysts, attempting to sort out these meanings, have distinguished between uncertainty and indeterminacy, the latter meaning a lack of knowledge rather than a precise probability of the estimated value being the true value.

The strength of the scientific approach lies in the interaction of theory and empirical research, the testing of theory by rigorous, quantifiable (if possible) investigation.

3.2.1 Why not just use quantitative data/model results?

In recent years researchers have built many computer-based models, integrated and not, that produce projections of various global change impacts. These models are not easy to build. Any model is a simplified representation of the world, so variables thought to be "key" or "critical" – driving forces of change, for example – are selected, and model design determines the modeled relationship among the variables. Assumptions, implicit and explicit, are inherent in the model design. For example, model builders typically assume that population will grow, social welfare will improve, and crops will become more productive. These qualitative assumptions become expressed as parameters and equations in the model. Quantitative data are located or gathered; sometimes little attention is paid to data quality in the first stages, when the objective is to get the model running. Furthermore, data (or high-quality data) do not always exist where theory, embodied in design, would like to have them, so compromises are necessary.

Often models represent the state of the art of scientific thinking about how variables are related to one another and how changes in one or several variables (population growth or increasing energy use, for example) cause changes in systems such as agriculture or forestry. Modeling is thus a way to explore the future based on what is known about past and present conditions. If the assumptions and processes of the model are transparent to users of the model and its results, modeling is a powerful tool for both researchers and policy makers.

However, the graphs, charts, sophisticated visualizations, and statistics produced by models are not enough. Qualitative theory and analysis breathes meaning into quantitative results. Without (qualitative) theory to guide the purpose for and methodology of gathering data, a study can be biased or meaningless. If the data just spoke for themselves, summary tables and model results would tend to converge. They do not, so researchers must discuss and rationalize their assumptions and give meaning to the data presented through sound analysis.

Furthermore, data may look hard but be soft. Data on land cover are best and most abundant for the United States, Canada, and Western Europe, but missing or of uncertain quality for many of the regions where it would be most advantageous to have good data. Data projections are by their nature tentative, becoming softer and softer the further out in time they go. Differing datasets and projections can be the basis for lively arguments that can only be resolved by examining assumptions and implicit or explicit theoretical frameworks (Halsnæs and Shukla 2007).

3.2.2 Why not just use qualitative case studies?

Data on the character and structure of an ecological or social system can be richly detailed and meaningful. But if comparative, order-of-magnitude, or quantified data support the qualitative analysis, the research gains credibility, objectivity, comparability, and generalizability. Moreover, the framework provided by quantitative approaches helps policy makers to make tradeoffs and prioritize as necessary. Finally, quantitative data in two or more domains can be "mapped" to gain insights (e.g., Yohe *et al.* 2006a, b).

Credibility and objectivity are particularly important in integrated regional assessment. The more scientific a study or analysis is, the more credibility it will have. As Black (2000: 351) points out, "Science is a matter of degree – scienticity. The scienticity of an idea increases with its testability, generality, simplicity, validity, and originality." These qualities are increased by distance from what is being studied, and quantification helps to distance scientists from their subjects by interposing objective-looking numbers. Empirical research, being close to a subject (as in ethnographic research) may yield messy results, with numerous caveats and qualifications. But if the data can be quantified, the research will look more objective because of the greater distance between the researcher and the subject.

Research also gains clarity if the findings can be discussed in quantitative terms. Verbal analysis may capture the nuances of institutional relationships, agricultural judgments, social networks, and so on, but providing summary data in quantitative terms allows users of the research to measure against numerical standards (e.g., percentage probabilities) and to compare, as necessary, for the purpose of making tradeoffs. Furthermore, the more closely detailed the studies are, the less they lend themselves to usable generalizations. Numerical and statistical analysis prods localized research out of itself to a more public realm for more general use and from a description of the present to possible scenarios of the future. Quantitative studies provide a more-or-less transparent basis for making projections of future situations.

Furthermore, quantitative data often yield good order-of-magnitude conclusions and recommendations, answering many policy relevant questions: What are the most important problems? What will happen if x policy is implemented as opposed to y policy? Will the effect of climate change be small or large?

The now-widespread use of quantification in general and cost–benefit analysis in particular originated as ways to make the arcane or professional knowledge of the "experts" available to the public who distrusted them. The insurance industry, accounting, and public works engineering, cases detailed by Porter (1995), relied on the implicit knowledge and career experience of professionals to set insurance premiums, keep corporate books, and decide on public investments in large infrastructure projects. When challenged, these experts tended to retreat to their privileged positions as experts rather than explaining how their conclusions were derived. Cost–benefit analysis, with its understandable logic, offered a way to make the expert knowledge of the US Army Corps of Engineers available to the public – even though in practice the recommendations of the Corps were often accepted without question in the US Congress.

So during the past two centuries quantification has, in industrialized countries, gradually come to mean public knowledge, as opposed to the esoteric knowledge possessed by professionals, including scientists. By a certain twist of irony, however, increasing information loads and sophisticated manipulation of numerical data have made information management and data visualization into specialized branches of knowledge in themselves. Furthermore, the overall seemingly common-sense logic of cost–benefit analysis can hide many expert judgments behind the figures. (See Halsnæs and Shukla 2007 for a summary of the issues.)

3.2.3 *The problem of integration*

The research study ideally involves theoretical analysis (and hypothesis formulation, if applicable), data gathering, statistical or other quantitative manipulation and presentation, and analysis. Qualitative and quantitative analysis are, in the ideal study, seamlessly integrated. In practice, however, studies tend to be characterized as case studies (qualitative), data-based (quantitative), or computer model-based (quantitative).

In part, the integration problem has to do with our failure to recognize and identify different kinds of data that are relevant to the regional study at hand. We often look only for quantitative data as the preferred grist for the research mill (following the standard definition of quantitative and qualitative at the beginning of this chapter) – or only for qualitative data, if we naturally tend toward "interpretive" scientific approaches (see Rayner and Malone 1998). The same may be said for aggregating lower-scale models into regional or global models, or for downscaling GCMs (Warrick, Chapter 14, this volume). Depending upon the natural bent of the research team, it tends to seek out numbers or narratives – and then to put its preferred data into pre-made Western forms. For example, the integrated assessment team at the US Pacific Northwest National Laboratory created its quantitative US module first and then defined a template for data from other countries and regions (see Edmonds *et al.* 1993, Fawcett and Sands 2005). Thus, the data categories from the United States became the model for other regions.

Another not-untypical approach to regional studies is to locate whatever quantitative data already exist within the study region and to build the project around analysis of those data. Collaborations often help to extend the amount and scope of discovered data and perhaps to develop the capability to collect other relevant data. However, such studies often reflect data availability more than a sound theoretical design.

Problems of scale also involve qualitative and quantitative issues that cannot be separated – not only global/regional/local scales, but also time and geographic scales. For ecosystem model building, these qualitative problems are treated as primarily technical problems of how to make data work together. This is understandable, since the qualitative problems involved, say, in scaling up field-level results on soil characteristics to a regional level may well be unresolvable (but see Izaurralde *et al.* (2001) for a review of methods that have been used). A more achievable objective is to scale up using a reasonable (if not elegant) method and then evaluate (or validate, if possible) the results. Polsky and Munroe (Chapter 4, this volume) discuss several simulation and fuzzy-logic-based methods for studying scalar dynamics.

The timescale differences among geologic (century-scale), ecological (short term to long term), and social systems (often year-to-year) have been identified and analyzed (see, for example, Folke *et al.* 1998; Folke 2006; Young *et al.* 2006; Polsky and Munroe, Chapter 4, this volume), but not systematically explored.[2] Social science researchers have not, by and large, focused on the specific geographic-scale problems of defining and analyzing a region over time; for example, the relevant social system may depend upon several different ecosystems, both contiguous with their settlements and at varying distances from those settlements. Similarly, people within a region may rely on far-flung support systems. If agricultural systems within a region provide crops for export, they depend on social institutional systems well outside the region. In what sense can researchers perform an integrated assessment under these circumstances? What are the region's boundaries?

Calls for more data/information are a standard part of the conclusions of collections of regional studies and to discussions of regional integrated assessment. Since new data are notoriously difficult to collect, such calls should be based not simply on number gaps or "it-would-be-nice-to-haves," but on theory-driven designs in which the data will be meaningful for regional assessment and policy input.

3.3 Diverse models, quantitative and qualitative

The qualitative–quantitative dichotomy has implications with respect to the tension between Western-style science and indigenous knowledge. The emphasis of the SysTem for Analysis, Research and Training (START) program on building collaborations and in-region capacity-building raises interesting questions about scientific (as opposed to economic) development. Like technology transfer, science transfer (including a preference for quantitative data and measurement, the

---

[2] Even the physical processes of the climate present scale problems for modelers; large- and small-scale processes affect each other. The most notable example is the difficulty in representing clouds.

"scientific method," economic cost–benefit analysis, and the accompanying rational actor paradigm) implies socio-cultural changes that may be at least as disruptive and harmful as they are beneficial. A country's GDP may increase but social support networks erode. Crop productivity may improve, while the poor become poorer.

For many actors in both scientific and policy realms, science means measurement, standardization, and objectivity – in practice, quantification. Quantification is social engineering in that it creates and often imposes new categories within social institutions (Scott 1998). Standard scientific measurements produce quantified data needed for regional assessment and for comparison with other regions and global analyses. The standardization in itself is a further distancing of scientific data from local measures. Standardization may also demand that models, for the sake of comparability, sacrifice some fidelity to the empirical world. Coordination of a regional integrated assessment may well call for such model standardization. The data and measures for models are often different from the localized measures for land use, social exchange, and economic activity.

For example, before forced villagization in Tanzania and Russia, peasants thought of land parcels in terms of what they would produce, not in terms of hectares or acres. Areas would be assigned to households to match production with need, and inequities might be dealt with by rotating assignments. In contrast, government-designed villages were laid out as a neat grid of same-size plots, regardless of the production potential of each. The imposed grid of the government-created village changed the basis for this way of life while simultaneously applying scientific principles of governance and rational utility.

In another example, every locality in France prior to the French Revolution had its own measure for a bushel. To be sure, the non-standard bushel sizes and shapes were often the basis for confusion and oppression: confusion in that trade among localities with different bushels was extremely difficult, oppression in that a lord was free to change the bushel that measured what his peasants owed him. After the Revolution, the peasantry clamored for a national standard bushel – and got a totally foreign, scientific weight measure. Because the peasants had no experience with the new measure, instituting it worked to the advantage of the elite classes.

Science brings not only benefits but also costs, often socio-cultural costs that are hard to quantify. Scientific quantification may measure a region's environmental ills – but lead to policies that bring poverty, debt, and further environmental harm. Quantification may allow comparative analyses but must be coupled with qualitative analysis before policies may be undertaken (Brenkert and Malone 2005). Integrated regional analysts should ideally be reflexive enough to understand their own cultural scientific models and the cultural models of the people in the region. Science in industrialized nations is apt to rely on quantitative analyses, but in less industrialized nations the metaphor and anecdote often carry the power that the data-driven graph has in the United States.

3.4 Strategies for integration

3.4.1 Turn qualitative data into quantitative data

One strategy to integrate quantitative and qualitative data is to extend the reach of quantification. Nominal data or ordinal data (e.g., "low," "medium," and "high") may become quantified so they can be used in computer-based models along with interval and ratio data. Ranges can be specified in quantitative terms (as the percentage ranges of confidence developed for the IPCC Third Assessment Report; see Moss and Schneider 2000). More elaborate integrative techniques include rational actor theory and Bayesian and other decision analytic frameworks.

One device used to integrate quantitative and qualitative data is the rational actor. The rational actor is a theoretical construct whose prime merit is that he (or she, but most often thought of as male – see Douglas and Ney 1998) is consistent. In market economics, the consumer is seen as a rational actor, consistently choosing to maximize utility by buying certain goods and not others. In decision making, the rational actor assesses the available options based upon a set of common criteria, then consistently chooses the best option to optimize the general welfare.

Originally the rational actor was a metaphor that helped in describing consumer preferences within models of the market. However, even economists recognized that it was a very limited view of human behavior, assuming "atomistic individuals free from complex interdependencies, the pursuit of pure self-interest (construed as happiness), farsighted rationality and accurate cost–benefit calculation . . ." (Zafirovski 2000: 455). The rational actor's usefulness rests on consistency, even though "the rarest of all human qualities is consistency" (Bentham, quoted in Zafirovski 2000: 457) and human life is "ontologically irrational" (Schumpeter, quoted in Zafirovski 2000: 456). In this chapter's terms, real human beings make qualitative choices; rational actors' choices are quantifiable.

The extension of the rational actor construct to models of decision making is similarly useful and similarly unrealistic. Consistency in making decisions can be quantified and made to operate, for example, in theoretic game models to achieve social optima and system equilibria. Again, the assumptions of wholly free and self-interested actors who maximize their utility (usually profit) are necessary to drive these models. And, again, many theorists recognize that the rational actor is an extremely limited model of human behavior. For instance, game theorists have a difficult time dealing with altruistic behavior (but see Panchanathan and Boyd 2005, Hauert *et al.* 2007).

Still, the gains in using the rational actor construct are substantial. It allows "clear and deductive theorizing despite the risk of substantial loss in realism" (Zafirovski 2000: 467). In other words, the construct allows quantification of human behavior, which can then be modeled along with crop productivities and water availability. Quantified human behavior is a narrow bridge, but a bridge nevertheless, between inconsistent but realistic human behavior and more consistent physical and chemical processes.

The rational actor is the usual approach taken in decision-making frameworks, again making it possible to extend quantification into social human processes. Probabilities are assigned to various potential outcomes, and the rational decision is calculated. The "benevolent planner," who is wholly rational, seeks to optimize social welfare (also often expressed in quantitative terms). Arrow *et al.* (1996) summarized the fairly restrictive assumptions of this approach: a unique decision maker, a limited number of alternatives, consistent valuation of alternatives, and rational choices. Although these are admittedly unrealistic assumptions, they allow formal decision analysis to proceed.

However, as Camerer and Fehr (2006: 47) point out (and as Arrow *et al.* 1996 admitted), "A large body of evidence accumulated over the last three decades shows that many people violate the rationality and preference assumptions." When people do not behave rationally – e.g., they act altruistically or do not assemble complete information – the results of the economistic model are unreliable.

In another approach, Bayesian procedures help to quantify decision analysis. Bayes' theorem provides a formula for updating any kind of belief when confronted by new evidence, and many different types of evidence can be weighed together. Furthermore, a Bayesian analysis allows analysts to weight evidence subjectively, although they must quantify the weights.

3.4.2 Use quantitative data in qualitative ways

Quantitative data may be summarized in a non-numerical framework to emphasize the relationships among elements rather than precise amounts. For example, a 2×2 matrix may map single/multiple threatening human activities and single/multiple threatened environmental components or resources. Or researchers may use model results to conclude that a projected impact will be generally large or small; the specific numbers will be de-emphasized. In computer simulation games, quantitative data make the game "work," but interest in the outcomes focused on a qualitative state. For example, the game *Power Play*, which focuses on energy efficiency, provides insights into the kinds of changes that households and markets would undergo as efficiencies result from higher energy prices (Bernier *et al.* 2004).

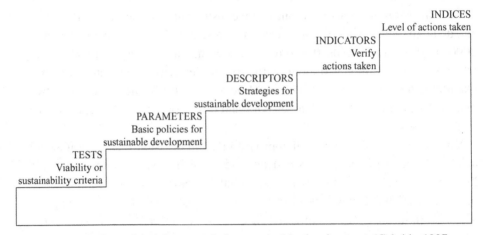

Figure 3.1. Operational framework for sustainable development (Cabrido 1997: 346; used by permission of John Wiley and Sons).

3.4.3 Present all data on their own terms

Gallopín (1997) has suggested approaches that do not involve aggregating but rather present a number of sustainability indicators simultaneously to give a picture of the quality being measured; he refers to these as "vector" indicators. These give users a more complete sense of the factors affecting the condition of a system of state variables, but one which is much more complex to manipulate and, to some degree, to compare. Use of vector indicators could be helpful for presentation of results that will not be compared across sectors or geographic units but are primarily intended for identification of priorities within a narrow domain.

Other approaches for presentation of aggregation are graphical. One potential approach would be to overlay results on geographic information systems or using other mapping methodologies. Cabrido (1997: 346; see Figure 3.1) develops a method for representing environmental quality indices using a Cartesian system of axes – the indices for the environmental components are graphed to indicate the relative level or strengths of stress among the environmental components.

The comparative case study method typically uses both quantitative and quali-tative data in a common framework. One broad approach is to identify variables-in-common across cases and, by comparison, determine which of the variables is crucial to the outcome being examined. Jared Diamond (1997, 2005) uses this method analyzing historical cases quantitatively, as do other scientists (e.g., Glantz 1988, Meyer et al. 1998) who explicitly reason by analogy. Kasperson et al. (1995) used this approach to characterize "regions at risk" of environmental collapse; and Abler et al. (2003) applied a common comparative methodology to study smaller areas within the United States to determine why greenhouse-gas emissions

	Quantitative	Qualitative
Interdisciplinary	Example: Integrated Assessment Models	Example: Collaboration among scientists to study impacts of climate change on agricultural productivity and the culture of farming
Disciplinary	Example: Measurement of soil carbon in the Earth's soils	Example: Ethnographic studies of African villages affected by drought in the Sahel

Figure 3.2. Four types of research approaches.

increase or decrease over time. A more formal and quantitative comparative method is Charles Ragin's (1987, 1994) comparative case study method, which allows for multiple pathways to determine similar outcomes. An innovative method is the identification of "syndromes" of global change – a set of patterns of co-evolution involving human–nature interaction (Lüdeke *et al.* 2004).

Perhaps the most powerful method of integrating quantitative and qualitative data lies in the process of collaboration, discussed in the next section.

3.5 Mechanisms for integrating qualitative and quantitative research in integrated regional assessment

We can continue the current practice of considering different kinds of data separately, using qualitative analyses of quantitative data, qualitative research methods, or theoretical studies. But perhaps a more fruitful strategy is to build on and develop other collaborative efforts in scientific research, either interdisciplinary or scientist–stakeholder collaborations. "Interdisciplinary" can mean a combination/ integration of qualitative and quantitative approaches, or it can mean various disciplinary quantitative approaches (as in integrated assessment models). Figure 3.2 shows this analytically.

Collaboration is important in establishing the balance between needs for quantification, standardization, and comparability in relation to established cultural measures appropriate to the region to be assessed. Even more, collaboration (conceived of broadly) can be itself an integrating mechanism. Case studies in the present volume illustrate how this works. For the Regional Integrated Sciences and Assessments (RISA, Pulwarty, Chapter 18, this volume), the Southeast Asian Regional Committee for START (SARCS, Lebel, Chapter 16, this volume) and the Inter-American (IAI, Liverman, Chapter 17, this volume), programs, scientific networks and collaborations have been the major products. Through interactions and negotiations, mutual learning and practical applications joined to computer-based models, these programs have created scientific spaces where quantitative and qualitative knowledge come together, thus opening the potential for integration.

The IAI lays great stress on international networks of scientists, collaborations, education/training, and free and open exchange of relevant data. All this is intended to yield better understanding of global environmental change in the Americas, although this goal is largely unrealized to date. Many political questions surrounding collaboration have been confronted and, in some cases, resolved; 14 countries ratified the IAI agreement and over 1000 scientists participate, and links with other international programs have been created. Workshops have been a principal activity, presumably to bring researchers together. The generality of the major themes may be a handicap in promoting interdisciplinary work; specific questions often galvanize collaborative research. Furthermore, the weaker representation of social science probably implies a dearth of qualitative research, and here again the point is that social scientists are working on such issues as urban pollution and forest concerns, and perhaps do not see or wish to make the effort to link their work to the broad IAI themes.

SARCS uses a network-based approach to address "overarching questions" about regional environmental change. The problem-and-place focus allows a high degree of integration of various kinds of knowledge.

The RISA program emphasizes collaborations and place-based, here-and-now research questions for regions within the United States. Highly qualitative activities characterize the first stages of effort, and both the problem focus and the requirement to include a spectrum of interests hold the potential for well-integrated activities. The National Research Council (NRC) evaluation of the US Climate Change Science Program (NRC 2007) counts the RISA program as a success in both approach and outcomes.

What of the integrated regional assessment programs that focus primarily on developing computer models? To the extent that getting the model to work predominates, the emphasis is on integrating quantitative information. The qualitative issues reside in the assumptions and model structure, and in this stage – often a recurring stage – the collaboration of scientists in defining the interactions among different "drivers" and factors can be a place where qualitative and quantitative research is integrated. Parson *et al.* (2007) discuss the tension in scenario-building exercises between qualitative storylines and quantitative realizations of scenarios. In the IPCC *Special Report on Emissions Scenarios* (SRES) (Nakićenović and Swart 2000) the qualitative storylines remained underdeveloped in favor of quantitative analyses. In the Millennium Ecosystem Assessment (MA) (2005) this tension was not resolved; instead, there are inconsistencies between the storylines and the quantitative scenarios.

The Asia–Pacific Integrated Model (AIM) (Kainuma *et al.*, Chapter 11, this volume) includes emissions, climate, and impact models; the overall goal is to evaluate policy responses. These incorporate assumptions about the socio-economic conditions of the future and embody those assumptions in numbers. The AIM collaborative team is diverse in institutions and disciplines.

The developers of CLIMPACTS (Warrick, Chapter 14, this volume) have chosen to focus on quantitative methods – both in the focus on biophysical sciences and in computer-based tools. The qualitative aspects of the program, besides the inescapable qualitative assumptions, reside in the intended purposes and user-friendliness of the model "family." Because the linked models have policy-relevant purposes, they have been developed to be easy to use, allow fast-running alternative analyses, and educate their users.

The ongoing SARCS is essentially network-oriented and development-focused. Capacity-building is an essential feature. In its first phase, the assessment provided the linkages among environmental change issues through the structures and processes of development in a relatively qualitative report. In the second phase, the assessment became more data-based and computer-model oriented. The "second generation" products aim to fill gaps identified in the overall report. The network is thus ready to attempt the second iteration of the integrated regional assessment. Throughout the process, members of the network have demonstrated awareness of the realities of the (qualitative) policy process. However, they recognize a "glaring weakness" in the lack of formal engagement of policy makers.

The IAI is attempting to establish a similar network of scientists across North and South America. Some progress has been made in engaging scientists, funding collaborative work, and agreeing to share data, but administrative and language barriers, inequalities in scientific education among countries, and a lack of interest on the part of social scientists have challenged IAI's efforts.

Policy relevance can be and is used as a factor that integrates qualitative and quantitative work. One example is RAINS, a model used as part of the process to address the acid rain issue in Europe (Alcamo *et al.* 1990, Hordijk 1991). The MacKenzie Basin study chose to engage the decision makers in research planning (Cohen 1993; see Box 2.1 in this volume). The ULYSSES project experimented with having citizen groups use computer models to make recommendations for their communities to address climate change, and the US National Assessment made stakeholder involvement central to the assessment process (Fisher and Kasemir, Chapter 8, this volume). The RAINS and MacKenzie Basin studies demonstrated the value of using quantitative tools to inform qualitative deliberations and judgments. Whether or not including qualitative stakeholder input is always a good strategy may depend on the purposes and methods used in obtaining the input.

3.6 Conclusion

This chapter began by suggesting the essential mutual dependence of so-called qualitative and so-called quantitative research. Quantitative research rests on judgmental, qualitative assumptions about how the world works, what suitable categories are for data, what constitute good data, and the validity of scientific procedures. For future projections, the role of qualitative assumptions is even more marked. Qualitative research, on the other hand, if it is to make sense of the world at all, must weigh and measure, at least in a comparative way, must judge what is important and what are critical variables in human development and change, and what constitute exhaustive and exclusive categories. Whether or not numbers are used, these are essentially quantitative tasks. Furthermore, the ability to quantitatively scope and assess issues provides the qualitative researcher with bases for framing problems.

Arguing the merits and demerits of qualitative and quantitative approaches is itself a wrongheaded and ultimately fruitless approach. The question is not "Which shall we use?" but "How can we use both to answer the question at hand fully and usefully?" In integrated regional assessment, this undoubtedly means explicit inclusion of qualitative dimensions such as building collaborations; including policy makers and other stakeholders in the research process; debating the starting assumptions; and being willing to re-examine categories, assumptions, and data of the research program as it goes along. Using both approaches also means defining numerical data sets that will be of use in scoping and defining the research question; undertaking the hard work of data gathering, checking, and validating/evaluating; performing adequate statistical analyses; and presenting results of quantitative analyses clearly and comprehensively.

The examples of integrated regional assessment in this volume suggest that ways to use both quantitative and qualitative research exist in the collaborations, networks, and interdisciplinary programs themselves. In such programs, theory and data are integrated, critiqued, and improved as researchers develop scientific knowledge.

References

Abler, R. *et al.* (eds.) 2003. *Global Change in Local Places: Estimating, Understanding, and Reducing Greenhouse Gases*. Cambridge: Cambridge University Press.

Alcamo, J., R. Shaw, and L. Hordijk (eds.) 1990. *The RAINS Model of Acidifications: Science and Strategies in Europe*. Dordrecht: Kluwer.

Arrow, K. J., J. Parikh, G. Pillet, *et al.* 1996. Decision-making frameworks for addressing climate change. In J. P. Bruce, H. Lee, and E. F. Haites (eds.), *Climate Change 1995: Economic and Social Dimensions of Climate Change*. Cambridge: Cambridge University Press.

Bernier, C., M. Ruth, A. Reeves, S. Laitner, and A. Meier 2004. *Power Play: The Energy Efficiency Game*. http://www.puaf.umd.edu/faculty/ruth/powerplay.htm

Best, J. 2001. *Damned Lies and Statistics: Untangling Numbers from the Media, Politicians, and Activists*. Berkeley, CA: University of California Press.

Best, J. 2004. *More Damned Lies and Statistics*. Berkeley, CA: University of California Press.

Black, D. 2000. Dreams of pure sociology. *Sociological Theory*, **18**(3), 343–368.

Brenkert, A. L. and E. L. Malone 2005. Modeling vulnerability and resilience to climate change: a case study of India and Indian states. *Climatic Change*, **72**(1–2), 57–102.

Cabrido, C. A., Jr. 1997. Sustainable development indicators: Philippine government initiatives. In B. Moldan and S. Billharz (eds.), *Sustainability Indicators: a Report on the Project on Indicators of Sustainable Development*. Chichester: Wiley.

Camerer, C. F. and E. Fehr 2006. When does "economic man" dominate social behavior? *Science*, **311**, 47–52.

Cohen, S. F. (ed.) 1993. *MacKenzie Basin Impact Study. Interim Report* 1, Toronto: Canadian Climate Center, Atmospheric Environment Service.

Diamond, J. 1997. *Guns, Germs, and Steel: the Fates of Human Societies*. New York: Norton, New York.

Diamond, J. 2005. *Collapse: How Societies Choose to Fail or Succeed*. New York: Viking.

Douglas, M. and S. Ney 1998. *Missing Persons: a Critique of Personhood in the Social Sciences*. Berkeley, CA: University of California Press.

Edmonds, J. A., H. M. Pitcher, D. Barns, R. Baron, and M. A. Wise 1993. Modeling future greenhouse gas emissions: the second generation model description. In L. R. Klein and Fu-chen Lo (eds.), *Modelling Global Change*. New York: United Nations University Press.

Fawcett, A. and R. D. Sands 2005. *The Second Generation Model: Model Description and Theory*. PNNL-15432. College Park, MD: Joint Global Change Research Institute. http://www.epa.gov/air/sgm_theory_final_1.pdf

Folke, C. 2006. Resilience: the emergence of a perspective for social-ecological systems analysis. *Global Environmental Change*, **16**, 253–267.

Folke, C., L. Pritchard, Jr., F. Berkes, J. Colding, and U. Svedin 1998. *The Problem of Fit between Ecosystems and Institutions. IHDP Working Paper* No. 2. Bonn: Human Dimensions Programme on Global Environmental Change.

Gallopin, G. C. 1997. Situational indicators. In B. Moldan and S. Billharz (eds.), *Sustainability Indicators: a Report on the Project on Indicators of Sustainable Development*. Chichester: Wiley.

Glantz, M. H. 1988. *Societal Responses to Regional Climate Change: Forecasting by Analogy*. Boulder, CO: Westview.

Halsnæs, K. and P. Shukla 2007. Framing issues. In *Mitigation of Climate Change. Contribution of Working Group III to the Intergovernmental Panel on Climate Change Fourth Assessment Report*. Cambridge: Cambridge University Press.

Hauert, C., A. Traulsen, H. Brandt, M. A. Nowak, and K. Sigmund 2007. Via freedom to coercion: the emergence of costly punishment. *Science*, **316**, 1905–1907.

Hordijk, L. 1991. Use of the RAINS model in acid rain negotiations in Europe. *Environmental Science and Technology*, **25**, 596–603.

Huff, D. 1993 (1954). *How to Lie with Statistics*. New York: Norton.

Izaurralde, R. C., K. H. Haugen-Kozyra, D. C. Jans, *et al.* 2001. Soil C dynamics: measurement, simulation and site-to-region scale-up. In R. Lal, J. M. Kimble, R. F. Follett, and B. A. Stewart (eds.), *Assessment Methods for Soil Carbon*. Boca Raton, FL: Lewis Publishers.

Kasperson, J. X., R. E. Kasperson, B. L. Turner, II (eds.) 1995. *Regions at Risk: International Comparisons of Threatened Environments*. Tokyo: UNU Press.

Likert, R. 1932. A technique for the measurement of attitudes. *Archives of Psychology*, **140**, 1–55.

Lüdeke, M. K. B., G. Petschel-Held, and H. J. Schellnhuber 2004. Syndromes of global change: the first panoramic view. *GAIA*, **1**, 42–49.

Merton, R. K. 1973. *The Sociology of Science: Theoretical and Empirical Investigations*. Chicago, IL: University of Chicago Press.

Meyer, W. B., W. N. Adger, K. Brown *et al.* 1998. Land and water use. In S. Rayner and E. L. Malone (eds.), *Human Choice and Climate Change, Volume 2: Resources and Technology*. Columbus, OH: Battelle Press.

Millennium Ecosystem Assessment (MA) 2005. *Ecosystems and Human Well-Being: Scenarios*. Washington, DC: Island Press.

Moss, R. H. and S. H. Schneider 2000. Uncertainties in the IPCC TAR: recommendations to lead authors for more consistent assessment and reporting. In R. Pachauri, R. Taniguchi, and K. Tanaka (eds.), *Guidance Papers on the Cross Cutting Issues of the Third Assessment Report of the IPCC*. Geneva: World Meteorological Organization.

Nakićenović, N. and R. Swart (eds.) 2000. *Special Report on Emissions Scenarios*. Cambridge: Cambridge University Press. http://www.grida.no/climate/ipcc/emission/

National Research Council (NRC) 2007. *Evaluating Progress of the U.S. Climate Change Science Program: Methods and Preliminary Results*. Washington, DC: National Academies Press. http://www.nap.edu/catalog.php?record_id=11934

Panchanathan, K. and R. Boyd 2005. Indirect reciprocity can stabilize cooperation without the second-order free rider problem. *Nature*, **432**, 499–502.

Parson, E., V. Burkett, K. Fisher-Vanden, *et al.* 2007. *Global Change Scenarios: Their Development and Use*. Sub-report 2.1B of Synthesis and Assessment Product 2.1 by the U.S. Climate Change Science Program and the Subcommittee on Global Change Research. Washington, DC: Department of Energy, Office of Biological & Environmental Research.

Porter, T. M. 1995. *Trust in Numbers: the Pursuit of Objectivity in Science and Public Life*. Princeton, NJ: Princeton University Press.

Ragin, C. 1987. *The Comparative Method: Moving beyond Qualitative and Quantitative Strategies*. Berkeley, CA: University of California Press.

Ragin, C. 1994. *Constructing Social Research: the Unity and Diversity of Method*. Thousand Oaks, CA: Pine Forge Press.

Rayner, S. and E. L. Malone 1998. The challenge of climate change to the social sciences. In S. Rayner and E. L. Malone (eds.), *Human Choice and Climate Change, Volume 4: What Have We Learned?* Columbus, OH: Battelle Press.

Scott, J. C. 1998. *Seeing Like a State: How Certain Schemes to Improve the Human Condition Have Failed*. Princeton, NJ: Yale University Press.

Thompson, M., M. Warburton, and T. Hatley 1986. *Uncertainty on a Himalayan Scale: an Institutional Theory of Environmental Perception and a Strategic Framework for the Sustainable Development of the Himalaya*. London: Ethnographica.

Yohe, G., E. L. Malone, A. L. Brenkert, *et al.* 2006a. A synthetic assessment of the global distribution of vulnerability to climate change from the IPCC perspective that reflects exposure and adaptive capacity. Palisades, NY: Center for International Earth Science Information Network, Columbia University. http://ciesin.columbia.edu/data/climate/

Yohe, G., E. L. Malone, A. L. Brenkert, *et al.* 2006b. Geographic distributions of vulnerability to climate change. *Integrated Assessment Journal*, **6**, 3.

Young, O. R., F. Berkhout, G. C. Gallopin, *et al.* 2006. The globalization of socio-ecological systems: an agenda for scientific research. *Global Environmental Change*, **16**, 304–316.

Zafirovski, Milan 2000. The rational choice generalization of neoclassical economics reconsidered: any theoretical legitimation for economic imperialism? *Sociological Theory*, **18**(3), 448–471.

4

Scale and scalar dynamics in integrated regional assessments

COLIN POLSKY AND DARLA K. MUNROE

4.1 Introduction

Integrated regional assessments (IRAs) are interdisciplinary research projects focusing on the regional implications of global environmental changes (GECs; see Yarnal, Chapter 2, this volume). A growing number of researchers are focusing on the region as a crucial – if not the most appropriate – unit of analysis for linking human and environmental processes across micro-, meso- and macro-spatial and temporal scales (Knight and Jäger, Chapter 1, this volume; Hewings 1986; Johnston et al. 1990; Abler et al. 2000; RISA 2001; Oinas and Malecki 2002). However, the importance of the regional scale does not mean that other scales should be ignored. IRA practitioners must analyze multiple scales simultaneously because GECs are characterized by *scalar dynamics* (Turner et al. 1995; Geoghegan et al. 1998). Given that the factors influencing a process typically operate on and interact across multiple scales, it follows that the results of any one single-scale analysis will be linked to the factors important to that scale of analysis (Stern et al. 1992; Wilbanks and Kates 1999). Thus a more holistic understanding of the regional system requires analysis of multiple scales.

Many chapters in this book identify scalar dynamics as an important issue for IRAs. The purpose of this chapter is to explore some methodological issues associated with studying scalar dynamics in IRAs. We begin with a general review of how the issue of scalar dynamics has been addressed in past IRAs. Then we discuss some methodological tools for examining scalar dynamics, with a particular emphasis on spatial statistics. We conclude by connecting the methodological discussion with the larger IRA research agenda, discussing future avenues for research.

4.2 Scale in integrated regional assessments: definitions
and theoretical background

We define an IRA as any study of human–environment interactions that satisfies Yarnal's definition. This definition contrasts with the common use of the term "integrated assessment" (and by extension, "integrated regional assessment") as synonymous with a relatively new class of computer models, such as the IMAGE 2.0 model of Alcamo *et al.* (1994). This relatively broad perspective allows us to engage theoretical perspectives often excluded from those computer models. We follow Gibson *et al.* (2000) in defining scale as the dimension(s) used to study a given process. In this way scale can refer to the spatial domain (e.g., whether tropical land-cover patterns should be studied at the tree, patch, stand, or forest level), the temporal domain (e.g., whether business cycle fluctuations occur at monthly or quarterly intervals), or even levels in a political hierarchy (e.g., municipal versus state versus Federal government).

There is no standard definition of the term region. In general, a region is large enough to reflect the average effects of macro-scale trends, and small enough to reflect the importance of micro-scale particularities (Easterling 1997). To promote regional-scale research is to deny that reductionist science – the idea that, in the current context, human–environment systems can be best understood by breaking them down into their smallest components – is necessarily the proper approach for IRAs (Turner and Meyer 1991). The reductionist approach is common, for example, in the field of economics, where it is often asserted that individual- (or household-) level circumstances are the most important factors for understanding human–environment interactions. The alternative to the reductionist approach, advocated by a growing number of scholars, is to acknowledge that there can be important actors and influences at all levels of a system (Peterson and Parker 1998; Moran *et al.* 2002), and that the effects of these various influences manifest clearly at the regional scale (Johnston *et al.* 1990).

Human–environment systems where outcomes at a given scale are influenced by factors operating on and interacting across multiple scales are said to exhibit *scalar dynamics* (Geoghegan *et al.* 1998). One of the first calls for research on scalar dynamics in IRAs (although the term IRA was not used) was Clark's (1985) paper on the importance of scale in understanding climate–society interactions. Clark argued that the broad range of characteristic spatial and temporal scales for natural and social processes provides important constraints and opportunities for the management of anthropogenic environmental problems. There is now a robust scholarly literature on scale and human–environment interactions, the theoretical basis for which derives principally from ecological thinking on systems, hierarchies, complexity, and thermodynamics (e.g., Root and Schneider 1995;

Kohn 1998; O'Neill 1988; Holling 2001; Easterling and Kok 2002). In recent years this research has also incorporated theoretical principles from landscape ecology (e.g., Turner *et al.* 1989; Wiens *et al.* 1993; Milne *et al.* 1996; Kok 2001); demography (e.g., Entwistle *et al.* 1998; Gutmann 2000), economics/economic geography (e.g., Walker and Solecki 1999; Abler *et al.* 2000; Hsieh 2000; Munroe 2000; Polsky and Easterling 2001; Polsky 2004), climate change impacts (e.g., Easterling *et al.* 1998a, b; Wilbanks and Kates 1999), political science (e.g., Cash and Moser 2000; Sprinz 2000), and political ecology (e.g., Grossman 1998; Turner 1999; Warren *et al.* 2001).

Space limitations here prevent a thorough review of these studies.[1] We make three brief observations about the literature on scalar dynamics in IRAs. First, although it is agreed that scalar dynamics should be important in both spatial and temporal domains, the bulk of the work to date has focused on spatial scale. The reason for this imbalance is not clear, especially since two disciplines so central to GEC studies – economics and climatology – have considerable expertise with temporal analytical techniques. Data availability may be a major concern. Although it is difficult to obtain a satisfactory spatial coverage for a set of variables, it is more challenging to do so consistently over a long time interval. More subtly, both the quantitative social sciences and ecology (e.g., Bisonette 1997) have traditionally heavily relied on the notion of "general equilibrium." This perspective assumes a priori that there is no important time trend in the relationships among variables (Picci 2001), or at least that such trends, if they exist, are theoretically and empirically uninteresting. As such, any observed temporal patterns are interpreted as reflecting either a short-term trend converging towards equilibrium, long-term market imperfections or policy interventions that prevent equilibrium, or statistical noise. In these cases, temporal effects are viewed more as a nuisance to be controlled and then forgotten than as a source of substantive information.[2] Whatever the reason for examining space without time (or vice versa), such an approach necessarily paints a partial picture of GECs. Explanations of GECs that do not adequately account for either spatial or temporal scale issues run the risk of misrepresenting key causal processes (see, e.g., Weinhold and Reis 2001).

Second, even though the field is characterized by a plethora of (sometimes conflicting) theoretical foundations, there is a developing consensus that human–environment systems are hierarchically organized. Such organization suggests that a regional perspective is needed to address *emergent properties*, or system

---

[1] In-depth theoretical reviews are provided by Gibson *et al.* (2000) and Evans *et al.* (2002).

[2] The same could be said for some studies that model spatial effects only for statistical reasons (Polsky 2004, Section 3.3).

characteristics not observable through single-scale analysis alone (Epstein and Axtell 1996; Jones and Duncan 1997; Peterson and Parker 1998; Easterling and Kok 2002). Emergence occurs when the "whole is not the sum of its parts" (Auyang 1998: 174), where critical local interactions and both local and cross-scale feed-backs lead to macroscopic patterns. For example, local-scale analysis of Chinese energy inefficiencies can be explained by, among other things, old equipment, but at larger scales other factors emerge as important, such as national-scale market distortions or the very nature of centrally planned economies at the global scale (Stern *et al.* 1992). Regional analysis may be useful to reflect how the effects of individual actions may be amplified or dampened after a critical threshold is exceeded and a larger-scale response materializes.

Third, spatial resolution is another important consideration for conducting and evaluating IRAs. The chosen resolution in IRAs may not match the ideal unit of analysis, for reasons of data availability, ignorance about processes, or both (Turner *et al.* 1995). Manson (2008) distinguishes between scale of *observation* (i.e., the fea-sible units of measurement) and scale of *explanation* (the level(s) at which processes occur). For example, the basic unit of observation for land-cover data is generally the pixel, which is an artifact of the technology of the remote sensing device, and as a division of space, may or may not fully represent the actual data in that space (Fisher 1997). NASA's Moderate Resolution Imaging Spectrometer's (MODIS) resolution of 1 km^2 would be unsuitable for studying garden plots of smaller dimension. When such a data/process mismatch occurs, the proposed model will be flawed, and any statistically significant associations estimated using those areal units suspect. This is the so-called Modifiable Areal Unit Problem (MAUP) (see, e.g., Openshaw and Taylor 1981; Green and Flowerdew 1997). The outstanding question is: By how much are such models flawed? Openshaw and Taylor (1979) demonstrated that for a given dataset, *any* correlation coefficient may be obtained between two variables, provided the boundaries of the areal units are sufficiently altered. This situation, which has been well studied in the spatial analysis literature is compounded in IRAs due to the multi-disciplinary nature of such analyses. Social and ecological systems do not necessarily coincide in terms of extent, resolution, or timescale (Bockstael and Bell 1998; Veldkamp *et al.* 2001). It is often the case in the practice of land-use and land-cover change (LUCC) modeling that distinct patterns or outcomes are evident at some scales and not others (Walsh *et al.* 1999; 2001). The integration of disparate social and ecological data leads to spatial error and additional aggregation problems (Rindfuss *et al.* 2004; Munroe and Müller 2007). As very-high resolu-tion land-cover data (e.g., Quickbird 0.6m imagery) become more accessible, the theoretical unit of analysis (e.g., household) may be of greater geographical extent than the imagery resolution (see Zhou *et al.* 2008, for a good example of such research).

Seminal work in spatial analysis has specifically focused on the impact of aggregation and zoning on multi-variate regression (Fotheringham and Wong 1991, Reynolds and Amrhein 1998). Administrative boundaries often represent an arbitrary division of space (Anselin 2001). However, in the spirit of Moran *et al.*, we argue that the boundaries of, for example, a county, do not simply obscure social organization and human behavior. For example, human activities often correspond to various non-nested hierarchical institutional arrangements: land use can be shaped by household, village, community, national or international effects. The last issue is thornier than a simple question of aggregation bias or data integration error, and warrants considerable attention. Over time policies and institutions develop that are specifically related to these boundaries, thereby justifying their use in spatial analysis (Gibson *et al.* 2000; Parker *et al.* 2003). In our view, cases where correlations vary with resolution and scale may indicate not only data/process mismatch but also additional information associated with scalar dynamics, in the form of previously omitted, yet important, dimensions of the process under study. Thus we do not examine the MAUP here (which is well treated elsewhere in the literature), and instead focus the remainder of this chapter on quantitative techniques for modeling scalar dynamics. Our assumption is that, notwithstanding legitimate concerns about the MAUP, a better understanding of scalar dynamics may improve our understanding of human–environment interactions.

4.3 Scalar dynamics in practice

4.3.1 Mathematical modeling

In light of the above review, a central challenge for IRA practitioners is to design studies that describe large-scale trends alongside localized deviations, in both space and time. There are analytical techniques that may help realize this goal, several of which are briefly reviewed here.[3] Simulation-based techniques (e.g., cellular automaton and agent-based models) are growing in popularity for examining scale effects in IRAs (Parker *et al.* 2003). These techniques represent a simulated social laboratory in which autonomous, potentially heterogeneous agents populate an environment, interact and communicate within that environment, and make decisions that tie behavior back to the environment. Thus, nature–society interactions are explicit (see, e.g., White and Engelen 1993; Batty and Longley 1994; Parker *et al.* 2002).

---

[3] In this paper we discuss quantitative techniques only, but we recognize that it is also necessary to employ qualitative techniques for understanding human–environment interactions. See Malone (Chapter 3, this volume) and Schröter *et al.* (2005) for related discussions.

One clear advantage of such techniques is the ability to model hierarchical relationships efficiently and explicitly, with interdependencies across scales, leading to the emergence of higher-level patterns from the collective result of individual actions (Epstein and Axtell 1996; Dean *et al.* 2000). For example, these models can describe how zoning laws (as a representative meso-level institution) emerge in response to many interactions among proximate individuals. These models are also potentially useful for examining the path (evolution) of a system in cases where the time required to reach equilibrium is beyond the time frame of interest to the researcher. In cases where these complexities are not present, simpler and more transparent modeling techniques may be appropriate (Parker *et al.* 2002).

Research using qualitative differential equations also shows promise for examining scalar dynamics in IRAs. In this approach, exemplified by the "Syndromes" work of (Petschel-Held *et al.* 1999), the large number of possible land-use classes (e.g., agriculture, urban) and associated influencing factors (e.g., climate, socio-economic characteristics) are grouped into a small set of archetypical patterns of GECs. For example, the so-called Sahel Syndrome describes the emergent property of food insecurity associated with the condition of intensively using agriculturally marginal lands in semiarid regions. The novelty of this analytical approach is that the regions identified need not be contiguous. The *functional* patterns of human–environment interactions that (may) give rise to the Sahel Syndrome exist not only in the Sahel proper but also in, for example, Central Asia, and may emerge anywhere the stated conditions are found (Lüdeke *et al.* 1999). Hence the potential for insight into scalar dynamics: the same process may be evaluated at local and regional scales independently, and inspected for what Turner *et al.* (1990) call the "cumulative" emergence of a global signature.

4.3.2 Statistical techniques: univariate spatial estimates

In the remainder of this paper we explore a few statistical techniques useful for studying scalar dynamics in IRAs. We examine spatial dimensions only because the multi-dimensional and multi-directional aspect of space yields an exceptionally complex conceptual and analytical issue to address. Tools to deal with scale patterns in the temporal domain are, in comparison, better known and substantially less complicated to implement. Statistical analysis is worth exploring in some detail here as it is perhaps the most commonly used quantitative technique in all of science, and as such is well understood by a wide audience. Even though the theory behind the statistical techniques equipped to examine spatial scalar dynamics dates from at least the 1970s (Cliff and Ord 1973a; b), these techniques are only now being widely used, as computing costs continue to decline and spatial data availability continues to increase (Anselin 1999; Goodchild *et al.* 2000). The calculations required for statistical analysis are greatly simplified if one assumes that the observations of analysis are statistically independent across geographic space. Sophisticated software (which is becoming increasingly commonplace and user-friendly) is required when there is statistical non-independence of observations by location, i.e., strong spatial autocorrelation.

Statistical independence by location implies that knowing the value for a variable for a given observation provides no additional statistical information about the value of that variable for another observation. For example, knowing the outcome of flipping a fair coin in Houston, Texas gives no statistical information about the outcome of flipping that coin in the nearby city of Galveston. The process is location-independent. By contrast, endemic to human–environment processes are spatial statistical dependencies, collectively termed *spatial effects*. For example, knowing the value of mean June rainfall in Houston means that one could make an educated guess about the value of that variable for Galveston – without having to measure it.

Geostatistical approaches (for more details see Cressie 1990, 1993) are designed to assess and predict the spatial distribution of a given process, such as the likely location of gold on a surface, given the locations of previous gold findings. An underlying assumption is that the expected pattern in a given variable is some function of distance. Geostatistical techniques, such as kriging and semivariogram analysis, aim both to predict a spatial surface at unsampled locations, and to assess overall spatial variability (i.e., to evaluate the strength of spatial predictions). Geostatistics, then, are often called a direct representation of space, since most approaches model spatial variability as a direct function of location (generally through x, y coordinates).

Two increasingly popular univariate statistical techniques designed to address spatial effects are the Moran's I and Local Indicator of Spatial Association (or LISA) statistics (Anselin 1995; 1996). For the Moran's I, which is a global statistic (i.e., it summarizes information from the entire study area), the difference between the value of one observation and the mean for that variable is compared to the difference between that variable's value at proximate observations and the global mean. A statistically insignificant Moran's I implies that the value of that variable is independent of its location, or that no visible spatial trend exists. Positive (negative) spatial autocorrelation occurs when observations with a certain value are close to other observations with similar (dissimilar) values. To identify localized spatial dependence or heterogeneity, the LISA statistic provides a measure of local instability. The LISA statistic yields a value for each observation as a function of its degree of deviation to the mean relative to its neighbors. The average of the LISA over all observations is the equivalent of the global Moran. Thus one can examine in a spatially explicit fashion each observation's contribution to the overall spatial pattern in the data set (Anselin 1995).

One IRA that uses the Moran's and LISA statistics to address scalar dynamics is a study of smallholder agriculture in Poland (Munroe 2000). Many researchers described the Polish agricultural structure at the time of transition as fragmented and inefficient (e.g., Adamowicz 1996; Noniewicz 1996; Florkowski 1997; Banski 1999), primarily the consequence of small average private farm sizes (just under seven hectares at the time of the 1996 agricultural census; GUS 1999a, b). Despite more than ten years of unregulated land markets since the end of the Socialist era, consolidation and/or abandonment of marginally productive land has not occurred. Therefore, more attention must be given to the spatial arrangement of the farm structure in Poland with regard to larger-scale (e.g., institutional) factors relating to market reform, as an attempt to forge more effective land-use policy (Munroe 2001).

In the Socialist era, Poland was divided into nine macro-regions, 49 *województwa* (or provinces), and some 3100 *gminy* (or municipalities). Land-use policy was largely forged at the macro-region. For example, the allowable maximum farm size depended upon migration incentives and other industrial policies at this level (Korbonski 1965). However, the provinces themselves did not have the authority to make or implement policy. The municipalities did have local authority and functioned as cohesive administrative bodies. There was no middle-tier decision-making body, and the planning economy operating at a regional level was often at odds with local-level efforts (Kosc 2000).

Data were collected from the 1996 Polish agricultural census on average private farm sizes by 3100 municipalities, the 49 provinces, and the nine macro-regions (GUS 1999a, b). Results from Moran's *I* and LISA analyses suggest a spatial scale-dependent relationship between the administrative structure and the distribution of agricultural land.[4] The value for Moran's *I* was high at the local level (0.61), increased at the provincial level (0.75), and decreased at the macro-region level (0.40) (Figure 4.1). This finding is contrary to most studies on the effect of aggregation on Moran's *I*, where the value of the Moran's *I* statistic typically decreases with aggregation, particularly when the variable in question exhibits significant spatial autocorrelation (Reynolds and Amrhein 1998). Therefore, the spatial patterns apparent in private farming may be a form of induced autocorrelation, i.e., the observable spatial autocorrelation is related to a second causal factor, how the land distribution was administrated, that is itself spatially autocorrelated. For example, because historically land policies differed sharply across regions, and economic

---

[4] Plots were created using the spdep package within R, available at http://www.cran.r-project.org/.

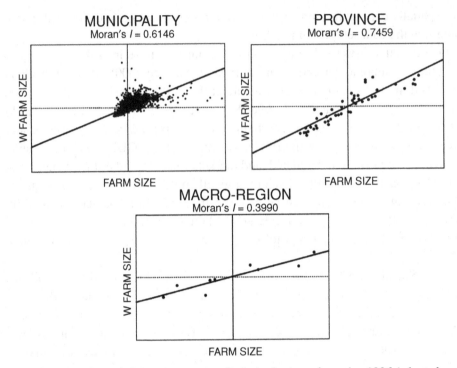

Figure 4.1. Moran's *I* for average area (in ha.) of private farm size 1996 (adapted from Munroe 2000).

recovery since the transition has been strongly influenced by local policies (Gorzelak 1998), it is not surprising that administrative policies shape the distribution of farm size.

Patterns in space are more similar at the local and the national level, whereas the middle (provincial) level shows less spatial similarity in the distribution of farm sizes. These results suggest that policy at the mid-tier provincial level could be used to link the local and the macro-regions more effectively in developing the necessary market infrastructure for more efficient land-use patterns. Since 1998, Poland has undergone a sweeping administrative reform, where the number of regions was consolidated from 49 to 17. It will be interesting to see what the ultimate effect the development of a more comprehensive middle-tier administrative structure, and its influence on land-use patterns, will be.

4.3.3 Statistical techniques: multi-variate spatial modeling

Multi-variate statistical techniques in IRAs and related studies are commonly embodied by regression modeling (e.g., Verburg and Chen 2000; Kok *et al.* 2001). For the purpose of studying spatial scalar dynamics, the basic Ordinary Least

Squares (OLS) regression model should be adapted to account for spatial effects. Spatial effects can be organized into two categories corresponding to two violations of OLS model assumptions. Residual spatial autocorrelation is understood in terms of *spatial dependence*, and heteroscedasticity with a spatial dimension in terms of *spatial heterogeneity* (Anselin 1992).[5] The former category is associated with micro-scale effects, and the latter, meso-scale, relative to the effects embodied for the study region (macro-scale) by the regression coefficients. The most common rationale for modeling these spatial effects is to improve the statistical properties of the OLS models that violate the assumptions of no residual spatial autocorrelation and/or homoscedasticity (e.g., Benirschka and Binkley 1994).

An additional, less common, and yet no less worthy goal of modeling spatial effects is to explore *why* the modeled spatial interactions yield an improved model fit over the original a-spatial OLS model (Irwin 2002). In this way evaluating regression models for spatial effects is a way of using pattern to suggest unspecified dimensions of the process under study. Statistical models can only go so far in this regard, so additional research (often of a qualitative nature) will likely be needed to elaborate on the importance of the unspecified factors (Schröter *et al.* 2005; Polsky *et al.* 2007).

Multi-level (or hierarchical) modeling is a multi-variate regression technique designed for gaining insight into the operation of hierarchically organized systems (see, e.g., Bryk and Raudenbush 1992; Goldstein 1995). Instead of specifying all independent variables at the same scale, as with OLS models, with this technique the scale(s) at which variables operate is explicit in how the variables are specified. Accordingly, this technique can be used to address heteroscedasticity with a spatial dimension. Multi-level modeling therefore permits the direct evaluation of scale-specific relationships as well as cross-scale interactions. Even though multi-level modeling is under-utilized in IRAs to date, there is a growing literature on the topic with respect to spatial effects in general and human–environment interactions in particular (Easterling and Polsky 2003; see, e.g., Gould *et al.* 1997; Westert and Verhoeff 1997; Raudenbush and Sampson 1999; Sampson *et al.* 1999; Hoshino 2001; Polsky and Easterling 2001). Given the adequate treatment of multi-level modeling elsewhere, we focus the remainder of discussion on multi-variate techniques on another regression-based approach, *spatial econometrics*. This technique boasts a rapidly growing literature (e.g., Fotheringham and Wong 1991; Bockstael 1996; Anselin 1988, 2001; Reynolds and Amrhein 1998; Paelinck 2000). However, the focus in this literature is often on small-scale interactions; issues of spatial scale and scalar dynamics are largely implicit.

---

[5] Technical discussions of spatial autocorrelation, heteroscedasticity, and satisfying model assumptions may be found in many previous discussions of regression modeling (Gould 1970; Anselin 1988; Kemedy 1998).

An IRA example that explicitly addresses scalar dynamics using spatial econometrics is a study of agricultural land values in the US Great Plains (Polsky 2004). This analysis uses Ricardian land-use theory to evaluate the importance of climate in the determination of agricultural land values[6] relative to other important factors (e.g., population density and soil quality). A spatial econometric regression model is used to estimate the statistical relationship between current climate and land values (i.e., the economic value of climate controlling for the other factors). The objective is to use the estimated relationships as a proxy for understanding the possible economic impacts of climate change, by applying a hypothetical climate change to the estimated historical relationships. For the study region of 446 counties, the model is estimated six times, once each for the years 1969, 1974, 1978, 1982, 1987, and 1992.

The spatial econometric framework is used to evaluate effects at three spatial scales, simultaneously. The *regional scale* – defined as the entire study area – is represented by the regional mean relationships for the independent variables. These relationships are embodied by the estimates of the regression coefficients. This part of the model is common to any regression model. The remaining two scales are represented by terms not traditionally included in regression models (the "spatial econometric" terms). The *sub-regional* scale is represented by the non-constant ("groupwise heteroscedastic" or GHET) error term, ε_i, for $i = 1, 2$, depending on whether or not a county overlies the Ogallala, the principal source of irrigation water in the Great Plains. The *local scale* – defined as sets of contiguous counties – is represented by the interfarmer communication term Wy and the associated estimate of the "spatial lag autocorrelation" coefficient ρ. The regional scale in this study encompasses one set of 446 counties, the sub-regional scale divides the region into two sets of counties (209 for the Ogallala, 237 for the rest of the Great Plains), and the local scale comprises many sets of small numbers of contiguous counties, about seven on average.

The three-scale structure to the model means that results can be associated with the three different scales. The regional-scale relationships are sufficiently obvious to not merit detailed discussion here. By construction, these estimates are the most representative values for the Great Plains considered as a whole. The local- and sub-regional-scale relationships deserve a more detailed explanation. Estimates for the coefficient of the local-scale interaction term ρ are significant for every year (Polsky 2002). As such, land values in a given county, for every year, must be explained in terms of the land values of neighboring counties, *even after accounting for the other independent variables*. Some form of interfarmer communication (be it the diffusion of a new farming technique, a tendency towards a bandwagon

---

[6] Data on county land values (per acre) are provided by the US Census of Agriculture (USDA 1995).

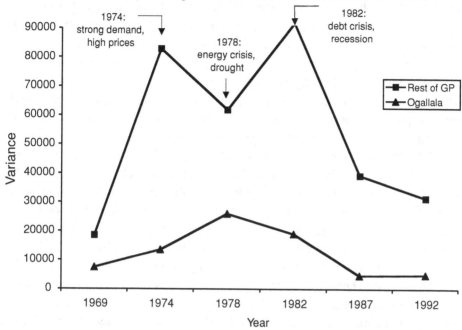

Figure 4.2. Unexplained variance from spatial regression models (used by permission from C. Polsky 2004).

effect, or any other social interaction-related factor) must be added to the model, to explain the observed patterns of land values. In the absence of specific variable(s) measuring such social interactions, the ρ term – which does not specify explicitly how, when, where, and why such important communications occur – is used as a placeholder to mark the need for further investigation into those processes. Studies that do not employ the spatial econometric framework are hard-pressed to model this dimension of human–environment interactions.

The sub-regional-scale GHET term provides a separate estimate for the variance in each of two sub-regions, for all years. This set of results can be understood as a measure of "poorness-of-fit," or the importance of all excluded variables (save the social interaction term ρ). Thus the closer the value of the variance for a given year (expressed as the sum of the squares of the residuals) is to zero, the better the model fit. Figure 4.2 graphs these results over time, and tells two dominant stories. The model fits significantly better within the Ogallala Aquifer than outside the Ogallala Aquifer, in all years, and the divergence in model fits between the two sub-regions differs significantly over time. These spatial and temporal differences in GHET values suggest that proximity to the Ogallala Aquifer is strongly associated with lesser land value sensitivity to variations in large-scale social and ecological factors omitted from the model (Polsky 2004). In this case, the omitted factors are associated with the dramatic swings in supply and demand (the "boom–bust" period) of the 1970s and 1980s in US agriculture, especially for grains. These factors are omitted from the original model because of problems with data availability, model stability, or both.

This finding does not suggest simply that in hard times it is better to be an irrigating farmer than not. The GHET values suggest a more general statement about the process of agricultural land use because the error terms are estimated *net of the effect of irrigation* on land values (irrigation is specified as an independent variable for the entire region). Thus the GHET analysis characterizes the importance of the Ogallala Aquifer in buffering land value fluctuations relative to the rest of the Great Plains not through the direct effects of irrigation but in terms of the network of human–environment relationships that have developed around the practice of irrigation. As with the large-scale (between-county) farmer communication term ρ, the specifics of these relationships are not explicit in the model. Instead, the GHET results act as a placeholder for future research by estimating the magnitude of the importance of this buffering function, expressed in terms of model poorness-of-fit. For example, in response to declining water tables over recent decades, many Ogallala communities have implemented innovative management programs that combine science and outreach to promote more conservative water use (Cash 2001; Emel and Roberts 1995). Yet it is unclear what makes one management regime more effective than others. Thus, following recent calls (Malone, Chapter 3, this volume; Schröter *et al.* 2005; Polsky *et al.* 2007) to blend quantitative and qualitative methods in IRAs, a key priority for spatial econometrics research is to unpack the GHET (meso) and ρ (micro) terms for greater insight into human–environment interactions. By construction, such qualitative analyses will also address scalar dynamics.

4.3.4 Moving from spatial analysis to spatio-temporal approaches

As noted in Section 2, there is an increasing interest in addressing scalar dynamics in time as well as in space. In this chapter we have emphasized statistical techniques for examining spatial scalar dynamics. Methods to model variations in both space and time in IRAs are emerging (see, e.g., Stern and Kaufmann 2000; Geoghegan *et al.* 2001; Kaufmann and Seto 2001; Serneels and Lambin 2001). However, these approaches, while spatially and temporally explicit, do not often model scalar dynamics. In the coming years we hope to see increasing attention to scalar dynamics in spatio-temporal models. In order to understand and model GECs, analytical techniques must increasingly account for complexity in each of these areas. This relatively small number of applications may be associated with practical matters such as data availability. However, given large-scale data collection efforts in recent decades (e.g., HERO 2002; SYPR 2002; AgTrans 2003), we believe there to be an increasing opportunity for space–time statistical models in the coming years.

That said, our optimism is tempered by technical challenges associated with estimating space–time statistical models. For example, identifying causality while accounting for both spatial and temporal dependence is exceedingly difficult. As Elhorst (2001) notes, alternative models for spatial-temporal specification testing are non-nested, so the usual practice of "general-to-specific" does not necessarily apply. Weinhold and Reis (2001) point out that lags in space and time relate very closely to the potential (mis)identification of causality, because identifying the temporal chain of causality in a given phenomenon is often complicated by spatial dependence, or vice versa. Thus although space–time modeling is a way to capture more of the real-world complexity of human–environment interactions than considering either space or time alone, adding realism often comes at the expense of interpretability.

Indeed, there is a general tradeoff between comprehensiveness and parsimony in statistical modeling. At present we believe this tension must be resolved on a case-by-case basis. There is a good argument to be made for both objectives, although in general we favor parsimony over comprehensiveness. At present, it is difficult to estimate models with more than a few points in time. Provided the sample size of cross-sectional dataset is relatively large to begin with, the comparative statics approach, where multiple points in time are evaluated independently and then compared post hoc, is an attractive option (Munroe *et al.* 2002; see, e.g., Polsky 2004). A set of cross-sectional observations may actually provide more statistical information regarding how a phenomenon evolved in prior time periods than sequential observations on one unit alone. An observation at a given time is the result of past processes, reflecting how the process has operated in the past (Nerlove 1999; Elhorst 2001; see, e.g., Polsky 2004).

4.4 Conclusions

In the recent literature on the human dimensions of global environmental change, the concept of scalar dynamics has been recognized as potentially important; the proposition that evidence for scalar dynamics represents more than a statistical nuisance merits continued investigation. Unfortunately, there is little guidance for operating and analyzing scalar dynamics in IRAs in general, abstract terms. An empirical literature on this topic is recently developing, the theoretical foundations of which are varied. Although many techniques are available for examining scalar dynamics in IRAs, we focus in this review on statistical methods as a relatively straightforward yet compelling set of techniques for this purpose. Examples of scalar dynamics insights to be gained by a statistical approach are provided by case studies of the area devoted to smallholder agriculture in Poland, and of agricultural land values in the US Great Plains. These empirical examples demonstrate how "spatial" violations of the common statistical assumption of independence of observations can be interpreted as evidence for substantively important (yet poorly understood) information about the process under study. To the extent that statistical models fail to capture system dynamics, examining "poorness-of-fit" in this way may highlight important factors missing from the original model, and that may be examined subsequently using qualitative and/or quantitative means.

In a recent, comprehensive review of modeling approaches to study GEC, Agarwal and colleagues (2002) developed a framework for evaluating the relative complexity of a wide variety of land-use models along three dimensions: time, space, and human decision making. The particular approach of a given study, and its chosen spatial and temporal focus, is determined by the degree of emphasis placed on the individual components of the system by the researchers. No single quantitative approach will ever likely be able to capture all the dynamics and dimensions of coupled human–environment processes. These systems are too complex to be completely understood in quantitative terms. More research and analysis is warranted to evaluate the intersection of multiple, competing sources of scale effects and scalar dynamics on these integrated efforts. To understand and design effective policy interventions addressing the most pressing issues of GEC, integrated assessment projects necessarily involve the merging and weaving of the three domains described by Agarwal *et al.* (2002). Best practice efforts should include careful consideration of the tradeoffs caused by confounding scalar dynamics corresponding to the most appropriate unit of analysis, the objectives of the study, and the corresponding relevant research questions. In this way IRA results will be more accurate and salient to a broader range of people. Consequently, we expect that IRAs will become increasingly policy relevant in the coming years, moving the field beyond the current state of theoretical disagreements towards a shared, interdisciplinary understanding of the causes and consequences of GECs.

Acknowledgements

This paper is based on research supported by the US National Science Foundation award BCS-0004236; the US National Oceanic and Atmospheric Administration's Office of Global Programs through the Research and Assessment Systems for Sustainability Program (Harvard University), and the Climate and Global Change Program; and the US National Science Foundation's award # SBR-9521918 as part of ongoing research at the Center for the Study of Institutions, Population, and Environmental Change (CIPEC) at Indiana University.

References

Abler, D., J. Shortle, A. Rose, and G. Oladosu 2000. Characterizing regional economic impacts and responses to climate change. *Global and Planetary Change*, **25**, 67–81.

Adamowicz, M. 1996. Main problems of the agribusiness sector development in Poland and ways of solution. *Annals of Warsaw Agricultural University: Agricultural Economics and Rural Sociology*, **32**, 89–94.

Agarwal, C., G. M. Green, J. M. Grove, T. Evans, and C. Schweik 2002. *A Review and Assessment of Land-Use Change Models: Dynamics of Space, Time, and Human Choice*. Gen. Tech. Rep. NE-297. Newton Square, PA: US Department of Agriculture, Forest Service, Northeastern Research Station. 61 p.

AgTrans 2003. Project description. Agrarian landscapes in transition: a cross-scale approach.
http://ces.asu.edu/agtrans (accessed February 14, 2003).

Alcamo, G. J., J. Kreileman, J. S. Krol, and G. Zuidema 1994. Modeling the global society–biosphere–climate system. Part I: model description and testing. *Water, Air and Soil Pollution*, **76**, 1–35.

Anselin, L. 1988. *Spatial Econometrics: Methods and Models*. Dordrecht: Kluwer.

Anselin, L. 1992. Spatial dependence and spatial heterogeneity: model specification issues in the spatial expansion paradigm. In J. P. Jones and E. Casetti (eds.), *Applications of the Expansion Method*. London: Routledge.

Anselin, L. 1995. Local indicators of spatial association – LISA. *Geographical Analysis*, **27**, 93–115.

Anselin, L. 1996. The Moran scatterplot as an ESDA tool to assess local instability in spatial association. In P. Longley, S. Brooks, R. McDonnell, and B. Macmillan (eds.), *Spatial Analytical Perspectives on GIS*. New York: Wiley, pp. 77–94.

Anselin, L. 1999. The future of spatial analysis in the social sciences. *Geographic Information Sciences*, **5**(2), 67–76.

Anselin, L. 2001. Spatial econometrics. In B. Baltagi (ed.), *Companion to Theoretical Econometrics*. Oxford: Blackwell, pp. 310–330.

Auyang, S. Y. 1998. *Foundations of complex-system theories in Economics, Evolutionary Biology, and Statistical Physics*. Cambridge: Cambridge University Press.

Banski, J. 1999. Obszary problemowe w rolnictwie Polski (Problem areas in Polish agriculture). *Prace Geograficzne*, **172**, 128–135.

Batty, M. and P. Longley 1994. From cells to cities. *Environment and Planning B*, **21**, S31–S48.

Benirschka, M. and J. K. Binkley 1994. Land Price Volatility in a Geographically Dispersed Market. *American Journal of Agricultural Economics*, **76**(May): 185–195.

Bisonette, J. A. 1997. Scale-sensitive ecological properties: historical context, current meaning. In J. A. Bisonette (ed.), *Wildlife and Landscape Ecology: Effects of Pattern and Scale*. New York Springer-Verlag.

Bockstael, N. 1996. Modeling economics and ecology: the importance of a spatial perspective. *American Journal of Agricultural Economics*, **78**, 1168–1180.

Bockstael, N. and K. Bell 1998. Land-use patterns and water quality: the effect of differential land management controls. In R. Just and S. Netanyahu (eds.), *Conflict and Cooperation on Trans-Boundary Water Resources*. Boston, MA: Kluwer, pp. 169–191.

Bryk, A. S. and S. W. Raudenbush 1992. *Hierarchical Linear Models: Applications and Data Analysis Methods*. Beverly Hills, CA: Sage.

Cash, D. 2001. Integrating information and decision-making in a multi-level world: cross-scale environmental science and management. Unpublished PhD Dissertation Thesis, Harvard University, Cambridge, MA.

Cash, D. and S. C. Moser 2000. Linking global and local scales: designing dynamic assessment and management processes. *Global Environmental Change*, **10**, 109–120.

Clark, W. C. 1985. Scales of climate impacts. *Climatic Change*, **7**, 5–27.

Cliff, A. D. and J. K. Ord 1973a. A note on statistical hypothesis testing. *Area*, **5**, 240.

Cliff, A. D. and J. K. Ord 1973b. *Spatial Autocorrelation*. London: Pion.

Cressie, N. 1990. The origins of kriging. *Mathematical Geology*, **22**(3), 239–252.

Cressie, N. 1993. *Statistics for Spatial Data*. New York: Wiley.

Dean, J. S. *et al.* 2000. Understanding Anasazi cultural change through agent-based modeling. In T. A. Kohler and G. J. Gumerman (eds.), *Dynamics in Human and Primate Societies*. Oxford: Oxford University.

Easterling, W. E. 1997. Why regional studies are needed in the development of full-scale integrated assessment modelling of global change processes. *Global Environmental Change*, **7**(4), 337–356.

Easterling, W. E. and K. Kok 2002. Emergent properties of scale in global environmental modeling: are there any? *International Journal of Integrated Assessment*, **3**(2–3), 2233–2246.

Easterling, W. E. and C. Polsky 2003. Crossing the complex divide: linking scales for understanding coupled human–environment systems. In R. McMaster and E. Sheppard (eds.), *Scale and Geographic Inquiry*, Oxford: Blackwell, pp. 55–64.

Easterling, W. E., C. Polsky, D. Goodin, M. Mayfield, W. A. Muraco, and B. Yarnal 1998a. Changing places, changing emissions: the cross-scale reliability of greenhouse gas emission inventories in the U.S. *Local Environment*, **3**(3), 247–262.

Easterling, W. E., A. Weiss, C. Hays, and L. Mearns 1998b. Spatial scales of climate information for simulating wheat and maize productivity: the case of the US Great Plains. *Agricultural and Forest Meteorology*, **90**, 51–63.

Elhorst, J. P. 2001. Dynamic models in the space and time. *Geographical Analysis*, **33**(2), 119–140.

Emel, J. and R. Roberts 1995. Institutional form and its effects on environmental change: the case of groundwater in the Southern High Plains. *Annals of the Association of American Geographers*, **85**(4), 664–683.

Entwistle, B., S. J. Walsh, R. R. Rindfuss, and A. Chamratrithirong 1998. Land-use/land-cover and population dynamics, Nang Rong, Thailand. In D. Liverman, E. F. Moran, R. R. Rindfuss, and P. C. Stern (eds.), *People and Pixels: Linking Remote Sensing and Social Science*. Washington, DC: National Academy Press.

Epstein, J. M. and R. Axtell 1996. *Growing Artificial Societies: Social Science from the Ground Up*. Washington, DC: Brookings Institution.

Evans, T. P., E. Ostrom, and C. Gibson 2002. Scale issues with social data in integrated assessment modeling. *International Journal of Integrated Assessment*, **3**, 135–150.

Fisher, P. 1997. The pixel: a snare and a delusion. *International Journal of Remote Sensing*, **18**(3), 679–685.

Florkowski, W. J., H. Szulce, and A. H. Elnagheeb 1997. Privatizing agricultural services in a transition economy. *Review of Agricultural Economics*, **19**(1), 45–57.

Fotheringham, A. S. and D. W. S. Wong 1991. The modifiable area unit problem in multivariate analysis. *Environment and Planning A*, **23**, 1025–1044.

Geoghegan, J., L. Pritchard, Y. Ogneva-Himmelberger, R. K. Chowdhury, S. Sanderson, and B. L. Turner 1998. "Socializing the pixel" and "pixelizing the social" in land-use and land-cover change. In D. Liverman, E. F. Moran, R. R. Rindfuss, and P. C. Stern (eds.) *People and Pixels: Linking Remote Sensing and Social Science*. Washington, DC: National Academy Press.

Geoghegan, J., S. C. Villar, P. Klepeis, P. M. Mendoza, Y. Ogneva-Himmelberger, R. R. Chowdhury, B. L. Turner and C. Vance 2001. Modeling tropical deforestation in the Southern Yucatan Peninsular Region: comparing survey and satellite data. *Agriculture, Ecosystems, and Environment*, **85**(1–3), 25–46.

Gibson, C., E. Ostrom, and T.-K. Ahn 2000. The concept of scale and the human dimensions of global change: a survey. *Ecological Economics*, **32**, 217–239.

Goldstein, H. 1995. Multilevel statistical models. In *Kendall's Library of Statistics*, vol. 3. London: Edward Arnold.

Goodchild, M., L. Anselin, R. P. Applebaum, and B. H. Harthorn 2000. Toward spatially integrated social science. *International Regional Science Review*, **23**(2), 139–159.

Gorzelak, G. 1998. *Regional and Local Potential for Transformation in Poland*. Warsaw: European Institute for Regional and Local Development.

Gould, M., K. Jones, and G. Moon 1997. Guest editorial: the scope of multilevel models. *Environment and Planning A*, **29**(4), 581–584.

Gould, P. 1970. Is *Statistix inferens* the geographical name for a wild goose? *Economic Geography*, **46**, 439–448.

Green, M. and R. Flowerdew 1997. New evidence on the modifiable areal unit problem. In P. Longley and M. Batty (eds.), *Spatial Analysis: Modelling in a GIS Environment*. Cambridge: Geoinformation International, pp. 41–54.

Grossman, L. S. 1998. *The Political Ecology of Bananas: Contract Farming, Peasants, and Agrarian Change in the Eastern Caribbean*. Chapel Hill, NC: University of North Carolina Press.

GUS 1999a. *Rocznik Gmin (Municipal Yearbook)*. ed. G. U. Statystyczny. Warsaw: GUS.

GUS 1999b. *Rocznik Wojewodztw (Provincial Yearbook)*, ed. G. U. Statystyczny. Warsaw: GUS.

Gutmann, M. P., 2000. Scaling and demographic issues in global change research: The Great Plains, 1880–1990. *Climatic Change*, **44**(3), 377–391.

HERO 2002. Project Description. HERO: Human-environment regional observatory. http://hero.geog.psu.edu/ (accessed 1 May 2002).

Hewings, G. J. D. 1986. Problems of integration in the modelling of regional systems. In M. Madden (ed.), *Integrated Analysis of Regional Systems*. London: Pion, pp. 37–53.

Holling, C. S. 2001. Understanding the complexity of economic, ecological, and social systems. *Ecosystems*, **4**, 390–405.

Hoshino, S. 2001. Multilevel modeling on farmland distribution in Japan. *Land Use Policy*, **18**, 75–90.

Hsieh, W. 2000. Spatial dependence among county-level land use changes. Unpublished PhD Dissertation Thesis, The Ohio State University, Columbus, OH.

Irwin, E. G. 2002. *Identifying Interaction Effects among Spatially Distributed Agents: an Application to Land Use Spillovers*. Working Paper, Department of Agricultural, Environmental, and Development Economics, the Ohio State University.

Johnston, R. J., J. Hauer, and G. Hoekveld (eds.) 1990. *Regional Geography: Current Developments and Future Prospects*. London: Routledge, pp. 1–10.

Jones, K. and C. Duncan 1997. People and places: the multilevel model as a general framework for the quantitative analysis of geographical data. In P. Longley and M. Batty (eds.), *Spatial Analysis: Modelling in a GIS Environment*. Geoinformation International.

Kaufmann, R. and K. Seto 2001. Change detection, accuracy, and bias in a sequential analysis of Landsat imagery in the Pearl River Delta, China: econometric techniques. *Agriculture, Ecosystems, and Environment*, **85**(1–3), 95–105.

Kohn, J. 1998. Thinking in terms of system hierarchies and velocities: what makes development sustainable? *Ecological Economics*, **26**, 173–187.

Kok, K. 2001. A modelling approach with case studies for Central America. Unpublished PhD Dissertation Thesis, Wageningen University, Wageningen, The Netherlands.

Kok, K., A. Farrow, A. Veldkamp, and P. H. Verburg 2001. A method and application of multi-scale validation in spatial land use models. *Agriculture, Ecosystems, and Environment*, **85**(1–3), 223–238.

Korbonski, A. 1965. *Politics of Socialist Agriculture in Poland: 1945–1960*. New York: Columbia University Press.

Kosc, W. 2000. After the reform: the effects of administrative change on Poland's cities. *Central Europe Review*, **2**(26).

Lüdeke, M., O. Moldenhauer, and G. Petschel-Held 1999. Rural poverty driven soil degradation under climate change: the sensitivity of the disposition towards the Sahel syndrome with respect to climate. *Environmental Modeling and Assessment*, **4**, 315–326.

Manson, S. M. 2008. Does scale exist? An epistemological scale continuum for complex human-environment systems. *Geoforum*, **39**(2), 776–788.

Milne, B. T., A. R. Johnson, T. H. Keitt, C. A. Hatfield, J. David, and P. T. Hraber 1996. Detection of critical densities associated with Pinon–Juniper woodland ecotones. *Ecology*, **77**, 805–821.

Moran, E. F., E. Ostrom, and J. C. Randolph 2002. *Ecological Systems and Multitier Human Organization, UNESCO Encyclopedia of Life Support Systems*. Oxford: EOLSS Publishers.

Munroe, D. 2000. Regional variations in Polish peasant farming: composed error, spatial econometric and spatial interaction techniques. Unpublished PhD Dissertation Thesis, The University of Illinois, Urbana-Champaign, IL.

Munroe, D. 2001. Peasant farm efficiency in Poland: an international perspective. *Regional Studies*, **35**(5), 459–469.

Munroe, D. K. and D. Müller 2007. Issues in spatially explicit statistical land-use/cover change (LUCC) models: examples from Western Honduras and the Central Highlands of Vietnam. *Land Use Policy*, **24**, 521–530.

Munroe, D., J. Southworth, and C. M. Tucker 2002. The dynamics of land-cover change in Western Honduras: spatial autocorrelation and temporal variation. *Agricultural Economics*, **27**, 355–369.

Nerlove, M. 1999. Properties of alternative estimators of dynamic panel models: an empirical analysis of cross-country data for the study of economic growth. In C. Hsiao, K. Lahiri, L.-F. Lee, and M. H. Pesaran (eds.), *Analysis of Panels and Limited Dependent Variable Models*. Cambridge: Cambridge University Press.

Noniewicz, C. 1996. Przestrzenne aspekty rozwoju gospodarki chlopskiej (Spatial aspects of development of peasant farms), Ekonomika i Polityka Rolna w Procesie Transformowania Gospodarki. Bialystok, Poland: University of Warsaw Press.

Oinas, P. and E. J. Malecki 2002. The evolution of technologies in time and space: from national and regional to spatial innovation systems. *International Regional Science Review*, **25**(1), 102–131.

O'Neill, R. V. 1988. Hierarchy theory and global change. In T. Rosswall, R. G. Woodmansee, and P. G. Risser (eds.), *SCOPE 35, Scales and Global Change: Spatial and Temporal Variability in Biospheric and Geospheric Processes*. Chichester: Wiley, pp. 29–45.

Openshaw, S. and P. J. Taylor 1979. A million or so correlation coefficients: three experiments on the modifiable areal unit problem. In N. Wrigley and R. J. Bennett (eds.), *Statistical Applications in the Spatial Sciences*. London: Pion.

Openshaw, S. and P. J. Taylor 1981. The modifiable areal unit problem. In N. Wrigley and R. J. Bennett (eds.), *Quantitative Geography: a British View*. London: Routledge & Kegan Paul.

Paelinck, J. H. P. 2000. On aggregation in spatial econometric modelling. *Journal of Geographical Systems*, **2**(2), 157–65.

Parker, D. C., J. Busemeyer, L. Carlson, *et al.* 2002. LUCIM: an agent-based model of rural landowner decision making in south-central Indiana, CIPEC Working Paper.

Parker, D. C., T. Berger, S. M. Manson *et al.* 2003. *Agent-Based Models of Land-Use and Land-Cover Change: Report and Review of an International Workshop. LUCC Report Series* No. 6, Irvine, CA.

Peterson, D. L. and V. T. Parker (eds.) 1998. Ecological scale: theory and applications. *Complexity in Ecological Systems*. New York: Columbia University Press.

Petschel-Held, G. *et al.* 1999. Syndromes of global change – a qualitative modelling approach to assist global environmental management. *Environmental Modeling and Assessment*, **4**(4), 295–314.

Picci, L. 2001. Explaining long- and short-run interactions in time series data. *Journal of Business and Economic Statistics*, **19**(1), 85–94.

Polsky, C. 2002. A spatio-temporal analysis of agricultural vulnerability to climate change: the U.S. Great Plains, 1969–1992. Unpublished PhD Dissertation Thesis, The Pennsylvania State University, University Park, PA.

Polsky, C. 2004. Putting space and time in Ricardian climate change impact studies: the case of agriculture in the U.S. Great Plains. *Annals of the Association of American Geographers*, **94**(3), 549–564.

Polsky, C. and W. E. Easterling 2001. Adaptation to climate variability and change in the US Great Plains: a multi-scale analysis of Ricardian climate sensitivities. *Agriculture, Ecosystems, and Environment*, **85**(1–3), 133–144.

Polsky, C., R. Neff, and B. Yarnal 2007. Building comparable global change vulnerability assessments: the vulnerability scoping diagram. *Global Environmental Change*, **17**, 472–485.

Raudenbush, S. W. and R. J. Sampson 1999. Ecometrics: toward a science of assessing ecological settings, with application to the systematic social observation of neighborhoods. In M. E. Sobel and M. P. Becker (eds.), *Sociological Methodology*. Oxford: Blackwell, pp. 1–41.

Reynolds, H. and C. Amrhein 1998. Some effects of spatial aggregation on multivariate regression parameters. In D. A. Griffith *et al.* (eds.), *Econometric Advances in Spatial Modelling and Methodology: Essays in Honour of Jean Paelinck*. Dordrecht: Kluwer, pp. 85–106.

Rindfuss, R. R., S. J. Walsh, B. L. Turner, J. Fox, and V. Mishra 2004. Developing a science of land change: challenges and methodological issues. *Proceedings of the National Academies of Science*, **101**(39), 13 976–13 981.

RISA 2001. Regional Integrated Sciences and Assessments Program. National Oceanic and Atmospheric Administration Office of Global Programs.

Root, T. L. and S. H. Schneider 1995. Ecology and climate: research strategies and implications. *Science*, **269**, 334–341.

Sampson, R. J., J. D. Morenoff, and F. Ealers 1999. Beyond social capital: spatial dynamics of collective efficacy for children. *American Sociological Review*, **64**(October), 633–660.

Schröter, D., C. Polsky, and A. Patt 2005. Assessing vulnerabilities to the effects of global change: an eight step approach. *Mitigation and Adaptation Strategies for Global Change*, **10**(4), 573–595.

Serneels, S. and E. F. Lambin 2001. Proximate causes of land-use change in Narok District, Kenya: a spatial statistical model. *Agriculture, Ecosystems, and Environment*, **85**(1–3), 65–81.

Sprinz, D. F. 2000. Cross-level inference in political science. *Climatic Change*, **44**, 393–408.

Stern, D. I. and R. Kaufmann 2000. Detecting a global warming signal in hemispheric temperature series: a structural time series analysis. *Climatic Change*, **47**, 411–438.

Stern, P. C., O. R. Young, and D. Druckman (eds.) 1992. *Global Environmental Change: Understanding the Human Dimensions*. Committee on the Human Dimensions of Global Change; Commission on the Behavioral and Social Sciences and Education, National Research Council. Washington, DC: National Academy Press.

SYPR 2002. Project Description. SYPR: Southern Yucatán Peninsular Region. http://www.clarku.edu/departments/geography/faculty/sypr.shtml (accessed 1 May 2002).

Turner, B. L. and W. B. Meyer 1991. Land use and land cover in global environmental change: considerations for study. *International Social Science Journal*, **130**, 669–679.

Turner, B. L., R. E. Kasperson, W. B. Meyer, K. M. Dow, D. Golding *et al.* 1990. Two types of environmental change: definitional and spatial-scale issues in their human dimensions. *Global Environmental Change* (December), 14–22.

Turner, B. L. D. L. Skole, S. Sanderson, G. Fischer, L. Fresco and R. Leemans (eds.) 1995. Land-use and Land-cover change: Science/Research Plan. *Joint publication of the International Geosphere–Biosphere Programme (Report No. 35) and the Human Dimensions of Global Environmental Change Programme (Report No. 7)*. Stockholm: Royal Swedish Academy of Sciences.

Turner, M. 1999. Merging local and regional analyses of land-use change: the case of livestock in the Sahel. *Annals of the Association of American Geographers*, **89**(2), 191–219.

Turner, M., V. Dale, and R. Gardner 1989. Predicting across scales: theory development and testing. *Landscape Ecology*, **3**(3/4), 245–252.

USDA 1995. Average value per acre of farm real estate. Economic Research Service, US Department of Agriculture Economics and Statistics System. http://usda. mannlib.cornell.edu/data-sets/land/87012/ (accessed 27 September 2000).

Veldkamp, A., K. Kok, G. H. J. De Koning, J. M. Schoorl, M. P. W. Sonneveld and P. H. Verburg 2001. Multi-scale system approaches in agronomic research at the landscape level. *Soil and Tillage Research*, **58**(3–4), 129–140.

Verburg, P. and Y. Chen 2000. Multiscale characterization of land-use patterns in China. *Ecosystems*, **3**, 369–385.

Walker, R. T. and W. D. Solecki 1999. Managing land use and land cover change: the New Jersey Pinelands biosphere reserve. *Annals of the Association of American Geographers*, **89**(2), 220–237.

Walsh, S. J., T. P. Evans, W. F. Welsh, B. Entwisle, and R. R. Rindfuss 1999. Scale-dependent relationships between population and environment in northeastern Thailand. *Photogrammetric Engineering and Remote Sensing*, **65**(1), 97–105.

Walsh, S. J., T. W. Crawford, W. F. Welsh, and K. A. Crews-Meyer 2001. A multiscale analysis of LULC and NDVI variation in Nang Rong district, northeast Thailand. *Agriculture, Ecosystems and Environment*, **85**(1–3), 47–64.

Warren, A., S. Batterbury, and H. Osbahr 2001. Sustainability and Sahelian soils: evidence from Niger. *The Geographical Journal*, **167**(4), 324–341.

Weinhold, D. and E. J. Reis 2001. Model evaluation and causality testing in short panels: the case of infrastructure provision and population growth in the Brazilian Amazon. *Journal of Regional Science*, **41**(4), 639–658.

Westert, G. P. and R. N. Verhoeff (eds.) 1997. *Places and People: Multilevel Modelling in Geographical Research. Nederlandse Geografische Studies*, vol. 227. Utrecht: The Royal Dutch Geographical Society.

White, R. and G. Engelen 1993. Cellular automata and fractal urban form: a cellular modelling approach to the evolution of urban land use patterns. *Environment and Planning A*, **25**, 1175–1199.

Wiens, J. A., N. C. Stenseth, B. V. Horne, and R. A. Ims 1993. Ecological mechanisms and landscape ecology. *Oikos*, **66**, 369–380.

Wilbanks, T. J. and R. Kates 1999. Global changes in local places: how scale matters. *Climatic Change*, **43**, 601–628.

Zhou, W., A. Troy, and J. M. Grove 2008. Modeling residential lawn fertilization practices: Integrating high resolution remote sensing with socioeconomic data. *Environmental Management*, **41**(5), 742–752.

5

Uncertainty management in integrated regional assessment

MARJOLEIN B. A. VAN ASSELT

5.1 Introduction

Uncertainty is not simply the absence of knowledge. Uncertainty can still prevail in situations where considerable information is available. For that reason, Funtowicz and Ravetz (1990) refer to uncertainty as inadequate information. More knowledge does not imply less uncertainty and vice versa: new information may reveal the presence of uncertainties that were previously unknown or were understated (van Asselt 2000; 2005). In this way, more knowledge highlights the fact that our understanding is more limited or that the processes are more complex than previously thought. Or as Shackle (1955) phrased it in his theory of "unknowledge": There would be no uncertainty if a question could be answered by seeking additional knowledge. The fundamental imperfection of knowledge is the essence of uncertainty. Heisenberg (1962) explained another dimension of the problematic relationship between knowledge and uncertainty. His uncertainty principle claims that we could not, in fact, obtain all the information we need, since the act of getting information often changes the phenomena being studied. In other words, scientists face inherent limitations to the reduction of uncertainty. However, that does not mean that acknowledging uncertainty necessarily corresponds to an extreme postmodern stand that nothing is truly knowable. In this chapter, I argue that it is possible to acknowledge uncertainty and explicitly to address salient uncertainties in integrated assessment, while maintaining classical scientific merits of systematic investigation, analysis, and interpretation.

An integrated assessment addresses complex problems that lie across, or at the intersection of, many disciplines; thus it also considers aspects not covered by any of the disciplines and thereby suffers from uncertainty due to lack of knowledge. Integrated assessments usually involve questions that are unanswerable due to inherent uncertainty, such as cases for which necessary historical records or monitoring systems are lacking, questions referring to human behavior, questions pertaining to the future and questions that involve value judgments. Integrated regional assessment furthermore has to deal with the fact that scientific knowledge is usually generic instead of context-specific, which implies that local conditions, circumstances and dynamics to be addressed in integrated regional assessment may vary.

From the above we can conclude that uncertainty is at the core of integrated regional assessment and that management of uncertainty is a key challenge. In this chapter, a typology of sources of uncertainty that can be used in the process of identifying and characterizing uncertainty is discussed. Understanding uncertainty helps to decide how uncertainty can be dealt with in a particular integrated regional assessment. Although guidelines and checklists for uncertainty analysis are under development (Penman *et al.* 2000; Risbey *et al.* 2001; Seebregts *et al.* 2001; Klinke and Renn 2002; van Aardenne 2002; Janssen *et al.* 2003; Petersen *et al.* 2003; van der Sluijs *et al.* 2003; Krayer von Kraus 2005; Petersen 2006), there is no standard recipe for uncertainty management in integrated assessment. Therefore, this chapter does not provide the framework for dealing with uncertainty in integrated assessment, but by means of two examples it indicates ways to accommodate uncertainty in regional integrated assessments. I first discuss the Dutch Environmental Outlooks as a case of uncertainty management in integrated regional assessment. This is an example of an expert assessment. The second example, an integrated assessment of the rivers Rhine and Meuse in view of a changing environment, is an example of a more participatory and model-based integrated assessment (for participatory integrated assessment, see van Asselt and Rijkens-Klomp 2002; Kasemir *et al.* 2003; and van de Kerkhof 2004).

5.2 Sources of uncertainty

Uncertainty can be seen as an aspect of the empirical domain that can be established through cognitive diagnosis or as a social construct that "exists" only in discourse. Uncertainty is usually defined through classification. One way to classify uncertainty is by investigating different sources of uncertainty. In this context, "source" is used to refer to the empirical and/or constructed origin of uncertainty. Analogous to investigating a physical phenomenon such as climate change, research on uncertainty would involve studying the underlying causes of uncertainty.

Generally speaking, a distinction can be made between uncertainty due to **variability** ("variability uncertainty") and uncertainty due to **limited knowledge** ("epistemic uncertainty"; Walker *et al.* 2003; Krayer von Kraus 2005; Petersen 2006; compare Hoffman and Hammonds 1994). In the first case, uncertainty is the result of relevant, but variable or even "random" system behavior. In the case of epistemic uncertainty, uncertainty is a property of the analyst(s) performing the study and the

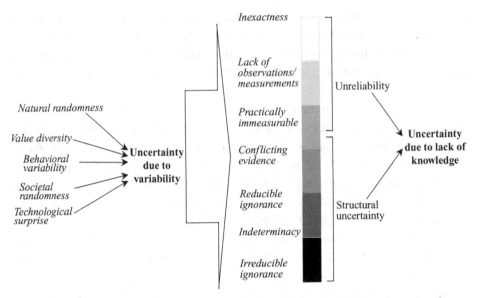

Figure 5.1. Sources of uncertainty according to van Asselt (2000) and van Asselt and Rotmans (2002). Used with kind permission of Springer Science and Business Media from M. B. A. van Asselt (2000) © Kluwer Academic Publishers.

collective state of knowledge. It should be noted that limited knowledge might be at least partly the result of variability uncertainty, as variability complicates analysis.

Different attempts have been made to further detail the various sources of uncertainty (see, for example, van der Sluijs 1997; van Asselt 2000; van Asselt and Rotmans 2002; Walker *et al.* 2003; Krayer von Kraus 2005; Petersen 2006). Such distinctions between, for example, inexactness, lack of observations, conflicting evidence and ignorance may help to further specify epistemic uncertainty. It may also help to specify whether variability uncertainty results from natural systems, from human behavior, from social, economic, and cultural dynamics, from value diversity or from technological surprises. Such categories may help to interpret uncertainty and to communicate about it, as it provides a basis to explain *why* something is uncertain. To illustrate various sources of uncertainty, we follow the vocabulary introduced in van Asselt (2000) and van Asselt and Rotmans (2002) (see Figure 5.1).

In this typology of sources of uncertainty, the following sources of variability uncertainty can be distinguished (compare Rowe 1994).

- *Inherent randomness of nature.* The non-linear, chaotic, and unpredictable nature of natural processes (see also Morgan and Henrion 1990). Also referred to as (unobserved) seasonalities (van Vlimmeren *et al.* 1991).

- *Value diversity.* Differences in people's mental maps, world views, and norms and values, due to which problem perceptions and definitions differ. Also referred to as subjective judgment and disagreement (Morgan and Henrion 1990) or moral uncertainties (de Marchi 1995).[1]
- *Human behavior (behavioral variability).* "Non-rational" behavior, discrepancies between what people say and what they actually do (cognitive dissonance), or deviations of "standard" behavioral patterns (micro-level behavior).
- *Social, economic, and cultural dynamics (societal variability).* The non-linear, chaotic and unpredictable nature of societal processes (macro-level behavior).
- *Technological surprises.* New developments or breakthroughs in technology or unexpected consequences ("side-effects") of technologies.

Due to variability in combination with limited resources to measure and obtain empirical information, variability contributes to limited knowledge. Limited knowledge is thus partly a result of variability, but knowledge with regard to deterministic processes can also be incomplete and uncertain. A continuum can be described that ranges from inexactness to irreducible ignorance.

- *Inexactness.* See also Funtowicz and Ravetz (1990) and Zimmermann (1996) also referred to as lack of precision, inaccuracy, metrical uncertainty (e.g. Rowe 1994), measurement errors (e.g. Beck 1987; van Vlimmeren *et al.* 1991), or precise uncertainties (e.g. Wallsten 1990). We roughly know.
- *Lack of observations/measurements.* Lacking data that could have been collected, but haven't been. We could have known.
- *Practically immeasurable.* Lacking data that in principle can be measured, but not in practice (too expensive, too lengthy, infeasible experiments). We know what we do not know.
- *Conflicting evidence.* Different data sets/observations are available, but allow room for competing interpretations (compare Zimmermann 1996). We don't know what we know.
- *Reducible ignorance.* Processes that we do not observe, nor theoretically imagine at this point in time, but may in the future (see also Funtowicz and Ravetz 1990 and Wynne 1992). We don't know what we do not know.
- *Indeterminacy (e.g. Wynne 1992).* Processes of which we understand the principles and laws, but which can never be fully predicted or determined. We will never know.
- *Irreducible ignorance.* There may be processes and interactions between processes that cannot be (or not unambiguously) determined by human capacities and capabilities. We cannot know.

The continuum thus ranges from unreliability to more fundamental uncertainty, the latter referred to as radical (Funtowicz and Ravetz 1993), structural (e.g. Rowe

---

[1] Klinke and Renn (2002) and Renn (2006) refer to this source of uncertainty as *ambiguity*, i.e., different interpretations of the available knowledge, different problem perceptions, and different definitions due to differences in values and framing. Building upon their conceptual analysis, van Asselt *et al.* (2007) propose to distinguish ambiguity as a separate class of uncertainty, next to variability and epistemic uncertainty.

1994) or systematic uncertainty (e.g. Henrion and Fischhoff 1986, Morgan and Henrion 1990). Uncertainties in the category of unreliability are usually measurable, or can be calculated, in the sense that they stem from well understood systems or processes. This implies that in principle either margins or patterns can be established, so that usually the uncertainty can be described quantitatively (either in terms of a domain or a stochastic equation). The other end of the continuum involves uncertainties that can at best be roughly estimated. Such radical uncertainty generally arises due to conflicting evidence, ignorance, indeterminacy, and uncertainty due to variability. It is even likely that the most salient uncertainties in an integrated assessment endeavor are radical.

Such a list of sources of uncertainty can be used in integrated regional assessment to communicate the idea of uncertainty (to raise uncertainty awareness (van Asselt 2004)), as a heuristic to systematically explore uncertainties (see van der Klis (2003) for an example) and to discuss specific uncertainties (vocabulary to share uncertainty information (van Asselt 2004)).

5.3 Approaches for uncertainty management in model-based assessments

Insight into sources of uncertainty is needed to determine *whether* and, if so, *how* the identified uncertainties could or should be addressed in quantitative uncertainty analysis (Walker *et al.* 2003). In van Asselt (2000) and van Asselt *et al.* (2001a), we have roughly explored the following classes of quantitative techniques (compare Kann and Weyant 2000; van der Sluijs *et al.* 2003).

- *Sensitivity analysis.* Study of the influence of variations in model parameters and initial values on model outcomes. Techniques: individual parameter variation, differential sensitivity analysis, response-surface method and meta-modeling.
- *Probability-based methods.* Probability distributions (usually based on expert judgments) for uncertain inputs/parameters are propagated through the model, with as outputs probability distributions or statistical measures as the 95-percentile. Techniques: usually Monte Carlo techniques are applied. Saltelli *et al.* (2000) provide a useful overview of recent developments in uncertainty and sensitivity analysis in a probabilistic context.
- *Formal scenario analysis.* Assessing sets of different assumptions of possible future states parameterized in the model.
- *Hedging-oriented methods.* The value of decision variables in the model is determined based on a joint distribution on the possible outcomes that may occur in the next period (Manne and Richels 1995).
- *Perspective-based model routes.* Different perspectives are reflected in choices concerning model inputs, parameter choices, model structure, and equations (van Asselt and Rotmans 1996; van Asselt *et al.* 1996; Rotmans and de Vries 1997; Hoekstra 1998b).

Each method has its strengths and weaknesses, so for understanding the scope and the relevance of uncertainty in an integrative study, a complementary use of various methods is advocated. Such combinations of uncertainty analysis methods are already applied. For example, hedging-oriented methods are combined with probability-based methods. Sensitivity analysis is quite often used to filter out those uncertain parameters that will be subjected to probability-based uncertainty analysis. Exploratory modeling (Bankes 1994; Lempert and Bonomo 1998; Lempert *et al.* 2003) is an example of an approach that explicitly aims to incorporate a combination of the above methods in order to address uncertainty explicitly. In its general form it combines sensitivity analysis with both quantitative and qualitative scenario approaches and it is usually applied in a participatory set-up. The ambition is not to be comprehensive with regard to uncertainty, as this is a mission impossible. The aim of uncertainty management in the context of integrated assessment is to specify the salient uncertainties and what these uncertainties imply or may imply for conclusions and recommendations. Salience in the context of integrated assessment means that the degree of uncertainty is significant and that the policy relevance is high.

Most methods for quantitative uncertainty analysis suffer from the fact that they only address uncertainties in model quantities and neglect the structure of the model itself. In doing so, significant uncertainties are "exogenized" and thereby become invisible (Wynne 1992). The PRIMA approach (van Asselt 2000; van Asselt and Rotmans 2002), a further extension of the idea of perspective-based model routes, aims to provide a structure for the process of uncertainty management, in which also uncertainties in model structure are addressed. The PRIMA approach served as a reference both in the Environmental Outlook case and the integrated water management case. We first discuss the main principles, before we turn to the two cases.

5.4 The PRIMA approach

The guiding principle in the PRIMA approach (van Asselt 2000; van Asselt and Rotmans 2002) is that uncertainty legitimates different perspectives and that as a consequence uncertainty management should consider various perspectives. Central in the PRIMA approach is the issue of disentangling controversies on complex issues in terms of uncertainties. The uncertainties are then "colored" according to various perspectives. Starting from these perspective-based interpretations, various legitimate and consistent narratives are developed, which serve as the basis for integrated analysis of developments as well as policy options. In this way, the relevance and impact of uncertainty associated with the underlying knowledge base can be made explicit, which facilitates the process of distilling robust insights relevant for decision making.

The PRIMA approach features five stages:

- definition of meta-perspective;
- uncertainties in perspective;
- scenarios in perspective;
- risks in perspective;
- quality assessment.

The first stage can be described as defining which perspective on plurality is adopted and in terms of the controversy or dilemma being assessed. In the second stage the scanning and selection of salient uncertainties and the perspective-based interpretation of these uncertainties are the central tasks. In the scenarios in perspective phase, various outlooks are assessed. This involves scanning the future from a wide variety of perspectives, which can be done both in a qualitative and quantitative manner. The next challenge is to assess future prospects, challenges, and risks, taking into account the variety of perspective-based assessments gathered in the previous phases. It is very important to test the quality of the associated robust insights by reflecting on the previous steps, i.e., by evaluating whether the uncertainties relevant for the conclusions have been considered in an adequate manner. This stage of quality assessment can be considered as closing the assessment cycle. Building upon these insights the process can be iterated with an adopted or new definition of the central controversy or dilemma, or with another perspective on plurality.

Applying different perspectives means taking a certain stand with regard to plurality.[2] There are four design choices crucial for setting-up the assessment dependent on the choice for the pluralistic stand, i.e.:

- the type of uncertainties to be included in the pluralistic analysis;
- whether a demand- or a supply-driven approach is advocated; in *supply-driven* studies, a group of scientists anticipates the societal relevance of a complex theme. The scientific problem definition scopes the assessment. In *demand-driven* studies the problem is defined in a participatory endeavor;
- the portfolio of methods for uncertainty analysis;
- the perspective framework used.

The first step in applying ideas from PRIMA to integrated regional assessment is to choose a meta-perspective on plurality that corresponds best with the attitude of both the analysts and the clientele, and then to be consistent with this perspective during the assessment process. To help practitioners to decide in choosing which perspective on plurality is most appropriate in their situation, I have tried to outline

---

[2] For a more comprehensive discussion on the issue of plurality, see van Asselt (2000), and van Asselt *et al.* (2003).

major differences in thinking about plurality, by means of a spectrum ranging from the rather positivist observation-in-perspective to the extreme social-constructivist reality-in-perspective (van Asselt 2000). The features associated with this spectrum of meta-perspectives are summarized in Table 5.1.

In the example studies described in the remainder of this chapter, the *science-in-perspective*-perspective is adopted. In this meta-perspective a typology of socio-cultural perspectives is welcomed, as this perspective holds that plurality in science derives from plurality in society. Such a typology can be found in cultural theory (key references are: Douglas and Wildavsky (1982), Funtowicz and Ravetz (1985), Rayner (1987), Rayner and Cantor (1987), Schwarz and Thompson (1990), Thompson *et al.* (1990), and Krimsky and Golding (1992)). The perspectives are summarized in heuristic rules (see Table 5.2) that can be used in applying these perspectives in integrated regional assessments.

These perspectives comprise a worldview (i.e., how people describe the world) and a management style (i.e., how they act upon it). Matching the management style of a perspective to its respective worldview is referred to as "utopia." A utopia is thus a world in which a certain perspective is dominant and the world functions according to its worldview. Utopias are characterized by the trust that the imagined future will be without problems. In contrast, "dystopias" either describe what would happen to the world if reality proved not to resemble the adopted worldview following the adoption of the favored strategy, or vice versa, i.e., where reality functions in line with one favored worldview, but opposite strategies are applied. Dystopias are thus scenarios involving mismatches between worldviews and management style.

5.5 PRIMA example I: the Dutch Environmental Outlooks

The National Environmental Outlooks (abbreviated to MV in Dutch) produced by the Dutch Institute for Public Health and Environment (RIVM) have the explicit purpose of informing and supporting Dutch environmental policy making. Approximately every four years, RIVM presents a long-term integrated assessment of the environment to the Dutch government. The Environmental Outlooks aim to provide an assessment of environmental and health impacts in the Netherlands associated with the future state of the environment that results from a particular development of societal pressures in terms of economic growth, demographic developments, and consumption and production patterns. The Environmental Outlooks are long-term assessments; generally with a time-horizon of 20–30 years. Their spatial focus is the Netherlands, but European and even global developments and trends are considered. The Environmental Outlooks can be considered as a series of integrated regional assessments.

Table 5.1. *Differences between meta-perspectives on plurality (van Asselt 2000). Used with kind permission of Springer Science and Business Media from M. B. A. van Asselt (2000), Perspectives on Uncertainty and Risk, p. 231 © 2000 Kluwer Academic Publishers.*

Meta-perspective design choice	Observation in perspective	Theory in perspective	Science in perspective	Reality in perspective
Uncertainties included	Radical uncertainties	Radical uncertainties	Radical and measurable uncertainties	Everything is uncertain
Type of assessment	Supply-driven	Supply-driven & participatory	Participatory	Demand-driven
Methods for uncertainty analysis	Standard methods with perspective-based as supportive	Perspective-based and standard methods as complementary approaches	Perspective-based with standard methods as supportive	Fully perspective-based
Perspective-framework	Hypotheses	Scientific paradigms	Socio-cultural perspectives	Perspectives result out of participation

Table 5.2. *Features of socio-cultural perspectives (van Asselt 2000). Used with kind permission of Springer Science and Business Media from M. B. A. van Asselt (2000),* Perspectives on Uncertainty and Risk, *p. 234 © 2000 Kluwer Academic Publishers.*

	Individualist (market-optimist)	*Egalitarian (environmental worrywart)*	*Hierarchist (controllist)*
Heuristic rule 1	Free market and anti-regulation; economic growth and technological development are progress.	Nature is vulnerable, and thus in need of protection from excessive exploitation; aversive to environmental risks; prevention is better than cure.	Societal stability through regulation, norms and hierarchy; acceptance of inequalities.
Heuristic rule 2	Individual development and material self-interest are the motives for action; success is a personal responsibility.	Equity.	Risk-aversive; anti-abrupt change; easy doing otherwise the line will break.
Heuristic rule 3	Nature is not fragile; it can stand rough handling.	The economy is a means and not an aim. Conscious consumption.	Reliance on expertise and experience of authoritative institutes experts.
Heuristic rule 4	Problems are solvable; risks are challenges and opportunities.	Solidarity is the leading principle and human beings act accordingly; collective interest.	Power and status are the motives for action.

The main message of the first Environmental Outlook (RIVM 1988), entitled "Concern for Tomorrow," was: the environment is getting worse at all scale levels despite all the efforts. This was a shock for the Dutch politicians, but also for society as a whole. This Environmental Outlook served as the scientific basis for the first National Environmental Policy Plan (VROM 1989). The second assessment, i.e., the "Environmental Outlook 2 1990–2010" (RIVM 1991/1992) was primarily an update of the first one. The third Environmental Outlook (RIVM 1993) indicated the effectiveness of policy plans in relation to the environmental quality objectives as formulated in the first Environmental Policy Plan. In addition the "Environmental Outlook 4 1997–2020" (RIVM 1997), the fifth National Environmental Outlook (RIVM 2000), and the sixth Environmental Outlook (MNP[3] 2006) were published.

The assessment process underlying the Environmental Outlooks is informed by monitoring, measuring, and modeling. Models are used to describe or explain environmental aspects in relation to other developments, to estimate future emissions, environmental quality, and impacts from economic and technological scenarios, and to assess possible futures in relation to objectives and targets. RIVM employs about thirty models in the assessment process (RIVM 1999). These models are as far as possible calibrated and validated against available monitoring data. RIVM does not possess one fully integrated model, but it uses the available models in cascades in order to assess the relevant environmental cause–effect chains.

To understand the practice of integrated regional assessment, how uncertainty was dealt with was explored in the Environmental Outlook reports (van Asselt 2000, 2004; van Asselt *et al.* 2001a). The argumentation analysis showed that only in cases of strong conclusions could uncertainties be discovered. Strong implies that, in principle, verifiable evidence/argumentation underpins the conclusion. To illustrate our argumentation analysis and how uncertainties are dealt with in the text, see Box 5.1 for an example.

Box 5.1
Example of uncertainty-oriented argumentation analysis

The number of people experiencing noise nuisance by road traffic is an example of an effect on human well-being, and is therefore considered to be one of the key outputs of RIVM's assessment. The following conclusion is found in the summary of the third Environmental Outlook:

Between 1985 and 1990, the percentage of people experiencing noise nuisance as a result of road traffic rose from 59% to 61% and the percentage of people suffering serious nuisance rose from 19–20% of the Dutch population. Despite the increase in car traffic and the expansion of the road network, the proposed policy will reduce the number of people experiencing serious nuisance to 10–15% of the population between 1990 and 2000. The number of people experiencing some nuisance will fall less sharply. The targets for the year 2000 will be easily achieved. However, the target for 2010 (a negligible level of serious nuisance) will not be achieved.

(MV-3, p.22)

The conclusion in the specific section about noise pollution in Chapter 4 of the third Environmental Outlook is more detailed than in the summary:

---

[3] The environmental policy unit of RIVM recently obtained independent status as the Dutch environmental and nature planning agency (Milieu en Natuurplanbureau (MNP)).

(. . .) the percentage of people experiencing noise nuisance or serious noise pollution between 1990 and 2000 will fall below the 1985 levels, which were 59% (nuisance) and 19% (serious pollution). In the ER scenario, these levels will drop to 56% and 15% and in the GS scenario to 51% and 12% (see figure 4.6.2a).

(MV-3, p.120)

The 12% and 15% mentioned here roughly match with the 10%–15% range in the overall conclusion. The phrasing "will fall" is here quantitatively argued by the estimates that the level of noise nuisance will decrease from 59% to 51% or 56% respectively.

No uncertainties regarding noise nuisance caused by road traffic are mentioned in the third Environmental Outlook itself. There is no indication of how the above numbers were generated and whether uncertainties are associated with these numbers. The background document on traffic and transport (van Wee *et al.* 1993) was checked for further justification. This background document presented the necessary information and argumentation in a transparent manner. It provided comprehensive and motivated information pertaining to noise nuisance by road traffic that matches with the conclusion on the anticipated effects of policy measures in relation to policy targets. Factors that determine noise pollution and nuisance are elaborated and assessed in a structured manner. Each step in the calculations, as well as assumptions and references, are made explicit. Although a specific reference in the third Environmental Outlook to this particular background document in the relevant section would have been appropriate, we argue that the conclusion on noise nuisance due to road traffic can be considered as an example of a strong conclusion. It can however be argued that argumentation that is judged as strong may hide uncertainty; it may be that only specialists are able to recognize that the underlying arguments involve uncertainties that are not considered in the assessment. The many assumptions in the background document on transport and traffic signify uncertainties. The sources of uncertainties can, in principle, be deduced from the assumptions, which also hint how the uncertainties are interpreted.

In the following we turn to a number of activities that have been organized in the assessment process of the fifth Environmental Outlook. The aim was to explore ideas put forward in the PRIMA approach in an actual integrated regional assessment case.

5.5.1 Uncertainty in perspective – workshop

Within the RIVM assessment process an uncertainty in perspective workshop was organised. The workshop involved 18 practitioners, all involved in the fifth Environmental Outlook process; it lasted half a day. The participants were selected and invited by the project management of the fifth Environmental Outlook. Prior to the workshop, the participants received a questionnaire and a discussion paper in which the PRIMA approach was summarized. This uncertainty in perspective workshop involved the following exercises (for the full workshop report see van Asten and van Asselt 1999):

- brainstorm on uncertainties relevant for the fifth Environmental Outlook;
- clustering of the uncertainties;
- interpretation of the uncertainties from the view point of different perspectives (working groups per perspective).

In the closing plenary, the sub-groups presented their results, and the workshop and its output were discussed in view of the fifth Environmental Outlook. The workshop yielded a wealth of information involving the notes of the workshop facilitators, audiotapes, pictures, and, most importantly, the material produced by the participants in the course of the workshop. A second questionnaire was completed after the workshop, in order to use the participants' evaluations, and feedback (for an analysis of the workshop data, see van Asselt 2000 and van Asselt *et al.* 2001a).

The objective of the brainstorm and clustering was to get an insight into uncertainties thought to be salient to the fifth Environmental Outlook. The exercise indicated that many of the articulated uncertainties involve variability and structural uncertainty. However, the analyses of the previous Environmental Outlooks (van Asselt 2000, van Asselt *et al.* 2001a) suggested that relevant sources of uncertainty primarily involve uncertainty due to unreliability (inaccurate measurement or calculations). This finding seems to suggest that the "articulation" of assessors' uncertainty knowledge in the way we did it in this PRIMA-workshop, reveals uncertainties that have been overlooked or neglected in traditional assessment processes. In the closing discussion as well as in 10 of the 15 questionnaires (out of the 18 participants) the participants argued that the workshop helped them to systematically consider uncertainty.

The aim of the second exercise (working groups per perspective) was to interpret the surfaced uncertainties. As discussed earlier, we used the three perspectives derived from cultural theory (i.e., the hierarchist/controllist, the individualist/market-optimist, and the egalitarian/environmental worrywart) as the perspective framework. The task for each working group was to interpret the surfaced uncertainties from the assigned perspectives. Facilitators aided the group work. They were asked explicitly not to interfere with the interpretations, but just to enhance discussion and group thinking. In each group, the heuristic statements – outlined earlier – were used to introduce the perspective. In order to put themselves in the perspective, each group brainstormed about what their perspective was associated with. The hierarchist/controllist group came forward with associations such as conservative, agricultural sector, water management authorities, and big brother is watching you. The individualist/market optimist group, for example, thought about the favored weekly magazine, preferred measures, and which Dutch ministries can be associated with this perspective. The egalitarian/environmental worrywart group came forward with associations such as lover of nature, economy as a means, think global – act locally, and be nice to people, strict with companies. The idea was that through this associative brainstorm the participants were helped to put themselves in the shoes of the perspective, from which "standpoint" they would then interpret the uncertainties.

The main task for the group was to attribute interpretations of uncertainties in line with the assigned perspective through discussion. The uncertainties that were put forward in the brainstorm, were copied and distributed. The group could go through this list in the way they wanted. Some preferred to go through the whole list, while others tried to select those uncertainties that were considered relevant for their perspective. In this way, each group discussed perspective-based interpretations of a number of uncertainties. For example the controllist group decided that instability, disasters, and global recession are risky uncertainties from the controllist's point of view. On the uncertainty consumption patterns, they agreed that in this perspective economic sectors would be regulated and consumption patterns controlled by levies and convenants. For the same uncertainty, the market optimist group concluded that this is an important uncertainty for their perspective, while the environmental worrywart group considered it a risky uncertainty. On technology as uncertainty, the controllist group agreed that no radical breakthroughs and innovations will take place, while the market optimist group expected technological innovations that could be stimulated by time and money. The latter group furthermore concluded that eco-technology will result out of consumer pressure. In the environmental worrywart group, it was argued that problems cannot be solved by technology and that eco-technology is a necessary evil. In this way, each group produced a list of perspective-based interpretations of uncertainties (for a full description, see van Asselt 2000).

With the perspective exercise, specific uncertain areas were highlighted. The workshop output allowed the identification of which issues and assumptions may turn out to be critical. It was argued that these uncertainties must be addressed, whether numbers are available or not, otherwise critical questions from societal actors can be expected. Exploring the PRIMA approach in practice furthermore yielded suggestions on how the uncertainties-in-perspective phase can be translated into concrete activities. It became clear that it is impossible to carry out the whole phase through one workshop. The experience taught us that a step-wise approach is needed. A flow-chart of the set-up of the uncertainties-in-perspective phase as derived from the experiences is summarized in Figure 5.2.

Uncertainties-in-perspective

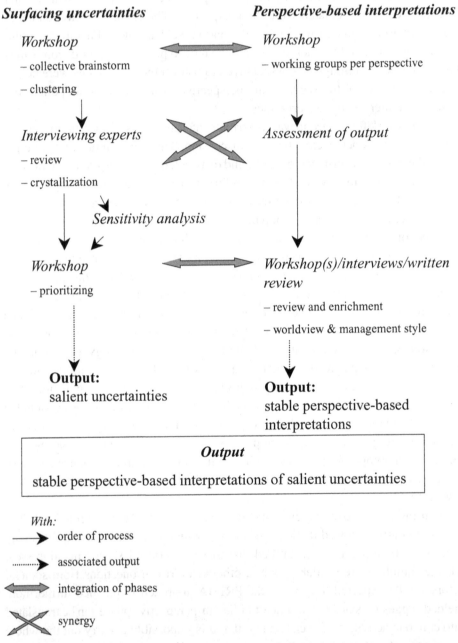

Figure 5.2. Proposed process for PRIMA uncertainties-in-perspective phase (van Asselt 2000). Used with kind permission of Springer Science and Business Media from van Asselt (2000) © Kluwer Academic Publishers.

5.6 PRIMA example II: integrated scenarios for the Rhine and Meuse basins

The floods of 1993 and 1995 illustrated that the threat of flooding is a recurring phenomenon in the Netherlands. The floods not only made clear that many people are dependent on the water, but also that the river seems to need more space. In this context the project "Integrated water management strategies for the Rhine and Meuse basins in a changing environment" was conducted (van Asselt *et al.* 2001b; Middelkoop *et al.* 2004), with the following aims:

- to develop a set of perspective-based internally consistent scenarios describing future developments with regard to economy, population, consumption patterns, transport, land use, climate change, and water management policies;
- to assess hydrological changes in the Dutch Rhine and Meuse basins as well as the consequences for various user functions associated with these scenarios by means of existing hydrological and impact models;
- to assess and evaluate the implications of different water management strategies as a basis for policy recommendations concerning the type of water management strategy needed in view of a changing environment and an uncertain future.

The PRIMA approach was taken as the basis for the process design. Using the steps proposed in PRIMA in a participatory manner (i.e., in collaboration with experts and stakeholders), we developed three families of utopian and dystopian scenarios. The hydrological changes and the consequences for user functions associated with these scenarios were assessed using existing hydrological and impact models. Different scenarios were represented in the model simulations by different model schemes (e.g. land-use maps), adaptation of model parameters (drainage coefficients), and different inputs (e.g. climate variables). A sensitivity analysis was performed to identify relevant parameters and inputs in the models. It was then decided which inputs and parameters could be used to represent the qualitatively formulated assumptions and trends of the perspective-based scenarios. The output of the model-based assessment of the future was used to underpin the potential range of hydrological changes and the consequences for user functions in a (semi-)quantitative manner. Finally, different water management strategies were evaluated through reflection on the water-specific management styles associated with the hierarchist, individualist, and egalitarian perspective.

5.6.1 Uncertainties in perspective process

For the identification of uncertainties with regard to the future of the Rhine and Meuse, three sources of information were used to articulate salient uncertainties, i.e., existing studies and policy reports, an expert meeting and a stakeholder workshop. These activities will be discussed below. The integration of this input into perspective-based scenario families is also described.

Salient uncertainties

For the purpose of the project we searched for studies that (a) deal with water management in a broad sense, which means that they must take into account economic, socio-cultural, institutional, as well as environmental aspects; (b) relate to the long term and, therefore, sketch at least one picture of the future and (c) deal with the Dutch parts of the Rhine and/or Meuse river basins. Through an iterative process involving water experts (both within and outside the project) about 20 relevant studies and policy reports were identified (van Asselt *et al.* 2001b). The selected reports cover a time period of nine years (1992–2000). The conclusion from an earlier evaluation study (Claessen and Dijkman 1999) was that starting points, assumptions, and the vision year are rather similar. Furthermore, it was concluded that spatial developments are almost ignored and that the time horizon (15–20 years) is relatively short.

The leading question in our review was which uncertainties, assumptions, and envisioned futures can be derived from the identified reports? The most often explicitly mentioned uncertainties concerned climate change, sea-level rise, and soil subsidence. In a number of studies, societal factors were identified as an important source of uncertainty, such as human behavior, public support, and the change of standards and values. Limited knowledge of the complex river system and the methods used were also mentioned as sources of uncertainty in two studies. Some studies questioned the reliability of the models.

Notwithstanding the acknowledgement of uncertainty, only one study stated that a way must be found to deal with future uncertainty. No study explicitly reasoned from uncertainty in their assessment of future water management. Partly as a consequence, the studies used just one set of assumptions and did not explore alternative interpretations of uncertainties. We concluded that uncertainties were not, or hardly, examined in the set of studies; they were mentioned, but not systematically addressed.

The selected studies assessed the future by means of snapshots of a future point in time. In the 1996 Watersystem assessment four scenarios are discussed. From our in-depth analysis of the various outlooks we concluded that these four can be considered as the core of four scenario clusters. The other studies and policy reports explicitly or implicitly, elaborate these outlooks:

- *current policy*: safety as an uncompromising priority;
- *use*: strong emphasis is put on economic growth;

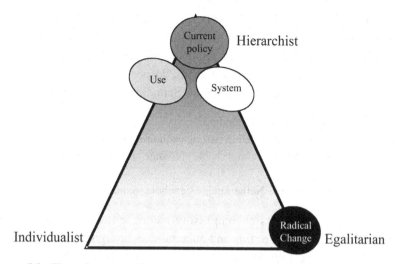

Figure 5.3. Clustering according to perspectives (source: van Asselt *et al.* 2001b).

- *system*: policy in which nature values have priority;
- *radical change*: assumes a radial reversal in the preferences of society, which is characterized by a strong environmental awareness (implying strong societal support for nature) and an altered production and consumption pattern.

In Figure 5.3 the four clusters are positioned in terms of the three perspectives. The individualistic domain is the least represented. This could partly be explained by the fact that water is not an important subject for individualists, since in the Dutch debate water management is mainly discussed in terms of safety (hierarchical) and the environment (egalitarian).

Except for the *radical change* scenario cluster, all scenario clusters can be categorized as primarily hierarchist. However, the scenario cluster *system* accommodates more egalitarian elements, such as the focus on nature values and sustainability and the scenario cluster *use* involves some individualistic assumptions, such as the focus on economic functions and technical solutions.

An *uncertainties-in-perspective*-brainstorming session to articulate expert knowledge was organized with the project researchers. The aim of this session was to obtain insight into their informed ideas about relevant uncertainties and various interpretations, that these water experts consider legitimate. The first part of the expert meeting consisted of a brainstorming session surfacing important uncertainties for the long-term future in view of the management of the Rhine and Meuse river basins.

A first stakeholder workshop was organized with 23 people from the following sectors: government, water management, agriculture, science, fishing, and recreation. A plenary session was held, in which the participants were requested to brainstorm freely about events and uncertainties related to the Rhine and Meuse in the Netherlands until the year 2050. For inspiration, the participants were provided with the list of uncertainties derived from the analysis of previous studies and the expert meeting. This session resulted in 294 *post-its* referring to events and uncertainties. The events included (compare Schneider *et al.* 1998):

- certain surprises (such as extreme flooding);
- imaginable surprises that are probable (strong economic fluctuations);
- imaginable surprises that are improbable (change in the Atlantic Thermohaline Circulation);
- "unimaginable" surprises (the Netherlands as a whole inundated).

The surprises included events encompassing nature, climate (flooding/drought), pollution, and terrorism. The events and uncertainties brought up by the stakeholders were subsequently clustered according to uncertainties in various domains.

The output from the analysis of studies and policy reports, the expert meeting and the stakeholder workshop was processed and clustered by means of the van Asselt taxonomy of sources of uncertainty (see Table 5.3).

Perspective-based interpretation of uncertainties

The next step involved interpreting the uncertainties according to the different perspectives. The project partners placed themselves in the shoes of the various perspectives during an expert meeting. Prior to this meeting, the experts studied the literature of perspective-based modeling, especially through the work of Hoekstra (1998a, b) and Rotmans and de Vries (1997) and thereby got acquainted with the various perspectives. The heuristic rules (Section 5.4) were used to facilitate the application of the scheme of perspectives in the expert meeting.

The next activity in interpreting the uncertainties consisted of a stakeholder workshop. Prior to the stakeholder workshop, the participants were asked to fill out a perspective questionnaire, which had been developed in the context of the TARGETS project (Rotmans and de Vries 1997) with input of Cultural Theory experts and which is available on CD-Rom (Hilderink *et al.* 1998). By means of this questionnaire we obtained some insights into the variety of perspectives held by the workshop participants. The participants were divided into three heterogeneous work groups with the assignment to develop storylines.

Storylines were developed by first selecting a *starting event* from the brainstorming session and then continuing the discussion on how this event may lead to developments up to the year 2050. It was emphasized that these storylines were not about what should happen, but what *might* happen. Each group developed one or more storylines according to this method. Next, the participants of a group were asked to develop a storyline from the viewpoint of the assigned perspective.

Table 5.3. *Classification of uncertainties according to sources of uncertainty.*

Limited knowledge
— Lack of observation/measurements   — Rainfall
— Practically immeasurable   — River flow
     — Sedimentation
     — Retention capacity
— Conflicting evidence   — Climate change
     — Sea-level rise
     — Resilience of ecosystems

Variability   **Uncertainty**
— Natural randomness   — Climate change
     — Sea-level rise
     — Soil subsidence
     — Rainfall
     — Flooding
— Value diversity   — Standards
     — Societal support for nature policy
     — Acceptance of risks
     — Safety standards
— Behavioral variability   — Lifestyle
     — Efficiency in water use
— Societal randomness   — Prosperity
     — Population growth
     — Land use
     — International relations
— Technological surprise   — Transport
     — Energy supply
     — Water technology

The perspective was explained to the groups using the heuristic statements. The workshop led to five concrete storylines (for a full report see van Asselt *et al.* 2001b).

The storylines represent various possible action–reaction patterns, illustrating different interpretations of uncertainty. After the workshop, the storylines were analyzed and characterized in terms of the perspectives by analyzing the underlying assumptions. This resulted in:

• perspective assumptions;
• insights into the combinations of worldview and management styles with which the stakeholders' storylines could be identified.

We synthesized the interpretation of uncertainties derived from the stakeholder workshop and the expert interpretation of perspectives into a water-specific worldview and management style for each perspective. In detailing the water-specific assumptions for each perspective, we also used the work of Hoekstra (1998a).

The next step involved the translation of the qualitative water perspectives to values for model parameters and inputs. To that end, utopian sets of assumptions were developed that provided the basis for quantitative estimates. In estimating values for socio-economic parameters and inputs, ranges derived from a broad range of existing scenario studies were used as benchmarks. In this way, for each of the models used, perspective-based input sets were defined. The models were run with these perspective-based inputs to perform the selected utopian and dystopian experiments.

The model outcomes and the qualitative descriptions of the perspectives on water were used to sketch utopian and dystopian images of 2050. The future out-looks were clustered into scenario families, i.e., clusters in which the worldview is shared. This implies that three clusters are distinguished, i.e., an egalitarian scenario family, a hierarchist scenario family, and an individualist one. The dystopian images can be considered as bifurcations from the utopian outlook. In doing so, the scenarios are presented as branching "what-if" assessments of the future (compare van Asselt *et al.* 2005). The pitfall of any scenario exercises is to classify one of the scenarios as the most likely scenario or best-guess scenario. In this way the output of the scenario analysis then masks inherent uncertainty. Through the proposed presentation of future outlooks through scenario families, this risk is recognized and counteracted.

A second stakeholder workshop was organized to review and evaluate the scenarios from the perspective of water use and users. The participants were asked to reflect on the scenarios in terms of coherency and relevance. The workshop started with a general introduction to the project, followed by a presentation of the three scenario families. Each scenario presentation was followed by a discussion. The egalitarian scenario family was discussed in a group session and the other two families were discussed in a plenary session. During the second half of the workshop, the scenarios were discussed in terms of policy strategies and favorable policy options.

The final step was to evaluate the three scenario families in terms of the hydro-logical situation (the Rhine and Meuse basins, the Ijsselmeer area and the terrestrial areas), the consequences for the user functions (agriculture, nature, and transport/shipping), and the characteristics of the water system (safety, reversibility, economic gains, and investments/costs). This was done through expert judgment informed by the stakeholder input. This perspective-based assessment of water-related issues pertaining to the Dutch Rhine and Meuse river basins was used to explore recommendations for policy.

From this integrated regional assessment study, in which uncertainty management in line with the PRIMA approach was central, a number of robust conclusions could be drawn. Robust in this context refers to the possibility of triggering a favorable future, while avoiding highly undesirable ones, in a way that is flexible enough to be changed or reversed if new insights emerge. The perspective-based assessment, for example, yielded that no water management strategy will be superior in all circumstances. In the Netherlands, no win–win situation can be guaranteed. The firm conclusion was that safety versus costs is a real policy dilemma that cannot be solved perfectly by an ingenious water management strategy. For example, irrespective of the uncertainty associated with the changing environment and an uncertain future, the egalitarian management style is associated with positive effects for nature and is safe in view of flood risks, but also has negative impacts on agriculture, both in terms of the area and the damage associated with high ground water levels, and with high investment costs and low economic benefits. This integrated regional assessment further indicated that in the Netherlands integrative water management cannot combine all functions at the same spot, but that it should be area-specific and involve differentiation of functions.

5.6.2 Lessons learned

During the stakeholder workshop three heterogeneous discussion groups consisting of stakeholders from various disciplines and professional backgrounds were formed: the hierarchist, the egalitarian, and the individualist group. As these groups were formed, it was felt that heterogeneity and confrontation between different perspectives would stimulate discussion in such a way that deviating standpoints would come to the fore. However, this choice turned out to have also three major drawbacks.

- First, there was a problem of identification as some participants were asked to empathize with, for them, unfamiliar assumptions and arguments. In some cases empathizing with a deviating perspective turned out to be problematic.
- Second, because some people were asked to empathize with an unfamiliar perspective, they felt that they were in a disadvantaged position and not able to defend their own interests in the discussion with other stakeholders in their group.
- Third, as a result of the heterogeneity of the group, the extremes were flattened out, rather than made more explicit.

From this experience, we learned that in stakeholder workshops with a heterogeneous group of perspectives such a role-play type of perspective exercise is not advisable. If such perspective role-playing games are used, it is better to create homogeneous groups, so that the members of the sub-group more naturally empathize with the assigned perspective or that they all need to think according to a perspective they don't like at first sight. Building upon our experiences from the first stakeholder workshop and building upon the questionnaire input, more or less homogeneous working groups were formed: the first representing the hierarchist perspective, the second the egalitarian, and the third the individualist perspective. The experiences in the second workshop indicate that a more homogeneous approach seems both possible and fruitful (compare van Notten 2005, who also did perspective-based imaging with homogeneous groups).

5.7 Conclusion

There is no recipe or best practice for uncertainty management in integrated regional assessment. However, concepts and methods are available that allow serious attention to be paid to uncertainty in regional assessment endeavors. One such approach, i.e., the PRIMA approach, has been illustrated in this chapter by means of two examples: the Dutch Environmental Outlooks and an integrated regional assessment for the rivers Rhine and Meuse. The two examples illustrate how management of uncertainty could look in practice, both in integrated assessments performed by experts and in more participatory regional assessments.

The experiments with PRIMA in regional assessment practice also yielded new topics for uncertainty research. Communication about uncertainty emerges as an important aspect of uncertainty management that needs more scholarly attention. Furthermore, it is clear that the most important role of methods for uncertainty management is that they can serve as heuristics for reflection in assessment processes and on assessment products (compare van Asselt and Petersen 2003).

The development of best practices for uncertainty management in the context of integrated regional assessments is an ongoing endeavor. Theoretical advances are being made, but, as the English say, "the proof of the pudding is in the eating." The current chapter provides some ingredients for further cooking and hopefully food for thought that inspires other assessors to propose new recipes and test them in practice.

Acknowledgements

This chapter builds upon two projects in which the author was involved, i.e., the uncertainty research on RIVM's Environmental Outlooks and the project Integrated water management strategies for the Rhine and Meuse basins in a changing environment. I would like to thank my colleagues in these two projects, i.e., Jan Rotmans, Rob Hoppe, Anton van der Giessen, Peter Janssen, Peter Heuberger, Rian Langendonck, and Frank van Asten for the first project and Hans Middelkoop,

Susan van 't Klooster, Willem van Deursen, Marjolijn Haasnoot, Jaap Kwadijk, Hendrik Buiteveld, Pieter Valkering, and Gunther Können with regard to the latter. Furthermore, I would like to thank all RIVM practitioners and the stakeholders who participated in the reported uncertainty management in integrated regional assessment activities. Writing of this book chapter took place in the context of the research program Perspectives on Uncertainty and Risk: an interdisciplinary methodology program on future studies, which is financed with a career grant to the author from the Dutch Science Foundation (NWO) – this support is gratefully acknowledged.

References

Bankes, S. 1994. Computational experiments and exploratory modeling. *CHANCE*, 7(1), 50–57.

Beck, M. B. 1987. Water quality modelling: a review of the analysis of uncertainty. *Water Resources Research*, **23**(8), 1393–1442.

Claessen, F. A. M. and J. Dijkman 1999. *Uitganspunten recente studies waterbeheer.* Delft: RIZA, WL Delft Hydraulics Lelystad.

de Marchi, B. 1995. Uncertainty in environmental emergencies: a diagnostic tool. *Journal of Contingencies and Crisis Management*, 3(2), 103–112.

Douglas, M. and A. Wildavsky 1982. *Risk and Culture: Essays on the Selection of Technical and Environmental Dangers.* Berkley, CA: University of California Press.

Funtowicz, S. O. and J. R. Ravetz 1985. Three types of risk assessment: a methodological analysis. In C. Whipple and V. T. Covello (eds.), *Risk Analysis in the Private Sector.* New York: Plenum Press.

Funtowicz, S. O. and J. R. Ravetz 1990. *Uncertainty and Quality in Science for Policy.* Dordrecht: Kluwer.

Funtowicz, S. O. and J. R. Ravetz 1993. The emergence of post-normal science. In R. von Schomberg (ed.), *Science, Politics and Morality: Scientific Uncertainty and Decision-making.* Dordrecht: Kluwer.

Heisenberg, W. 1962. *Physics and Philosophy: the Revolution in Modern Science.* New York: Harper and Row.

Henrion, M. and B. Fischhoff 1986. Assessing uncertainty in physical constants. *Annual Journal of Physics*, **54**(9), 791–797.

Hilderink, H. B. M., E. Mosselman, A. H. W. Beusen, *et al.* 1998. *Targets* 1.0 (CD-ROM), ESIAM. Bussum: Baltzer Science Publishers.

Hoekstra, A. Y. 1998a. Appreciation of water: four perspectives. *Water Policy*, **1**, 605–622.

Hoekstra, A. Y. 1998b. *Perspectives on Water: an Integrated Model-based Exploration of the Future.* Utrecht: International Books.

Hoffman, F. O. and J. S. Hammonds 1994. Propagation of uncertainty in risk assessment: the need to distinguish between uncertainty due to lack of knowledge and uncertainty due to variability. *Risk Analysis*, **14**(5), 707–712.

Janssen, P. H. M., A. C. Petersen, J. P. Van Der Sluijs, J. S. Risbey, and J. R. Ravetz 2003. *Quickscan Hints & Actions List.* Bilthoven: RIVM.

Kann, A. and J. P. Weyant 2000. Approaches for performing uncertainty analysis in large-scale energy/economic policy models. *Enviromental Modeling and Assessment*, 5(1), 29–46.

Kasemir, B., J. Jäger, C. Jaeger, and M. T. Gardner 2003. *Public Participation in Sustainability Science*. Cambridge: Cambridge University Press.

Klinke, A. and O. Renn 2002. A new approach to risk evaluation and management: risk-based, precaution-based, and discourse-based strategies. *Risk Analysis*, **22**(6), 1071–1094.

Krayer von Kraus, M. P. 2005. *Uncertainty in Policy Relevant Sciences*. Copenhagen: Technical University of Denmark.

Krimsky, S. and D. Golding 1992. *Social theories of risk*. Westport, CT: Praeger.

Lempert, R. J. and J. L. Bonomo 1998. *New methods for robust science and technology planning. DB-238-DARPA*. Santa Monica, CA: RAND.

Lempert, R. J., S. W. Popper, and S. C. Bankes 2003. *Shaping the Next One Hundred Years: New Methods for Quantitative Long-term Policy Analysis*. Santa Monica, CA: RAND.

Manne, A. and R. Richels 1995. *The Greenhouse Debate: Economic Efficiency, Burden Sharing and Hedging Strategies*. Stanford, CA: Department of Operations Research, Stanford University.

Middelkoop, H., M. B. A. van Asselt, S. A. van 't Klooster, *et al*. 2004. Perspectives on flood management in the Rhine and Meuse rivers. *River Research and Applications*, **20**, 327–342.

MNP 2006. *Nationale Milieuverkenning* 6: 2006–2040. MNP report 500085001. Bilthoven: MNP and SDU uitgevers.

Morgan, G. M. and M. Henrion 1990. *Uncertainty – a Guide to Dealing with Uncertainty in Quantitative Risk and Policy Analysis*. Cambridge: Cambridge University Press.

Penman, J., D. Kruger, I. Galbally, *et al*. 2000. *Good Practice Guidance and Uncertainty Management in National Greenhouse Gas Inventories*. IPCC National Greenhouse Gas Inventories Programme. Tokyo: Institute for Global Environmental Strategies.

Petersen, A. C. 2006. *Simulating Nature: a Philosophical Study of Computer-simulation Uncertainties and Their Role in Climate Science and Policy Advice*. Apeldoorn: Het Spinhuis.

Petersen, A. C., P. H. M. Janssen, J. P. Van Der Sluijs, J. S. Risbey, and J. R. Ravetz 2003. *Mini-checklist & Quickscan Questionnaire*. Bilthoven: RIVM.

Rayner, S. F. 1987. Risk and relativism in science for policy. In B. B. Johnson and V. T. Covello (eds.), *The Social and Cultural Construction of Risk*. Dordrecht: Reidel.

Rayner, S. and R. Cantor 1987. How fair is safe enough? The cultural approach to societal technology choice. *Risk Analysis*, **7**(1), 3–9.

Renn, O. 2006. Risk governance: towards an integrative approach. *IRGC White Paper No. 1*. Geneva: International Risk Governance Council (IRGC).

Risbey, J., J. Van Der Sluijs, and J. Ravetz 2001. Protocol for the assessment of uncertainty and strength of emissions data. *NW&S report E-2001–10*. Utrecht: Department of Science, Technology and Society, Utrecht University.

RIVM 1988. *Concern for Tomorrow*, ed. H. D. Samson. Alphen aan de Rijn: Tjeenk Willink.

RIVM 1991/1992. *National Environmental Outlook 2 1990–2010*. Bilthoven: RIVM.

RIVM 1993. *National Environmental Outlook 3 1993–2015*, ed. H. D. Samsom. Alphen aan den Rijn: Tjeenk Willink.

RIVM 1997. *National Environmental Outlook 4 1997–2020*, ed. H. D. Samsom. Bilthoven: Tjeenk Willink (in Dutch).

RIVM 1999. *Addendum Measure, Calculate, and Uncertainties: the Working Method of RIVM's Environmental Research*. Bilthoven: RIVM (in Dutch).

RIVM 2000. *National Environmental Outlook 5 2000–2030*. Alphen aan de Rijn: Samson (in Dutch).

Rotmans, J. and B. de Vries 1997. *Perspectives on Global Change: the TARGETS Approach*. Cambridge: Cambridge University Press.

Rowe, W. D. 1994. Understanding uncertainty. *Risk analysis*, **14**(5), 743–750.

Saltelli, A., K. Chan, and E. M. Scott 2000. *Sensitivity Analysis*. London: Wiley.

Schneider, S. H., B. L. Turner, II, and H. Morehouse Garriga 1998. Imaginable surprise in global change science. *Journal of Risk Research*, **1**(2), 165–185.

Schwarz, M. and M. Thompson 1990. *Divided We Stand: Redefining Politics, Technology and Social Choice*. New York: Harvester Wheatsheaf.

Seebregts, A. J., B. W. Daniels, P. C. Van Der Laag, and S. Spoelstra 2001. To be sure! *First Initiatives to a More Conscious Management of Uncertainty in Policy Relevant Studies of ECN*. Petten/Amsterdam: Energy Centre of the Netherlands (ECN) (in Dutch).

Shackle, G. L. S. 1955. *Uncertainty in Economics and Other Reflections*. Cambridge: Cambridge University Press.

Thompson, M., R. Ellis, and A. Wildavsky 1990. *Cultural Theory*. Boulder, CO: Westview Press.

van Aardenne, J. A. 2002. *Uncertainty in Emission Inventories*. Wageningen: Wageningen University.

van Asselt, M. B. A. 2000. *Perspectives on Uncertainty and Risk: the PRIMA Approach to Decision Support*. Dordrecht: Kluwer.

van Asselt, M. B. A. 2004. Foresight and the art of uncertainty communication. *Beleidswetenschap*, **18**(2), 137–168 (in Dutch).

van Asselt, M. B. A. 2005. The complex significance of uncertainty in a risk era: logics, manners and strategies in use. *Internal Journal for Risk Assessment and Management*, **5**(2/3/4), 125–158.

van Asselt, M. B. A. and J. Rotmans 1996. Uncertainty in perspective. *Global Environmental Change*, **6**(2), 121–157.

van Asselt, M. B. A. and N. Rijkens-Klomp 2002. A look in the mirror: reflection on participation in integrated assessment from a methodological perspective. *Global Environmental Change*, **12**(3), 107–180.

van Asselt, M. B. A. and J. Rotmans 2002. Uncertainty in integrated assessment modelling: from positivism to pluralism. *Climatic Change*, **54**, 75–105.

van Asselt, M. B. A. and A. P. Petersen 2003. *Not Afraid of Uncertainty*. The Hague: Lemma/RMNO (in Dutch).

van Asselt, M. B. A., A. H. W. Beusen, and H. B. M. Hilderink 1996. Uncertainty in integrated assessment: a social scientific approach. *Environmental Modelling and Assessment*, **1**(1/2), 71–90.

van Asselt, M. B. A., R. Langendonck, F. van Asten, *et al.* 2001a. *Uncertainty and RIVM's Environmental Outlooks. Documenting a Learning Process*. Maastricht/Bilthoven: ICIS/RIVM.

van Asselt, M. B. A., H. Middelkoop, S. A. van 't, Klooster, *et al.* 2001b. Development of flood management strategies for the Rhine and Meuse basins in the context of integrated river management. *Report of the IRMA-SPONGE project* 3/NL/1/164/99 15 183 01, Maastricht/Utrecht.

van Asselt, M. B. A., S. Huijs, and S. A. van 't Klooster 2003. The intriguing relationship between uncertainty and normativity: the need for pluralistic assessment. In N. Gottschalk-Mazouz and N. Mazouz (eds.), *Nachhaltigkeit und globaler Wandel*.

Integrative Forschung zwischen Normativität und Unsicherheit. Stuttgart: Campus, University of Stuttgart.

van Asselt, M. B. A., J. Rotmans, and D. S. Rothman 2005. *Scenario Innovation: Experiences from a European Experimental Garden.* London: Taylor & Francis.

van Asselt, M. B. A., W. F. Passchier, and M. P. Krayer von Kraus 2007. Uncertainty assessment: an analysis of regulatory science on wireless communication technology, RF EMF and cancer risks. *Report for the IMBA project – workpackage 1.* Maastricht: Maastricht University.

van Asten, F. and M. B. A. van Asselt 1999. *Uncertainty and the fifth Environmental Outlook: Workshop Report.* Maastricht: ICIS (in Dutch).

van de Kerkhof, M. 2004. *Debating Climate Change. A Study on Stakeholder Participation in an Integrated Assessment of Long-term Climate Policy in the Netherlands.* Utrecht: Lemma.

Van Der Klis, H. 2003. *Uncertainty Analysis in Numerical River-morphological Models.* Delft: Delft University.

Van Der Sluijs, J. P. 1997. *Anchoring Amid Uncertainty: on the Management of Uncertainties in Risk Assessment of Anthropogenic Climate Change.* Utrecht: Univesiteit Utrecht.

Van Der Sluijs, J. P., J. S. Risbey, A. C. Petersen, P. H. M. Janssen, and J. R. Ravetz 2003. *Tool Catalogue for Uncertainty Management.* Bilthoven: RIVM.

van Notten, P. W. F. 2005. *Writing on the wall: scenario-development in times of discontinuity*, Maastricht University – Thela Thesis & Dissertation.com.

van Vlimmeren, J. C. G., F. J. H. Don, and V. R. Okker 1991. *Composition and pattern of forecast uncertainty due to unreliable data: further results, 81.* The Hague: Central Planning Bureau.

van Wee, G. P., J. van der Waard, M. J. van Doesburg, *et al.* 1993. *Traffic and Transport in the National Environmental Outlook 3 and the SVV Outlook 1993.* 251701014. Bilthoven: RIVM, AVV (in Dutch).

VROM 1989. *National Environmental Policy Plan.* The Hague: SDU.

Walker, W. E., P. Harremoës, J. Rotmans, *et al.* 2003. Defining uncertainty: a conceptual basis for uncertainty management in model-based decision-support. *Integrated Assessment*, **4**(1), 5–17.

Wallsten, T. S. 1990. Measuring vague uncertainties and understanding their use in decision making. In G. M. von Furstenberg (ed.), *Acting Under Uncertainty: Multidisciplinary Conception.* Dordrecht: Kluwer.

Wynne, B. 1992. Uncertainty and environmental learning: reconceiving science and policy in the preventive paradigm. *Global Environmental Change*, **2**(June), 111–127.

Zimmermann, H. J. 1996. Uncertainty modelling and fuzzy sets. In H. G. Natke and Y. Ben-Haim (eds.), *Uncertainty: Models and Measures.* Berlin: Akademie Verlag.

6

Vulnerability of people, places, and systems to environmental change

NEIL LEARY AND SARA BERESFORD

6.1 Introduction

The consequences of environmental change are not uniform. They vary for different people, places, and times. This is the clear picture that emerges from studies of the impacts of global environmental change as well as observations of the distribution of impacts of natural hazards, the incidence of hunger and famine, and problems of land-use and land-cover change.[1] The responses to the ensuing risks will also differ among people, places, and times.

These differences have given impetus to *integrated regional assessment* of global environmental change, as noted by Yarnal (Chapter 2, this volume). An important strand of research within regionally specific studies of global environmental change is the exploration of the causes of differential consequences, which is the focus of *vulnerability assessment*. Vulnerability assessment seeks answers to questions such as: who and what are vulnerable to the multiple environmental and human changes underway, and where? How are these changes and their consequences attenuated or amplified by different human and environmental conditions? What can be done to reduce vulnerability to change? How may more resilient and adaptive communities and societies be built? (Turner *et al.* 2003a). The relevance and importance of this line of research is the necessity of understanding the patterns and causes of differential vulnerabilities in order to formulate effective responses that can lessen the potential harm. In this chapter we examine frameworks for answering these and other questions about vulnerability to environmental change. The assessment of adaptation responses, which is closely related to vulnerability assessment, is explored by Dickinson, Bizikova, and Burton (Chapter 7, this volume).

---

[1] See, for example, Burton *et al.* (1993), FAO (1999), IFRC (2002), Kasperson *et al.* (1995), Leary *et al.* (2008a), McCarthy *et al.* (2001), and Meyer and Turner (1994).

Vulnerability is context dependent and changing, being shaped by the dynamics and interactions of local environment, society, and environment–society processes, as well as by the interactions of local processes with processes at larger spatial scales. In consequence, vulnerability assessment is reliant on case study approaches that provide the necessary context of place, time, society, and scale for understanding the complexity of environmental risks and feasible and effective responses to them (Kasperson *et al.* 2001: 5).

However, case studies give rise to a potential difficulty for vulnerability assessment: the causes of and remedies for vulnerability may appear to be specific and unique to a particular group, place, or time and hence inapplicable to other contexts. The challenge for vulnerability assessment is to find explanations of the causes of differential consequences of environmental change from studies of specific contexts that are nonetheless robust and applicable to a wider set of contexts. Finding robust explanations can be facilitated by working at regional scales. This intermediate scale between the local and the global permits (1) horizontal integration across different local places and systems within the region and (2) vertical integration of interactions and feed-backs from local-to-regional and regional-to-global scales (Kasperson *et al.* 1995: 23). As argued by Wilbanks and Kates (1999: 608), focus on a single scale can lead to misunderstanding and so research at multiple scales is needed.

This chapter introduces and defines vulnerability and related concepts, compares vulnerability and impact assessment approaches, and gives an overview of selected frameworks for vulnerability assessment. The review of vulnerability assessment draws from and complements previous reviews by Adger (2006), Cutter (1996), Kasperson *et al.* (2001, 2005, 2006), and Liverman (2001). Some general conclusions about *who* and *what* are vulnerable to environmental change, and the causes of vulnerability, are drawn from recent case studies of the project Assessments of Impacts and Adaptation to Climate Change (AIACC; Leary *et al.* 2008a), as well as other recent literature. The chapter closes with suggestions for future directions for vulnerability assessment.

6.2 Concepts: vulnerability, exposure, sensitivity, and resilience

Numerous definitions of vulnerability have been offered, a sample of which are presented in Box 6.1.[2] While definitions differ in their emphases and details, common to most definitions of vulnerability is the potential to suffer harm. This very simple definition is the one that is adopted in this chapter.

---

[2] A more comprehensive collection of definitions of vulnerability is provided by Cutter (1996) and updated and expanded by Kasperson *et al.* (2006).

Box 6.1
Definitions of vulnerability

Kates (1985)
Vulnerability is the capacity to suffer harm and react adversely.

Blaikie *et al.* (1994)
By vulnerability we mean the characteristics of a person or group in terms of their capacity to anticipate, cope with, resist, and recover from the impact of a natural hazard. It involves a combination of factors that determine the degree to which someone's life and livelihood are put at risk by a discrete and identifiable event in nature or in society.

Bohle *et al.* (1994)
Vulnerability is best described as an aggregate measure of human welfare that integrates environmental, social, economic, and political exposure to a range of potential harmful perturbations. Vulnerability is a multi-layered and multi-dimensional social space defined by the determinate, political, economic, and institutional capabilities of people in specific places at specific times.

Cutter (1996)
Vulnerability is broadly defined as the potential for loss. Vulnerability is conceived as both a biophysical risk as well as a social response, but with a specific areal or geographic domain. This can be geographic space, where vulnerable people and places are located, or social space – who in those places is most vulnerable.

Kelly and Adger (2000)
The ability or inability of individuals and social groupings to respond to, in the sense of cope with, recover from or adapt to, any external stress placed on their livelihoods and well-being.

Kasperson *et al.* (2001)
The differential susceptibility to loss from a given insult.

McCarthy *et al.* (2001)
Vulnerability (to climate change) is the degree to which a system is susceptible to, or unable to cope with, adverse effects of climate change, including climate variability and extremes. Vulnerability is a function of the character, magnitude, and rate of climate change, and variation to which a system is exposed, its sensitivity, and its adaptive capacity.

Kasperson *et al.* (2006)
The degree to which a person, system, or unit (such as a human group or place) is likely to experience harm due to exposure to perturbations or stresses.

People, places, and systems can be vulnerable to a wide range of insults. Vulnerability has been looked at as the susceptibility of a group, place, or system to harm from a particular perturbation or stress, such as climate change, or a suite of multiple perturbations and stresses that might include, for example, climate change, land degradation, demographic change, urbanization, technological change, and economic globalization. Vulnerability has also been looked at from the perspective of the potential for a particular outcome or harm, for example poverty, hunger, or dislocation, that is determined by multiple and interacting forces.[3]

Liverman (2001: 204–205) and Cutter (1996: 530) distinguish two general strands in studies of the vulnerability of people and places, one biophysical and the other social. The biophysical approach has its roots in the natural hazards field. Primary attention is given to characterizing exposure to a hazard in biophysical terms. The spatial distribution of some hazardous condition is identified; human occupancy of the hazardous zone (e.g., floodplain, coastal area, seismic zone) is estimated; the magnitude, duration, and frequency of the hazard (e.g., flood, hurricane, earthquake) is determined; and the potential loss of life and property associated with particular events are estimated.

The other strand gives primary attention to the social determinants of vulnerability. The causes of vulnerability are sought in the social processes and conditions that place people in harm's way and shape their capacities to absorb stresses, cope with and adapt to change, and recover from harm. Similar to the "wounded soldier" from the Roman use of *vulnerabilis*, whose risk is primarily determined by his wounded state, Kelly and Adger (2000) make the analogy that it is primarily the existent state of the exposed that determines their vulnerability.

These two strands have been integrated during the past decade or more by researchers trying to achieve a more holistic and complete theory of the causes of vulnerability.[4] Chambers (1989) described vulnerability as having an external side, comprised of external stresses to which a system is exposed, and an internal side, or the characteristics of the exposed system. The internal characteristics have been grouped into system sensitivity and resilience.

Building on Chambers' synthesis of biophysical and social conceptions of vulnerability, Kasperson *et al.* (2006: 9) characterize vulnerability as having three dimensions: *exposure* to stresses, perturbations, and shocks; *sensitivity* of people, places, and ecosystems to stress or perturbation, including their capacity to anticipate and cope with the stress; and *resilience* of exposed people, places, and ecosystems in terms of their capacity to absorb shocks and perturbations while

---

[3] See Clark *et al.* (2000) for a brief discussion of different perspectives in vulnerability assessments.

[4] Adger (2006, p. 271) observes that the natural hazards research tradition attempted to incorporate physical science, engineering, and social sciences since its inception. In application, however, the emphasis has tended to be more on physical science and engineering than on the social aspects.

maintaining function. Exposure, sensitivity, and resilience, defined in Box 6.2, are seen as the three major dimensions of vulnerability. Other authors decompose vulnerability into the components *exposure, sensitivity,* and *adaptive capacity* (e.g. Adger, 2006: 270). While different authors have applied slightly different decompositions, the basic concepts of exposure, sensitivity, and resilience or adaptive capacity have provided structure for organizing the many different causes of vulnerability and to draw together the biophysical and social strands of vulnerability research.

Box 6.2
The three dimensions of vulnerability
(from Kasperson *et al.* 2006)

Exposure is the contact between a system and a perturbation or stress.
Sensitivity is the degree to which a group, place, or system is affected by exposure to a perturbation or stress.
Resilience is the ability of a group, place, or system to absorb perturbations or stresses without changes in its fundamental structure or function that would drive it into a different state (or extinction).

Figure 6.1 shows the dimensions of vulnerability and their linkages to causal processes of vulnerability and relates them to the assessment process and societal responses. Demographic, social, economic, biophysical, and other processes of the coupled human–environment system drive environmental changes that, together with natural variability, expose people, places, and systems to stresses. The processes also shape the internal system characteristics that determine system sensitivities, capacities, and resilience to cope with and respond to exposures. Vulnerability can be lessened by interventions of stakeholders at different points in the causal structure. Assessment of the causes, nature, and distribution of vulnerability can yield information for more effective interventions.

A probabilistic evaluation of the exposures, system interactions, and consequences establishes the attendant risks, while the realization of a specific scenario of exposures, sensitivities, capacities, and resilience results in a set of impacts. The impacts can also feed back to bring additional environmental and social changes and exposures and to alter the conditions of the human–environment system.

The causal structure of vulnerability depicted in the schematic figure suggests that responses can lessen vulnerabilities and risks of harmful impacts by operating on the dimensions of vulnerability. Vulnerability can be lessened by measures that lessen exposure to perturbations and stresses; lessen sensitivities to exposures;

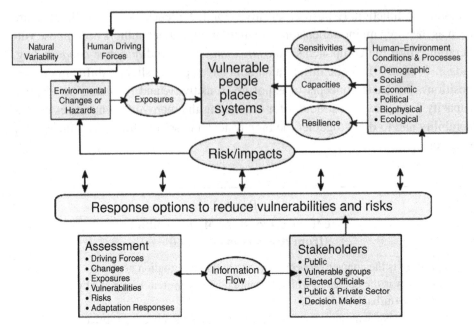

Figure 6.1. Vulnerability dimensions, processes, assessment, and responses. Processes of the coupled human–environment system drive environmental changes that, together with natural variability, expose people, places, and systems to stresses. The processes also shape the internal system characteristics that determine system sensitivities, capacities, and resilience to cope with and respond to exposures. Vulnerability can be lessened by interventions of stakeholders at different points in the causal structure. Assessment of the causal structure of vulnerability can yield information for more effective interventions (adapted from Kasperson *et al.* 2001 and Kasperson *et al.* 2006).

increase capacities to cope and adapt; and increase resilience and recovery potential. Examples of responses to lessen vulnerability are listed in Table 6.1 and the dimensions of vulnerability on which they primarily act are identified.[5]

Responses can be taken at many different scales, ranging from the individual to the local, community, nation, region, and globe. Whether and how public sector responses to vulnerabilities and risks are made are controlled by the political economy of the system in which a range of stakeholders participate that include public and private sector decision makers, elected officials, experts, members of vulnerable groups, and the general public. The generation and flow of information, sometimes from formal assessment of the vulnerabilities, risks, and response options,

---

[5] Adaptation responses to climate hazards and climate change have been investigated by the AIACC studies and are explored in Leary *et al.* (2008b).

Table 6.1. *Responses to lessen vulnerability.*

Response measure	Dimension of vulnerability acted upon			
	Exposure	Sensitivity	Capacities	Resilience
• Dampen driving forces of change (e.g. reduce greenhouse-gas emissions)	×			
• Anticipate exposures and prepare for potential effects	×	×		
• Migrate from or limit development in exposed place	×	×		
• Divest from exposed and sensitive activity	×	×		
• Build dykes	×			
• Eliminate or suppress disease vectors	×			
• Control damages (e.g. disaster relief)		×		
• Transfer water to areas/activities of priority need		×		
• Switch to more robust crop varieties		×		
• Develop new, more robust crop varieties		×	×	×
• Diversify sources of household income and livelihoods	×	×	×	×
• Establish mechanisms and rules for transfer of water (e.g. water markets, water courts)			×	×
• Expand/create mechanisms to share and spread risks (e.g. insurance)			×	×
• Expand health care infrastructure		×	×	×
• Accumulate stocks of human, physical, natural, financial, and social capital			×	×
• Reform land tenure rules	×	×	×	×
• Create incentives and support for soil conservation	×	×	×	×
• Promote universal education and enfranchizement			×	×
• Promote family planning	×		×	×

occasionally from explicitly participatory and recursive processes of assessment and deliberation, but most often from informal assessment of "signals," influence the responses made to environmental hazards. Kasperson *et al.* (2001: 19) also emphasize the potential for failures to address environmental degradation that possibly feed back into the driving forces to aggravate problems.

6.3 Comparison of vulnerability and impact assessment approaches

The dominant analytic framework for research on global environmental risks has taken an impacts assessment approach that is often focused on biophysical impacts (Liverman 2001: 206). However, questions about the vulnerability of social and ecological systems have emerged as a central focus, particularly in assessments that are driven by policy needs (Clark *et al.* 2000: 1).

Impact and vulnerability assessment approaches have different, yet potentially complementary motives. In impact assessments, analysts seek to understand the overall severity of consequences from an environmental change or stress so as to provide a basis for making decisions about measures to curtail the change. In contrast, analysts performing vulnerability assessments seek to understand the causes of potential consequences so as to inform decisions about measures to limit damaging consequences that include strategies to cope with or adapt to change, as well as measures to curtail the offending change.

An impact assessment begins with selection of an environmental stress of concern. Scenarios are developed for the evolution through time of the selected stress, and possibly other factors thought to influence the scale of the exposed system or its sensitivity to the chosen stress. Models of the exposed system or sub-system components are used to simulate impacts for the different scenarios. At the end of the chain of analyses, selected adaptive responses are incorporated into the system models and the simulations are performed again. The simulated residual or net impacts provide an indicator of the susceptibility of the system or sub-system to harm from the environmental stress, or its vulnerability. But vulnerability itself is not explicitly analyzed.[6]

In a vulnerability assessment, the characteristics of exposed people, places, and systems are examined to understand not only how they might be harmed by exposure to stresses, but also, and importantly, why. Interacting, multiple stresses are emphasized, though in practice it is common to give more detailed treatment to a particular environmental stress. Historical analyses figure prominently in vulnerability assessments as a tool to investigate the social, political, and economic processes that formed past and present vulnerabilities (Ribot 1995; Adger 1999; Kelly and Adger 2000). Scenarios for analysis of future vulnerability can be constructed, although the permutations can become overwhelming when treating multiple stresses and multiple causal factors of vulnerability. But understanding the present causes of vulnerability can provide useful guidance for generating a parsimonious set of scenarios to explore future vulnerabilities. Also, an inverse approach to scenarios has been proposed for vulnerability assessments that would

---

[6] See Carter *et al.* (1994) for a description of impact assessment methods.

attempt to determine the combinations of environmental and social stresses that would significantly increase the likelihood of specified adverse outcomes (Clark *et al.* 2000: 2–3).

The growing emphasis being given to vulnerability assessment is driven by a number of trends in policy and science discourses. Increasing attention is being given to differential consequences of global environmental change, causes of these differences, and their equity implications. A second contributing trend is a growing appreciation of coping and adaptation measures as critical components of strategies to reduce environmental risks. In the context of climate change, widening recognition that efforts such as the Kyoto Protocol cannot halt human-caused climate change has led to the realization that adaptive strategies to address the stresses from climate change also will be needed as a complement to mitigation efforts (see Parry *et al.* 1998: 741; McCarthy *et al.* 2001: 6–8; Dickinson *et al.*, Chapter 7, this volume). A third trend is the growing recognition that *multiple* environmental and social stresses act on people, places, and systems, that the consequences will depend on interactions among the multiple stresses, and that responses need to take account of the multiple stresses if they are to be effective, efficient, and sustainable (see, for example, Watson *et al.* 2001: 123–134). A fourth trend is a growing belief that many of the fundamental causes of vulnerability to one environmental stress are likely common to other stresses, both environmental and social, that common remedies may exist, and that remedies need to be integrated into broader development planning if they are to be successful.

Vulnerability assessment is designed to examine differential consequences of multiple stresses, and their causes. For these reasons a number of authors make the point that a vulnerability approach provides a framework that is better suited than an impacts approach for evaluating responses to risks from global environmental change (see Ribot 1995; Kelly and Adger 2000; Liverman 2001). A vulnerability approach is also well suited for placing environmental change in context with other social and development problems and goals and exploring common causes of risks and common remedies. Consequently, the trends in policy and science discourses have made vulnerability assessment an essential part of integrated regional assessment.

6.4 Approaches to vulnerability assessment

A variety of approaches or frameworks have been developed to conceptualize the factors and linkages that cause vulnerability and to guide the assessment of vulnerability. Kasperson and colleagues (2006) trace the evolution of approaches to vulnerability from those of the natural hazards field to political economy, human ecology, and the most recent and more integrative approaches. In this chapter, selected approaches are described to illustrate some of the main features of vulnerability assessment and their evolution. These are the entitlements theory approach of Sen (1981), the human ecology approach of Bohle–Downing–Watts (1994), and the coupled human–environment system approach developed at Clark University and Stockholm Environment Institute (Kasperson and Kasperson 2001; Turner *et al.* 2003a; Kasperson *et al.* 2006).

6.4.1 Entitlements theory

Amartya Sen opens his well-known essay on poverty and famines noting that: "Starvation is the characteristic of some people not *having* enough food to eat. It is not the characteristic of there *being* not enough food to eat. While the latter can cause the former, it is but one of many *possible* causes" (Sen 1981: 1, emphasis in the original). In the essay, Sen presents his theory of entitlements, an approach to understanding starvation, famine, and poverty that focuses on ownership patterns and factors that influence the processes and rules governing what bundles of commodities a person can command through, for example, production or exchange. The set of potential bundles a person can command represents his or her entitlements and determines whether or not the person will have enough to eat as well as provide for other basic needs.

The approach shifted analysts away from the then prevailing explanation of famines as a problem of insufficient food supply and toward causes embedded in economic and social structures and processes of ownership and exchange. A key implication of the approach for vulnerability assessment is that much of vulnerability is socially controllable (Kasperson *et al.* 2006: 12). In this framework, an individual starts with an endowment, or ownership bundle, that consists of his or her own labor power, land, and other resources. The individual has a range of options for transforming the initial endowment into alternative bundles of commodities. He or she might, for example, produce food for direct consumption, produce other commodities for sale in the market, or sell his or her labor for wages. The income from the sale of commodities and labor can then be exchanged for food and other commodities. The full set of commodity bundles that can be commanded through these and other processes from an initial endowment point is referred to by Sen as the entitlement set. It is synonymous with the opportunity set or budget constraint in the economic theory of household production and consumption.

What commodity bundles can be commanded by the individual will depend on the legal, political, economic, and social rules and processes of the society and the person's position in it (Sen 1981: 46). These include, for example, market structure and regulation, rights of entrepreneurs to the profits of an enterprise, rights of a peasant family member to family and communal output, social security provisions, employment guarantees, subsidies, and tax policy.

The rules and processes for transforming endowments into entitlements provide a system or function termed an "exchange entitlement mapping" by Sen, and the "architecture of entitlements" by Adger (1999). Using this mapping function, one can partition endowments into those that map into entitlement sets that encompass a minimum food requirement, and hence allow the individual to avoid starvation, and those that do not, and in consequence do not prevent starvation (Sen, 1981: 48).

In applying this framework in the analysis of global environmental change, there are two pathways by which people may be made more or less vulnerable to hunger or poverty. One is by collapsing or expanding endowments. For example, climate change may reduce (or enhance) the productivity of a peasant's land, thereby contracting (expanding) the endowment and the entitlement set. The second is by shifting the entitlement mapping in ways that can strongly affect those living near the margin of hunger and poverty. For example, the stresses of environmental change may result in higher food prices and place minimum food requirements or basic needs beyond reach for some. The framework can also be used to identify processes that amplify or dampen vulnerability to environmental change and the groups being made more or less vulnerable, operating again through either changing endowments or shifting the entitlement mapping.

Adger (1999) and Kelly and Adger (2000) apply entitlements theory to examine the vulnerability of coastal Vietnam to adverse outcomes of present and historical exposures to tropical cyclones and of future climate change. The context for their analyses is the process of *doi moi*, which has been transforming the collectivized system of Vietnam toward a private property, market-oriented system since the late 1980s. They examine how the reforms of *doi moi* have altered the structure of entitlements and trace the implications for capacities and vulnerability in coastal communities. Some of the consequences of *doi moi* include decreased collective action for maintenance of dykes and diversion of resources from coastal defenses into other infrastructure investments, privatization and conversion of communal mangroves to agriculture, increased aquaculture, development of informal credit systems, and greater average incomes accompanied by more pronounced inequality.

These changes have brought substantial and complex changes in entitlements. Some have expanded the bundles of commodities that can be commanded by households, others have contracted them. The net effects vary for coastal and inland communities, and vary across livelihood groups and households within communities. Kelly and Adger conclude that it is unwise to attempt to determine the overall impact of a broad-based change such as *doi moi* on vulnerability.

However, it is possible to determine the tendencies of different aspects of *doi moi* to amplify or dampen vulnerability. For example, vulnerability is increased by diminished coastal defenses, loss of access to goods and services from mangroves, and increased investment in aquaculture that is highly exposed to storm damage. Vulnerability is decreased by, for example, increased average incomes, greater diversity of incomes, and the expanded availability of credit. An understanding of these tendencies can help to inform policies to counteract trends that are leading to greater vulnerability to climate hazards.

6.4.2 The human ecology approach

Human ecologists (e.g. Hewitt 1983, 1997; Watts 1983) attempt to explain why the poor, women, and other marginalized groups experience the greatest risks from natural hazards by looking at the influences of economic development and political economy on class structure, governance, and economic dependency (Adger 2006). Bohle and colleagues (1994) explain vulnerability in terms of the individual's position within a causal structure of vulnerability that has three dimensions: human ecology, expanded entitlements, and political economy. They define *human ecology* as the relations between nature and society, which encompass the means by which humans interact with and transform nature to derive goods, services, and livelihoods; the risks and threats from the interactions and transformations; and the properties of ecosystems and society that govern the interactions and transformations. *Expanded entitlements* includes the commodity bundles of Sen (1981) plus a wider set of social entitlements, including those such as empowerment and enfranchizement by which access to entitlements are secured, fought over, and contested. *Political economy* refers to the macro-scale processes of accumulation and distribution of the society in which entitlements are embedded.

In their framework, vulnerability is a function of exposure to stress and shocks, the capacity to cope with stress and shocks, and the potentiality to recover from the effects. According to Bohle *et al.* (1994: 41), exposure is determined by human ecology and expanded entitlements, while coping capacity is primarily shaped by entitlements and political economy. Entitlements determine the resources that can be drawn on to help cope with environmental stress and shocks. Political economy plays a role by determining whether and how entitlements can be claimed, contested, defended, and lost. Political economy and human ecology combine to determine recovery potential.

Bohle *et al.* applied their framework to examine the vulnerability of households and communities in Zimbabwe to hunger from climate change. Downing (1992) characterized the biophysical exposure by estimating changes in agricultural area and maize yields for scenarios of temperature and precipitation change.

Sensitivity of the system was characterized by estimating the impacts on household income and food security, which varied depending on household size and composition, caloric energy requirements, size of land holdings, area cultivated, average maize yield, seed and maize prices, cost of hired labor, and off-farm income. Food availability for a representative household was estimated to be reduced by roughly 60–70%. Bohle *et al.* (1994) examined the resilience and vulnerability of the system with respect to the exposures and system sensitivities estimated by Downing and found that food insecurity in Zimbabwe was exacerbated by weak macro-economic performance, inequitable land distribution, and misdirected social policy.

6.4.3 Coupled human–environment systems

Drawing on the work of Kasperson *et al.* (2001, 2006) and others, participants in the Research and Assessment Systems for Sustainability Program[7] produced a comprehensive, highly integrative framework for analyzing vulnerability. The framework is presented in Turner *et al.* (2003a). There are two basic parts to the framework: perturbations/stresses and the coupled human–environment system. Some features of the framework that distinguish it from earlier frameworks include treatment of human systems and the natural environment as a coupled system; a shift in focus from a single perturbation or stress to multiple and continuous perturbations and stresses; consideration of internally generated as well as external perturbations and stresses; integration of human responses to reduce exposure, cope, adapt, and build resilience into the analysis; and nesting of spatial and temporal scales to capture cross-scale interactions and dynamics.

The coupled human–environment framework was not put forward by its creators with the expectation that it would be fully implemented in all vulnerability assessments. This would exceed the capacity and often the needs of most vulnerability analysts according to Kasperson and colleagues. More limited applications that might focus on sub-systems of the framework and simplify other parts are envisioned. But the framework "serves as a reminder of what is missing in assessments based on such simplifications" (Kasperson *et al.* 2006: 45).

Application of the coupled human–environment framework to the southern Yucatan highlights the complex dynamics of human and environmental systems of the region that give rise to pressures on the region's farmers, its forests, and a vision of future "green" development that would be based on ecological and archaeological tourism (Turner *et al.* 2003b). *Ejido* farmers of the region cultivate maize for subsistence and chili for commercial sale. The farmers are exposed

---

[7] See http://sust.harvard.edu.

to environmental hazards from water stress and hurricanes that damage crops and to market hazards from highly volatile chili prices that impact incomes. They are also exposed to hazards that result from their own farming practices, which are deforesting and fragmenting the landscape, causing greater exposure to severe winds and fire hazards, depleting soil nutrients, increasing crop pests and disease, and enabling invasion by bracken fern that arrests forest regrowth for decades.

Responding to the deforestation in the region, and with the goal of shifting the regional economy to one based on ecological and archaeological tourism, the government of Mexico created the Calakmul Biosphere Reserve in the center of the region. This has reduced the availability of new lands for *ejidos*, which, with an increase in landless migrants, has increased land pressures on the *ejido* farms and resulted in illegal squatting on reserve lands. Meanwhile, federal farm programs provide payments to farmers that, as an unintended consequence, promote the clearing of forests for pasture. Use of the coupled human–environment framework potentially could help guide interventions that would avoid such unintended outcomes.

6.5 Measures of vulnerability

Measures of the vulnerability of different people, places, and systems have been developed to compare the degree of vulnerability of different exposure units and to examine changes in vulnerability through time. Because vulnerability is a complex, multi-dimensional concept that is not directly observable (Moss *et al.* 2001: 8), researchers have experimented with a variety of methods to develop proxy indicators. These indicators are used in attempts to quantify the dimensions of vulnerability: exposure, sensitivity, coping and adaptive capacity, and resilience. One source of interest in vulnerability indicators is their potential application for identifying areas of priority for aid for adaptation, enhancement of adaptive capacity, and more intensive analysis of vulnerabilities and adaptation needs. Caution is needed, however, because measures of vulnerability are rarely validated with empirical evidence (Kasperson *et al.* 2005: 150). Downing *et al.* (2001) provided a useful summary of vulnerability indices.

In this chapter we present three examples of vulnerability indices to illustrate some of the approaches. All three are constructed to measure vulnerability to climate hazards. Hurd *et al.* (1999) construct indices of present day vulnerability of water users at a watershed scale. Brooks and Adger (2003) construct national scale indices of vulnerability using data on present risks from climate-related disasters. Moss *et al.* (2001) also develop a national scale indicator of vulnerability and use models to simulate how vulnerability may change in the future.

6.5.1 *Watershed scale vulnerabilities in the United States*

Hurd *et al.* (1999) present a set of indicators and alternative aggregations of indicators for measuring the vulnerability to climate change of water users in different watersheds of the United States. They build on the work of Gleick (1990) who measured the vulnerability of water users in 18 major water basins of the United States using five different indicators related to water supply and water withdrawals for consumptive uses (e.g., irrigation, industrial, and domestic water consumption). Hurd *et al.* (1999) extend the number of indicators to 12, including indicators related to instream water uses (e.g. water quality, ecosystem health, and navigation), flood risks, and flexibility in both water use and institutions for allocating water. They also increase the spatial resolution to provide a more spatially detailed picture of vulnerability by developing and mapping their indicators for each of the 204 watersheds of the contiguous United States (the scale of analysis is the 4-digit hydrologic unit classification of the US Geological Survey).

The indicators, selected through a consultative process that included fourteen water resource experts, focus largely on biophysical aspects of vulnerability. There are indicators that measure exposure to scarcity and variability of water supply, poor water quality, and stresses on aquatic habitats and the sensitivity of water users to these exposures. Adaptive capacity is incorporated by indicators that measure water storage capacity, flexibility of water use, and flexibility of institutions for water allocation.

Hurd *et al.* (1999) present two possible aggregations of the indicators, each of which gives a very different picture of the geographic distribution of vulnerabilities in the United States. The first index aggregates the six indicators that are most closely related to exposures, sensitivities, and adaptive capacities in consumptive uses of water. A second index aggregates six indicators most closely associated with instream water uses and flood risk. In each case the individual indicators are given equal weight in constructing the index and the resulting scores are classified into three different rankings of vulnerability: low, medium, and high.

The results for the two different vulnerability indices are mapped in Figures 6.2a and b. Inspection of Figure 6.2a reveals a spatially cohesive pattern in which vulnerabilities of consumptive uses of water are greatest in the southwestern and high plains watersheds of the United States, and lower in the east and Great Lakes region. These results generally agree with the more aggregate scale findings of Gleick (1990) and reflect the general aridity, high water withdrawals relative to mean runoff, high variability of runoff, and groundwater depletion that characterize many parts of the western United States. In contrast, the map of vulnerabilities of instream water uses and flood risks in Figure 6.2b shows a very different geographic distribution of vulnerability across the United States that is

(a)

(b)

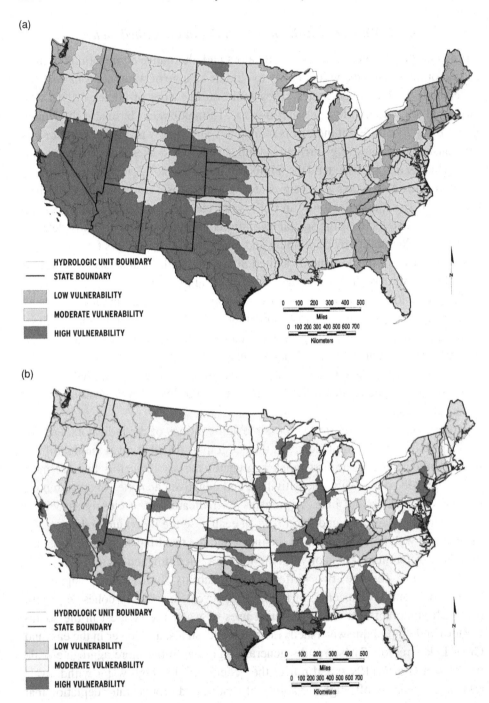

Figure 6.2. Vulnerability of water users in the United States. Used by permission of Blackwell Publishing from Hurd *et al.* (1999).

much less spatially cohesive. The maps of the two indices illustrate the sensitivity of results to the weighting of different factors thought to shape vulnerability. And the lack of spatial cohesion for the second index is an indication of the importance of spatial scale when considering vulnerabilities of instream water uses and flood risks.

6.5.2 Country level vulnerability to climate related disasters

Brooks and Adger (2003) employed country level data for observed outcomes of climate related disasters to measure vulnerability to climate hazards and use their results to infer vulnerability to climate change. Data on deaths and numbers of persons affected by climate-related disasters integrate exposure to climate hazards and social, economic, institutional, and political factors that shape vulnerability to hazards. The rationale for this approach is that countries that have experienced relatively severe and widespread impacts from climate hazards in recent decades are those that are presently less able to cope with climate hazards and that many of these countries are likely to continue to lag in their abilities to cope with hazards in the near future. Also, climate change is expected to include changes in the frequency, severity, and geographic distribution of some types of extreme events such as flooding, drought, heat waves, and coastal storm surges (IPCC 2001: 15). While the changes in extremes can lead to changes in who is exposed, Brooks and Adger argued that it is likely that many of the changes will exacerbate vulnerability in countries that are presently highly exposed to climate extremes and are deficient in their abilities to cope, as evidenced by the numbers of persons killed or affected by such events (Brooks and Adger 2003: 20). An attractive feature of the approach is that vulnerability is measured by observable outcomes and so is capable of being empirically verified.

Five proxy measures of vulnerability are developed for three different time periods: 1971–1980, 1981–1990, and 1991–2000. The chosen proxies are the percentage of population killed or affected by climate-related disasters during the period, the percentage of population killed, the absolute number of persons killed, and the ratio of the number killed to number affected, the latter calculated by two different methods. The source of the data is the Emergency Events Database of the US Office of Foreign Disaster Assistance and the Centre for Research into the Epidemiology of Disasters at the Universite Catholique de Louvain in Brussels. An event is classified as a disaster and included in these datasets if it results in 10 or more deaths, 100 or more persons affected, a call for international assistance, or the declaration of a state of emergency. Brooks and Adger screen out events that are judged to be unrelated to climate. The types of disasters that are included in their analysis include drought, epidemics of diseases that exhibit strong seasonal variability, extreme temperature, famine, flood, insect infestation, land- and mud-slides, wave and surge, wild fire, and windstorm.

Brooks and Adger used the percentage of the population either killed or affected by climate related disasters for the period 1991–2000, their preferred proxy for vulnerability, to rank 167 countries from most to least vulnerable. For all 20 of the most vulnerable countries, the number of persons either killed or affected during 1991–2000 is greater than 50% of its 1995 population. All but one of the twenty most vulnerable countries during this period are developing countries.[8] The same indicator was constructed for the period 1981–1990. A comparison of the indicator for the two periods show some changes in vulnerability rankings from one decade to the next, but about half of the countries listed as most vulnerable in 1991–2000 also rank highly for earlier decades. Changes in rankings are likely due primarily to differences in the number of events to which the countries were exposed in the two decades and not to changes in their sensitivities to these events.

6.5.3 *Vulnerability in the future*

Moss *et al.* (2001) present a methodology for constructing a national-scale indicator of vulnerability to climate change and projecting changes in vulnerability through time. Their indicator of vulnerability is a composite of indicators of sensitivity to climate change in five different sectors and adaptive capacity in three different dimensions. The sectors in which they measure sensitivity are human settlements, food security, human health, ecosystems, and water resources. The dimensions of adaptive capacity that are measured are economic capacity, human and civic resources, and environmental capacity. Table 6.2 identifies the proxy variables that were selected to measure sectoral sensitivities and adaptive capacity.

Present values of their vulnerability indicator are constructed for 38 countries using 1990 data from a variety of sources. Each of the individual indicators listed in their table 5 is given equal weight in constructing the composite indicator of vulnerability. Values are normalized such that the average world value in 1990 is equal to zero; a positive value for the composite indicator of a country indicates lesser vulnerability (greater resilience) than the world average in 1990; a negative value indicates greater vulnerability than the world average. Their analysis for 1990 finds that, while a few of the developing countries rank above the world average, all countries that are found to be more vulnerable than the world average are developing countries.

---

[8] The exception, Australia, is ranked high by this proxy due to the way in which that country reports on events that affect a large number of persons (Brooks and Adger, 2003, p. 17).

Table 6.2. *Indicators of sectoral sensitivity to climate change and adaptive capacity. (Adapted from Moss et al. 2001: 11–12).*

Dimension of vulnerability	Sector/Category	Indicator	Proxy for
Sensitivity	Settlements, Infrastructure	Population at risk from sea-level rise	Potential extent of disruption from sea-level rise and storm surges
		% population with access to safe water	Access to basic services to buffer against climate variability and change
		% population with access to sanitation	
	Food security	Cereal production per unit land area	Degree of modernization in agriculture; access of farmers to inputs
		Animal protein consumption per capita	Access of population to markets and other mechanisms to compensate for production shortfalls
	Human health	Birth rate	Composite of conditions that affect human health
		Life expectancy	
	Ecosystems	% land managed	Degree of human intrusion in natural landscape; land fragmentation
		Fertilizer use/crop land area	Nitrogen and phosphorous loadings to ecosystems and stresses from pollution
	Water resources	Ratio of water withdrawals to supply of water	Sensitivity of water system to meet current or future needs
Adaptive and coping capacity	Economic capacity	GDP per capita	Distribution of access to markets, technology, and other resources useful for adaptation
		Gini coefficient	
	Human and civic resources	% population in workforce	Social and economic resources available for adaptation after meeting other present needs
		Illiteracy rate	Human capital and adaptability of labor force
	Environmental capacity	% land not managed	Landscape fragmentation and ease of ecosystem migration
		SO_2 emissions per unit land area	Air quality and other stresses on ecosystems
		Population density	Population pressure and stresses on ecosystems

Moss and colleagues also simulated changes in their selected indicators to the year 2095 using an integrated assessment model and related models. These models provide projections of, for example, population, gross domestic product, crop production, numbers of people at risk from sea-level rise, water withdrawals for agriculture, and other variables at the scale of 11 world regions. Values of their indicators of sectoral sensitivity and adaptive capacity for the future were constructed using assumed relationships between the indicator variables and outputs of the integrated assessment model. The projections were performed for three different scenarios: rapid growth, local sustainability, and delayed development.

The results projected for 2095 include mixed changes in sectoral sensitivities, increased adaptive capacities, and reduced vulnerabilities (Moss *et al.* 2001, pp. 39–40, Tables 11 and 12). Projected increases in adaptive capacities and reductions in vulnerability are generally greater for developing countries than for developed countries. By 2095, almost all countries are estimated to be less vulnerable than the global average for 1990. But of the 19 countries ranked as most vulnerable in 1990 out of the 38 countries included in the analysis, all of which are developing countries, 15 are still among the 19 most vulnerable countries in 2095.

The declining vulnerabilities yielded by the analyses of Moss and colleagues are driven by assumptions of generally rising welfare and rising per capita incomes throughout the world. Even in the delayed development scenario, global per capita income rises almost four-fold from 1990 to 2095. Whether or not these assumptions prove true will have important implications for future vulnerability. Also important will be the extent to which there continue to be large inequities within and between countries in the distribution of wealth, resources, and other determinants of adaptive capacity.

6.6 Who and what are vulnerable?

In a review of the state of vulnerability assessment, Clark and colleagues found that conclusions about who and what are vulnerable to global environmental change is hampered by conflicting conceptual frameworks, incomplete and incomparable data, and inadequate models (Clark *et al.* 2000: 1). More recently, Kasperson and colleagues (2005: 145) found that knowledge about the sources and patterns of vulnerability is still incomplete and that the use of different conceptual frameworks, applied to different types of threats, yield different patterns of where vulnerability is found and who is vulnerable. Considerably more work needs to be done to advance understanding of the causes of vulnerability, knowledge of the exposures, sensitivities, capacities, and resilience of different people, places, and systems, and methods to anticipate and explore future dynamics of the dimensions of vulnerability. Yet, with these limitations in mind, some general conclusions emerge from available studies.

6.6.1 Vulnerable households and livelihoods

Households in the same community can experience broadly similar environmental conditions and yet have very different vulnerability to environmental hazards due to different exposures, sensitivities, and capacities. For example, exposures to floods and mudslides vary depending on the location of a household relative to ravines, steep slopes, and floodplains. Such locations are often occupied by poorer and marginalized households (Diaz 1992; Barros *et al.* 2008; Pulhin *et al.* 2008).

Households whose livelihoods are directly and strongly dependent on natural resources of variable and fragile productivity can be highly sensitive to climate and other environmental stresses, particularly if the household is not well diversified in the resources upon which it relies for its production activities. Examples of environment sensitive livelihoods include subsistence crop farming (Chinvanno *et al.* 2008; Nyong *et al.* 2008; Osman-Elasha and Sanjak 2008; Ziervogel *et al.* 2008), small-scale commercial farming and livestock production (Chinvanno *et al.* 2008; Eakin *et al.* 2008), plantation agriculture (Ratnisiri *et al.* 2008), pastoral herding (Batimaa *et al.* 2008), and fishing (Nagy *et al.* 2008). Sensitivity can also extend to households that supply inputs to such producers, use their outputs in the household's own production of other goods and services, or that spend a large portion of the household budget to purchase food and other necessities from such producers.

The capacities of households to resist, cope, recover, and adapt are determined to a large degree by their entitlements to market and non-market commodities and the social entitlements emphasized by Bohle *et al.* (1994). A household's or individual's entitlements determine what resources may be called upon to respond to the stresses of environmental change and the resulting impacts. Some of the entitlements found by the AIACC case studies to be important determinants of capacity include access to safe water and sanitation (Adejuwon 2008; Pulhin *et al.* 2008), security of water rights (Osman-Elasha and Sanjak 2008; Yin *et al.* 2008), land-tenure status (Batimaa *et al.* 2008; Eakin *et al.* 2008; Pulhin *et al.* 2008), quantity and quality of land, and livestock holdings (Chinvanno *et al.* 2008; Nyong *et al.* 2008; Ziervogel *et al.* 2008), quantity and quality of household labor supply (Nyong *et al.* 2008; Ziervogel *et al.* 2008), and financial assets such as monetary income, financial savings, and access to credit (Chinvanno *et al.* 2008; Eakin *et al.* 2008; Nyong *et al.* 2008; Osman-Elasha and Sanjak 2008; Pulhin *et al.* 2008; Ziervogel *et al.* 2008). Access to formal and informal social safety nets are also an important determinant of capacity of households (Eakin *et al.* 2008; Osman-Elasha and Sanjak 2008; Ziervogel *et al.* 2008).

As the above suggests, there is a strong relationship between poverty and vulnerability to environmental pressures. Households living in poverty are more likely to occupy hazardous locations, to be highly dependent on environment sensitive livelihoods and to lack the entitlements necessary to access resources and support systems that are critical for responding to hazards. Within the household, differences in entitlements and the division of labor can result in greater vulnerability of women and children than for men and greater vulnerability for the elderly, infirm, and handicapped (Bohle *et al.* 1994; Cutter 1995; Nyong *et al.* 2008; Pulhin *et al.* 2008). Victims of conflict, migrant workers and their families, marginal populations in urban areas, ethnic minorities, and members of low income households within vulnerable livelihood systems, among others, are found to be vulnerable to food insecurity (Food and Agriculture Organization 1999). Indigenous peoples are often highly vulnerable because their traditional livelihoods are highly sensitive to climate and environmental changes and they are often marginalized from political and social support systems (Scott *et al.* 2001: 392; Kasperson *et al.* 2005: 149).

6.6.2 *Vulnerable communities*

The vulnerability of households can be dampened or amplified by processes at the community and higher levels. Several of the AIACC studies demonstrate that communities with poorly diversified rural economies and high proportions of households engaged in subsistence or small-scale farming or herding on marginal lands typically lack the resources and institutions to provide sufficient infrastructure and services to sustain resiliency and capacity to adapt to environmental change (Leary *et al.* 2008c). Poor management of resources and poor local governance can degrade the natural resource base, decrease local livelihood opportunities and incomes, widen inequality, and increase out-migration of able-bodied workers, all of which add to the vulnerability of the community (Batimaa *et al.* 2008; Nyong *et al.* 2008; Osman-Elasha and Sanjak 2008; Pulhin *et al.* 2008). Communities that are highly dependent on degraded resources, that are experiencing multiple and accumulating stresses, and in which human systems are in or near a failed state, and so incapable of effective response, are at greatest risk of worst-case outcomes such as collapse of rural livelihoods, deepening and widening poverty, displacement of population, hunger, epidemics, and violent conflict.

In many of the AIACC case studies, vulnerability is heightened because key functions of risk management are inadequate or absent in communities due to weaknesses in supporting institutions (Batimaa *et al.* 2008; Chinvanno *et al.* 2008; Dube and Sekhwela 2008; Mataki *et al.* 2008; Osman-Elasha and Sanjak 2008; Pulhin *et al.* 2008). The institutions are often poorly resourced, lacking in human capacity, overloaded with multiple responsibilities, and overwhelmed by the demands of their communities. In some instances, traditional institutions have been diminished in role by socio-economic changes and government policies, which reduces the capacity of the community to respond and increases vulnerability, as has happened in Botswana's Limpopo Basin (Dube and Sekhwela 2008). In Nigeria and Sudan, competition for scarce water and productive land contribute to marginalization of people, conflict and violence, which erode the social institutions necessary for coping with stress and responding to shocks (Nyong *et al.* 2008; Osman-Elasha and Sanjak 2008). Economic globalization brings a variety of changes that impact at the community level, both positively and negatively, including increased market competition that can weaken local livelihoods and decrease the ability of communities to provide support services (Eakin *et al.* 2008; Mataki *et al.* 2008).

Communities of low-lying coasts and small islands are highly exposed to a variety of climate hazards that are expected to be amplified by global climate change and sea-level rise. These climatic hazards converge with growing concentrations of people and infrastructure in harm's way, poor erosion control and management of coastal development, transformations of land cover in upland watersheds, scarce and sensitive water supplies, destruction of coastal wetlands and damage to coral reefs to create conditions of heightened vulnerability (Barros *et al.* 2008; Mataki *et al.* 2008). Because individual events can adversely impact a large portion of the population and resources of a small island state, their ability to respond to disasters and to recover can be severely compromised. High dependence of island economies on tourism, which can be sensitive to climate and environmental variations, also adds to the vulnerability of small islands (Payet 2008).

In the Third Assessment Report of the Intergovernmental Panel on Climate Change (IPCC), Scott *et al.* (2001) evaluated the vulnerabilities of stylized settlement types to different stresses anticipated to arise from climate change. Vulnerabilities of the different settlement types to flooding, landslides, tropical cyclones, changes in primary resource productivity, water shortage, fires, and other stresses were rated as low, moderate, or high depending upon the severity of likely impacts, degree of disruption to the settlement, and the ease or difficulty and cost for a settlement to overcome or adapt to the impacts. One of their main conclusions is that vulnerability to flooding and landslides is significant and widespread across all the settlement types they considered. Resource-dependent settlements are most vulnerable to stresses from changes in the productivity of primary resources, while coastal–riverine–steepland settlements are more vulnerable to floods and/or landslides, as expected. Rural settlements generally are judged by Scott and his colleagues to be more vulnerable than their urban counterparts, with exceptions for air quality and heat island effects. The reason for this is the relative isolation of rural settlements and typically limited infrastructure.

6.6.3 Vulnerable regions

Observed impacts of natural hazards display a pattern of greater impacts in developing country regions than in developed countries. More than one million deaths are attributed to tropical cyclones, floods, and droughts over the period 1980–2000, of which roughly 90% are in developing countries (Pelling *et al.* 2004). Asia has been disproportionately impacted, experiencing 43% of the natural disasters and 70% of the deaths from natural disasters. Countries with low human development represent roughly 10% of the world population exposed to natural disasters but account for more than 50% of the deaths. In comparison, countries with high human development, representing 15% of the exposed population, account for less than 2% of the deaths. Developed countries experience greater economic losses from natural disasters in absolute terms, but losses in developing countries are larger relative to their gross domestic product.

While vulnerability has multiple causes and is a complex, multi-dimensional and dynamic phenomenon, these observations show a clear link between the level of development, which is closely associated with poverty, and vulnerability to natural hazards. The association of poverty and low levels of development with vulnerability is also found in the national scale indicators of Brooks and Adger (2003) and Moss *et al.* (2001) and patterns found in the AIACC studies of vulnerability to climate change (Leary *et al.* 2008a) and is consistent with the findings of the IPCC's Third Assessment Report (McCarthy *et al.* 2001).

The IPCC examined the vulnerability to climate change of continental-scale regions of the world, the results of which are summarized in Table 6.3. While continental-scale regions are not the ideal unit for an assessment of vulnerability, some general features do emerge. Not surprisingly, a central finding of the IPCC is that there are substantial differences in expected impacts, adaptive capacities, and vulnerabilities among populations and systems within each of the continental-scale regions. However, there is a consistent pattern of low adaptive capacity and high vulnerability in the developing country regions of the world. Within the developed world, adaptive capacity is higher and vulnerability is generally lower, though even in the developed countries there are marginalized populations that are considered to be highly vulnerable. The high vulnerability of developing country regions is attributed primarily to the low levels of human, financial, natural, and physical resources, limited institutional and technological capabilities, and failures of political-economies that characterize these regions (McCarthy *et al.* 2001: 13–15). Another contributing factor is the large share of economic activity

Table 6.3. *Regional adaptive capacity, vulnerability, and key concerns related to climate change (Adapted from McCarthy et al. 2001: 14).*

	Adaptive capacity	Vulnerability	Key concerns	Human–environment conditions
Africa	Very low	High	• Food security • Water availability • Infectious disease • Desertification • Extreme weather • Biodiversity • Coastal settlements	• Lack of human, financial, natural, and physical resources • High proportion of population in conditions of poverty, low food security, poor health status • Low HDI and low food security, except North Africa where these are medium/low • One-third of incomes from farming; 70% of population earn livelihood from farming • High reliance on rainfed agriculture; highly variable rainfall
Asia	Varied	Varied	• Extreme weather • Food security • Water availability • Infectious disease and heat stress • Coastal settlements • Biodiversity • Infrastructure in permafrost zones	• Wide range of development levels across Asia; HDI ranges from low in south Asia, medium in southeast, and high in other parts • Adaptive capacity is low and vulnerability high in developing countries of Asia; developed countries have high adaptive capacity and low vulnerability • Large population living in poverty • Two-thirds of world's undernourished people live in Asia, particularly in south, southeast, and China; low to medium/low food security in Asia • Many countries characterized by rapid population growth
Australia/ New Zealand	High	Low to high	• Agricultural productivity • Water availability • Extreme weather • Ecosystem change and loss • Biodiversity	• Some groups have low capacity to adapt and high vulnerability (e.g., indigenous peoples) • High HDI and high food security • Competition for water is intense in arid and semiarid areas • Agricultural productivity very sensitive to rainfall changes; ENSO related droughts cause significant distress • Growing population and investments near coasts are increasing exposures to cyclones and storm surge • Coral reefs, arid and semiarid habitats, alpine systems, and freshwater wetlands in coastal zone are particularly vulnerable

(cont.)

Table 6.3. (*cont.*)

	Adaptive capacity	Vulnerability	Key concerns	Human–environment conditions
Europe	High	Varied	• Water availability • Alpine glaciers, permafrost environments • River flood hazard • Agricultural productivity • Ecosystem change and loss • Tourism	• Vulnerability varies from low to high; highest in less developed parts, e.g. Southern Europe and European Arctic • High HDI and medium/high food security • Diversified economy with low proportion of GDP derived from primary resources • Reduced water availability anticipated in drought prone south; increased in wetter north • Agricultural productivity expected to decrease in south and east, increase in north • Greater risks of flooding, erosion, and wetland loss anticipated
Latin America	Low	High	• Extreme weather • Water availability and quality • Agricultural productivity • Infectious diseases • Coastal settlements	• Low HDI and food security in Central America; high HDI and medium/low food security in South America; • Weather extremes are common in Central America, southern Mexico, and other areas and capacity to cope is low • ENSO drives large part of interannual climate variability; ENSO related droughts and floods have had substantial impacts • Populations dependent on glacial melt for water are vulnerable • Crop yields projected to decrease at many sites; subsistence farming threatened
North America	High	Low	• Agricultural productivity • Water availability • Ecosystem change and loss • Coastal settlements • Extreme weather and insurance losses • Human health	• High HDI and high food security • Some communities are vulnerable; e.g., indigenous peoples and those dependent on climate-sensitive resources • Mixed effects on crop yields; potential for negative effects rises with greater warming • Populations dependent on snowmelt for water exposed to changes in seasonal availability • Unique ecosystems such as prairie wetlands, alpine tundra, and cold-water ecosystems at risk; effective adaptation unlikely

Region			Key vulnerabilities	Description
Polar regions	Low to high	Low to high	• Natural systems • Coastal erosion • Ecosystem change • Species distribution and abundance • Feedbacks to climate system	• Natural systems have low capacity and high vulnerability; Indigenous, traditional peoples have little capacity and high vulnerability; technologically developed communities have high capacity and low vulnerability • Climate change in polar regions expected to be greater, more rapid than elsewhere • Some changes already evident in sea ice, ice sheets and shelves, permafrost, and coastal erosion, species distribution and abundance • Ice edge ecosystems threatened • Processes that feed back to climate, once triggered, may continue for centuries
Small Island States	Low	High	• Sea-level rise • Coastal settlements and resources • Water availability and quality (salinity) • Ecosystem change and loss • Biodiversity • Fisheries • Agricultural productivity • Tourism	• Low to high HDI and low to high food security for different island groups • High dependence on coastal resources, many of which are in fragile state • Little economic diversification; lack of infrastructure and human and financial capital • Small open economies that are highly sensitive to external shocks • Rapid population growth and high population densities • Limited water supplies • Limited arable land • High exposure to storm surge and sea-level rise

in developing countries that is accounted for by climate-sensitive primary resource sectors, and the even larger share of population who earn their livelihoods from these sectors. Also, some of the biophysical exposures appear to be harsher. For example, most studies of climate change impacts on crop yields suggest that there is greater potential for productivity declines in the tropics and sub-tropics, home to most of the developing countries, than in temperate and cooler climates of the mid-latitudes that are common for much of the developed world (McCarthy *et al.* 2001: 5).

6.7 Conclusions and future directions

A rich literature of vulnerability case studies has been accumulating and adding to the body of knowledge about vulnerability to environmental change. An overarching and robust result from this literature is that vulnerability is not a simple function of exposure but is caused and shaped by multiple interacting environmental and human processes. The state and dynamics of these processes vary from place to place and from person to person, generating conditions of vulnerability that differ in both character and degree. Consequently, people exposed to similar environmental stresses are not impacted to the same extent.

The most severe outcomes from exposure to environmental stress are not expected to arise where a single stress acts alone but where multiple stresses, environmental and other, are at work to create conditions of high vulnerability. The specific interacting stresses and their relative importance vary across contexts but often include rapid population growth, growing numbers of people living in poverty, urbanization that concentrates people and infrastructure in harm's way, economic globalization, conflict, human transformations of the environment, unsustainable intensification of land uses, and climate variability and change. These processes operate at spatial scales from local to global and cross-scale interactions shape vulnerability in important ways. Vulnerability is often associated with poverty and can be particularly high where cumulative stresses have severely degraded the natural resource base of livelihoods and regional economies and eroded the capacities of institutions to respond effectively.

However, the state of knowledge about vulnerability to environmental change and methods for assessment of vulnerability are incomplete. Assessments have been based on varied concepts and frameworks and have applied varied methods. This variety and experimentation has been a necessary stage in the development of vulnerability assessment as a system of inquiry into the differential consequences of environmental change. But comparing and synthesizing results across case studies, and finding robust explanations of the causes of vulnerability, have been hampered by the diversity of approaches.

There is a need for a robust conceptual framework and development of analytic methods and tools to make further advances in understanding that can inform vulnerability-reducing actions (Kasperson *et al.* 2005: 145). Recently there has been convergence on a core set of ideas, which we have summarized in this chapter. From these ideas, the coupled human–environment system of Turner *et al.* (2003a) has emerged as a promising framework for vulnerability assessment. The coupled human–environment system is a broad and flexible conceptual framework for relating how different components and processes of a system interact to give rise to vulnerabilities. A number of elements have been identified as essential for vulnerability analysis (Turner *et al.* 2003a: 8075) and adoption of these elements in future studies will help to unify knowledge about vulnerability. These elements include investigation of system exposure, sensitivity and resilience, consideration of multiple, interacting stresses, and integration across nested scales.

Future assessments of vulnerability need to address a number of critical gaps in our knowledge and methods. Exposures to environmental change, climate change for example, are often characterized at very coarse spatial scales, while vulnerabilities and response options need to be understood at much finer spatial scales that correspond to those at which decision making takes place. Methods are needed to realistically and credibly downscale environmental changes to better match the spatial scales relevant to decision making. Cross-scale interactions among different components of coupled human–environment systems need to be explored to better understand how they shape vulnerabilities, capacities to respond, and the availability and feasibility of response options. Evaluations of the cumulative effects of multiple stresses operating over time and interacting with each other are a pressing need. This includes richer examination of the human driving forces of vulnerability, the relationship between poverty and vulnerability, and the roles of institutions and governance. Identification of entry points for interventions to reduce vulnerability and evaluation of the effectiveness and consequences of response options are needed. The measurement of vulnerability needs to move in directions that allow for empirical validation.

An important motivation for vulnerability assessment is to inform more effective responses for lessening the potential harm from environmental change. This has led to assessment approaches that are user-oriented and that engage stakeholders in the assessment process. These important developments will be a central feature of future vulnerability work. To facilitate advances in this direction, research is needed to better understand decision processes of different classes of actors for managing environmental risks, the roles of institutions, the knowledge needed to make good decisions, the transformation of information into relevant and usable knowledge, and the design of assessment processes that can contribute to deliberation and decision making.

References

Adejuwon, J. O. 2008. Vulnerability in Nigeria, a national level assessment. In N. Leary, C. Conde, J. Kulkarni, *et al.* (eds.), *Climate Change and Vulnerability*. London: Earthscan.

Adger, W. N. 1999. Social vulnerability to climate change and extremes in coastal Vietnam. *World Development*, **27**(2), 249–269.

Adger, W. N. 2006. Vulnerability. *Global Environmental Change*, **16**, 268–281.

Barros, V., A. Menéndez, C. Natenzon, *et al.* 2008. Climate change vulnerability to floods in the metropolitan region of Buenos Aires. In N. Leary, C. Conde, J. Kulkarni, *et al.* (eds.), *Climate Change and Vulnerability*. London: Earthscan.

Batimaa, P., L. Natsagdorj, and N. Batnasan 2008. Vulnerability of Mongolia's pastoralists to climate extremes and change. In N. Leary, C. Conde, J. Kulkarni, A. Nyong, and J. Pulhin (eds.), *Climate Change and Vulnerability*. London: Earthscan.

Blaikie, P., T. Cannon, I. Davis, and B. Wisner 1994. *At risk: natural hazards, people's vulnerability and disasters*. London: Routledge.

Bohle, H. G., T. E. Downing, and M. J. Watts 1994. Climate change and social vulnerability. *Global Environmental Change*, **4**(1), 37–48.

Brooks, N. and W. N. Adger 2003. Country level risk measures of climate related natural disasters and implications for adaptation to climate change. Tyndall Centre for Climate Change Research, *Working Paper* 26.

Burton, I., R. W. Kates, and G. F. White 1993. *The Environment as Hazard*. New York: Guildford Press.

Carter, T. R., M. L. Parry, H. Harasawa, and S. Nishioka 1994. *IPCC Technical Guidelines for Assessing Climate Change Impacts and Adaptations*. London: University College London, Centre for Global Environmental Research and Tsukuba: National Institute for Environmental Studies.

Chambers, R. 1989. Vulnerability, coping and policy. *IDS Bulletin*, **20**, 1–7.

Chinvanno, S., S. Boulidam, T. Inthavong, *et al.* 2008. Climate risks and rice farming in the lower Mekong River basin. In N. Leary, C. Conde, J. Kulkarni, A. Nyong, and J. Pulhin (eds.), *Climate Change and Vulnerability*. London: Earthscan.

Clark, W. C., J. Jäger, R. Corell, *et al.* 2000. Assessing vulnerability to global environmental risks. *Report of the Workshop on Vulnerability to Global Environmental Change: Challenges for Research, Assessment and Decision Making*. May 22–25, 2000. *Discussion Paper* 2000–12. Warrenton, VA: Airlie House, Cambridge, MA: Belfer Center for Science and International Affairs.

Cutter, S. L. 1995. The forgotten casualties: women, children and environmental change. *Global Environmental Change*, **5**, 181–194.

Cutter, S. L. 1996. Vulnerability to environmental hazards. *Progress in Human Geography*, **20**(4), 529–539.

Diaz, V. J. 1992. Landslides in the squatter settlements of Caracas: towards a better understanding of causative factors. *Environment and Urbanization*, **4**(2), 80–89.

Downing, T. E. 1992. *Climate Change and Vulnerable Places: Global Food Security and Country Studies in Zimbabwe, Kenya, Senegal, and Chile*. Oxford: Environmental Change Unit, University of Oxford.

Downing, T. E., R. Butterfield, S. Cohen, *et al.* 2001. *Vulnerability Indices: Climate Change Impacts and Adaptation. UNEP Policy Series*. Nairobi: United Nations Environment Programme.

Dube, P. and M. Sekhwela 2008. Indigenous knowledge, institutions and practices for coping with variable climate in the Limpopo basin of Botswana. In N. Leary, J. Adejuwon, V. Barros, *et al.* (eds.), *Climate Change and Adaptation*. London: Earthscan.

Eakin, H., M. Wehbe, C. Avila, G. S. Torres, and L. A. Bojorquez-Tapia 2008. Social vulnerability of farmers in Mexico and Argentina. In N. Leary, C. Conde, J. Kulkarni, *et al.* (eds.), *Climate Change and Vulnerability*. London: Earthscan.

Food and Agriculture Organization 1999. *The State of Food Insecurity in the World 1999*. Rome: Food and Agriculture Organization of the United Nations.

Gleick, P. H. 1990. Vulnerability of water systems. In P. E. Waggoner (ed.), *Climate Change and U.S. Water Resources*. New York: Wiley, pp. 223–240.

Hewitt, K. (ed.) 1983. The idea of calamity in a technocratic age. In *Interpretations of Calamity from the Viewpoint of Human Ecology*. Boston, MA: Allen and Unwin, pp. 3–32.

Hewitt, K. 1997. *Regions at risk: a geographical introduction to disasters*. Harlow: Longman.

Hurd, B., N. Leary, R. Jones, and J. Smith 1999. Relative regional vulnerability of water resources to climate change. *Journal of the American Water Resources Association*, **35**(6), 1399–1409.

Intergovernmental Panel on Climate Change 2001. Summary for policymakers. In J. T. Houghton, Y. Ding, D. J. Griggs, *et al.* (eds.), *Climate Change 2001: The Scientific Basis*. Cambridge: Cambridge University Press.

International Federation of Red Cross and Red Crescent Societies 2002. *World Disasters Report 2002*.

Kasperson, J. X. and R. E. Kasperson 2001. *International Workshop on Vulnerability and Global Environmental Change, a Workshop Summary*, 17–19 May 2001. Stockholm: Stockholm Environment Institute.

Kasperson, J. X., R. E. Kasperson, and B. L. Turner, II (eds.) 1995. *Regions at Risk, Comparisons of Threatened Environments*. Tokyo: United Nations University Press.

Kasperson, J. X., R. E. Kasperson, B. L. Turner, II, W. Hsieh, and A. Schiller 2006. Vulnerability to global environmental change. In E. Rosa, A. Diekmann, T. Dietz, and C. Jaeger (eds.), *The Human Dimensions of Global Environmental Change*. Cambridge, MA: MIT Press.

Kasperson, R. E., J. X. Kasperson, B. L. Turner, II, K. Dow, and W. B. Meyer 1995. Critical environmental regions: concepts, distinctions, and issues. In J. X. Kasperson, R. E. Kasperson, and B. L. Turner, II (eds.), *Regions at Risk, Comparisons of Threatened Environments*. Tokyo: United Nations University Press.

Kasperson, R. E., J. X. Kasperson, and K. Dow 2001. Introduction: global environmental risk and society. In J. X. Kasperson and R. E. Kasperson (eds.), *Global Environmental Risk*. Tokyo: United Nations University Press, London: Earthscan, pp. 1–48.

Kasperson, R. E., K. Dow, E. Archer, *et al.* 2005. Vulnerable peoples and places. In R. Hassan, R. Scholes, and N. Ash (eds.), *Ecosystems and Human Wellbeing: Current State and Trends*, vol. 1. Washington, DC: Island Press, Washington, pp. 143–164.

Kates, R. W. 1985. The interaction of climate and society. In R. W. Kates, J. H. Ausubel, and M. Berberian (eds.), *SCOPE 27*. New York: Wiley.

Kelly, P. M. and W. N. Adger 2000. Theory and practice in assessing vulnerability to climate change and facilitating adaptation. *Climatic Change*, **47**, 325–352.

Leary, N., C. Conde, J. Kulkarni, A. Nyong, and J. Pulhin (eds.) 2008a. *Climate Change and Vulnerability*. London: Earthscan.

Leary, N., J. Adejuwon, V. Barros, I. Burton, J. Kulkarni, and R. Lasco (eds.) 2008b.
    Climate Change and Adaptation. London: Earthscan.
Leary, N., J. Adejuwon, W. Bailey, *et al.* 2008c. For whom the bell tolls, vulnerabilities in
    a changing climate. In N. Leary, C. Conde, J. Kulkarni, A. Nyong, and J. Pulhin
    (eds.), *Climate Change and Vulnerability*. London: Earthscan.
Liverman, D. M. 2001. Vulnerability to global environmental change. In J. X. Kasperson
    and R. E. Kasperson (eds.), *Global Environmental Risk*. Tokyo: United Nation
    University Press, London: Earthscan, pp. 201–216.
Mataki, M., K. Koshy, and V. Nair 2008. Top-down, bottom-up: mainstreaming
    adaptation in Pacific Island townships. In N. Leary, J. Adejuwon, V. Barros, *et al.*
    (eds.), *Climate Change and Adaptation*. London: Earthscan.
McCarthy, J. J., O. F. Canziani, N. A. Leary, D. J. Dokken, and K. S. White (eds.) 2001.
    *Climate Change 2001: Impacts, Adaptation, and Vulnerability. Contribution of
    Working Group II to the Third Assessment Report of the Intergovernmental Panel on
    Climate Change*. Cambridge: Cambridge University Press.
Meyer, W. B. and B. L. Turner, II (eds.) 1994. *Changes in Land Use and Land Cover: a
    Global Perspective*. Cambridge: Cambridge University Press.
Moss, R. H., A. L. Brenkert, and E. L. Malone 2001. Vulnerability to climate change, a
    quantitative approach. *Report to US Department of Energy, PNNL-SA-33642*.
    http://www.pnl.gov/globalchange/projects/vul/index.htm
Nagy, G. J., M. Bidegain, R. M. Caffera, *et al.* 2008. Climate and water quality in the
    estuarine and coastal fisheries of the Rio de la Plata. In N. Leary, C. Conde, J.
    Kulkarni, *et al.* (eds.), *Climate Change and Vulnerability*. London: Earthscan.
Nyong, A., D. Dabi, A. Adepetu, A. Berthe, and V. C. Ihemegbulem 2008. Vulnerability
    in the Sahelian zone of northern Nigeria, a household level assessment. In N. Leary,
    C. Conde, J. Kulkarni, A. Nyong, and J. Pulhin (eds.), *Climate Change and
    Vulnerability*. London: Earthscan.
Osman-Elasha, B. and A. Sanjak 2008. Livelihoods and drought in Sudan. In N. Leary,
    C. Conde, J. Kulkarni, *et al.* (eds.), *Climate Change and Vulnerability*. London:
    Earthscan.
Parry, M., N. Arnell, M. Hulme, R. Nicolls, and M. Livermore 1998. Adapting to the
    inevitable. *Nature*, **395**(22 October 1998), 741.
Payet, R. A. 2008. Climate change and the tourism dependent economy of the Seychelles.
    In N. Leary, C. Conde, J. Kulkarni, A. Nyong, and J. Pulhin (eds.), *Climate Change
    and Vulnerability*. London: Earthscan.
Pelling, M., A. Maskrey, P. Ruiz, and L. Hall 2004. *Reducing Disaster Risk: a Challenge
    for Development*. New York: United Nations Development Bank, Bureau for Crisis
    Prevention and Recovery.
Pulhin, J. M., R. Peras, R. Cruz, *et al.* 2008. Climate variability and extremes in the
    Pantabangan–Carranglan watershed of the Philippines, an assessment of
    vulnerability. In N. Leary, C. Conde, J. Kulkarni, *et al.* (eds.), *Climate Change and
    Vulnerability*. London: Earthscan.
Ratnisiri, J., A. Anandacoomaraswamy, M. Wijeratne, *et al.* 2008. Vulnerability of Sri
    Lankan tea plantations to climate change. In N. Leary, C. Conde, J. Kulkarni, *et al.*
    (eds.), *Climate Change and Vulnerability*. London: Earthscan.
Ribot, J. C. 1995. The causal structure of vulnerability: its application to climate impact
    analysis. *Geo-Journal*, **35**(2), 19–122.
Scott, M., S. Gupta, E. Jauregui, *et al.* 2001. Human settlements, energy, and industry. In
    J. J. McCarthy, O. F. Canziani, N. A. Leary, *et al.* (eds.), *Climate Change 2001:
    Impacts, Adaptation, and Vulnerability. Contribution of Working Group II to the*

Third Assessment Report of the Intergovernmental Panel on Climate Change. Cambridge: Cambridge University Press.

Sen, A. 1981. *Poverty and Famines, An Essay on Entitlement and Deprivation.* Oxford: Oxford University Press.

Turner, B. L., II, R. E. Kasperson, P. A. Matson, *et al.* 2003a. A framework for vulnerability analysis in sustainability science. *Proc. Nat. Acad. Sci.,* **100**(14), 8074–8079.

Turner, B. L., II, P. A. Matson, J. J. McCarthy, R. W. Corell, L. Christensen, N. Eckley, G. K. Hovelsrud-Broda, J. X. Kasperson, R. E. Kasperson, A. Luers, M. L. Martello, S. Mathiesen, R. Naylor, C. Polsky, A. Pulsipher, A. Schiller, H. Selin, and N. Tyler. 2003b. Illustrating the coupled human–environment system for vulnerability analysis: three case studies. *PNAS,* **100**(14), 8080–8085.

Watson, R. T. and the Core Writing Team 2001. *Climate Change 2001: Synthesis Report. Contributions of Working Groups I, II and III to the Third Assessment Report of the Intergovernmental Panel on Climate Change.* Cambridge: Cambridge University Press.

Watts, M. 1983. On the poverty of theory: natural hazards research in context. In K. Hewitt (ed.), *Interpretations of calamity from the viewpoint of human ecology.* Boston, MA: Allen and Unwin, pp. 231–262.

Wilbanks, T. J. and R. W. Kates 1999. Global change in local places: how scale matters. *Climatic Change,* **43**, 601–628.

Yin, Y., N. Clinton, B. Luo, and L. Song 2008. Resource system vulnerability to climate stresses in the Heihe River basin of Western China. In N. Leary, C. Conde, J. Kulkarni, *et al.* (eds.), *Climate Change and Vulnerability.* London: Earthscan.

Ziervogel, G., A. Nyong, B. Osman, *et al.* 2008. Household food security and climate change: comparisons from Nigeria, Sudan, South Africa and Mexico. In N. Leary, C. Conde, J. Kulkarni, *et al.* (eds.), *Climate Change and Vulnerability.* London: Earthscan.

7

Integrating climate change adaptation into sustainable development

THEA DICKINSON, LIVIA BIZIKOVA, AND IAN BURTON

7.1 The past neglect of adaptation

One of the first principles of integration is that there are always some relevant things necessarily left out. It is impractical to manage or model the environment in an all-encompassing way. As scientific understanding grows, and as risk perceptions evolve, it often happens that some of the omitted dimensions are seen to be much more important than originally supposed. A case in point is the issue of climate change. It was initially constructed as a pollution problem, and now is coming to be understood as a problem in human adjustment (coping or adaptation), to inevitable climate change, and therefore also as a problem in sustainable development and global economic and social equity. Furthermore, adaptation has been seen primarily as a problem for developing countries, since it is the poorer communities and those with least adaptive capacity that will be most affected at least in the near term. Increasingly, however, adaptation is also receiving more attention in developed countries as they realize that the inevitable impacts can no longer be ignored. All nations and the international community are now becoming engaged in a game of "catch-up" to take into account that which was initially left out or neglected.

In this chapter we first offer a brief explanation for the initial narrow formulation of the problem and how it has failed to produce results. We next show why adaptation has become an imperative. This leads to a discussion of the ways in which it might be integrated into the international sustainable development agenda. This can be approached in two ways. First, adaptation can be directly integrated into sustainable development as part of international development assistance. Second, adaptation can be better supported through the United Nations Framework Convention on Climate Change (UNFCCC). These two tracks are sometimes seen to be in conflict. We argue that they can be complementary and that both are needed. A further question relates to the integration of adaptation and mitigation. The promotion of adaptation in isolation from the management of the greenhouse gas emissions (that are causing climate change) can have perverse consequences. We conclude with a discussion of the content of on-going negotiations and its relation to research needs for integration.

7.2 Climate change as a pollution issue

Anthropogenic climate change was initially seen predominantly as a pollution issue. This view has recently been espoused by the film "An Inconvenient Truth" and has been adopted by the vast majority of climate activists. It is widely shared by government agencies and the media. The construction is not entirely wrong, but it is incomplete, and has a distorting effect on policy. The emergence of climate change onto the international environmental agenda followed hard on the heels of the issues of acid precipitation and stratospheric ozone depletion. Both of these problems were addressed through international agreements to control emissions: sulfur dioxide in the case of acid rain and chlorofluorocarbons in the case of the ozone layer. Greenhouse gas emissions, especially carbon dioxide, were identified as the relevant pollutants causing "global warming." It was appreciated that climate change was (and is) a much more complicated problem given the diverse impacts and benefits of climate including its variability and extremes, and the heavy reliance of the global economy on fossil fuels. However, given an understandable urgency to start tackling the problem it was the pollution perspective that dominated at the expense of the dimensions of adaptation, development, and equity. The Framework Convention (UNFCCC) was rapidly negotiated and prepared for signature at the Rio de Janeiro "Earth Summit" in 1992, closely followed by a rush to move to emission reductions in the Kyoto Protocol (KP) to the Convention, which was agreed in principle in 1997. Complaints about the lack of serious attention to adaptation, and development issues coming largely from developing countries, have steadily increased in volume and in sense of urgency (Burton *et al.* 2007a).

Meanwhile the developed world has mostly focused on mitigation, but without much success. Major reasons for the failure to date include the lack of effective technology to control emissions or produce energy with less reliance on fossil fuels. Closely linked to the technology constraint, is concern about costs and loss of lifestyle amenities. That technology which does exist is considered to be too expensive and/or its deployment is not in harmony with dominant public preferences. Also climate change is seen as a global pollution problem that requires broad international agreement on common action if "free-riding" is to be avoided. In the face of the widely diverse interests of nation-states, it has not been possible to reach agreement on more than token gestures towards the serious reduction of greenhouse gas emissions.

At the time of writing (December 2007) it is more than 15 years since the UNFCCC was first opened for signature in Rio, and 10 years since the Kyoto agreement. In that time only very small steps have been taken at the international level. Greenhouse gas emissions continue to rise at an increasing rate and the damage from extreme climate-related events continues to mount. Events with disastrous consequences such as the European heat wave of 2003 and Hurricane Katrina in New Orleans in 2005 are becoming more intense and more frequent. Although these events have been highly publicized in the developed world, it is the poorer and least developed communities and countries that will bear the brunt of the impacts of climate change largely because they have less adaptive capacity or ability to cope (see Leary and Beresford, Chapter 6, this volume). It is widely accepted that losses in these countries are already an increasing threat to development in general and in particular to the attainment of the Millennium Development Goals (UN 2007). The reduction of greenhouse gas emissions and the stabilization of greenhouse gas concentrations in the atmosphere are essential actions, as too is adaptation; it is necessary to adapt to the inevitable climate change which it is now too late to prevent.

7.3 The adaptation imperative

People have always had to cope with climate. The story of human evolution and development includes a very successful record of adjustment and adaptation to a wide variety of climates. Humans, as one of the most adaptable species, have spread widely over the Earth and created flourishing societies from the sub-arctic to the margins of hot deserts, from high mountains to low coasts. The climate of this planet varies as much or more over space than over time. In theory at least, there is a strong case to be made for adaptation to anthropogenic climate change. Human societies can and do adapt to, and benefit from, climate variety as demonstrated over millennia with much less advanced technology than is available today. To be sure there are constraints on adaptation, such as the cost of some measures (Dutch-style dykes are probably beyond the financial reach of Bangladesh) and the acceptability of some policies (restricting the use of flood plains or coasts or accommodating large numbers of climatically displaced refugees), will be neither popular nor easily achieved, but given the political will there is little doubt that the financial resources and the technology could be made available to substantially reduce the impacts of climate change through adaptation in both developed and developing countries. Given the more extreme and long-term projections of climate change perhaps there are ultimate limits to adaptation, but if so they are still far out of sight for human societies, (unmanaged ecosystems and their species are another matter). For such a change to be achieved, better understanding of the meaning and

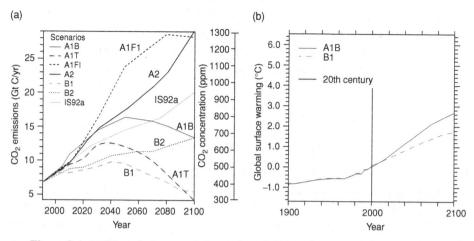

Figure 7.1. IPCC emission scenarios and modeled surface temperature increases (IPCC 2000).

value of adaptation has to be realized and communicated effectively to the policy process. Now adaptation is not only practicable and desirable, it has become an imperative.

Four main sets of global emission scenarios (storylines) were adopted in the work of the IPCC (Figure 7.1). The *A1* scenarios are of an increasingly integrated world with rapid economic growth, a global population that reaches nine billion by 2050, and new and efficient technologies. The *A2* scenarios reflect more independently operating and self-reliant nations, still with increasing population, but with slower and more fragmented technological advancements. The family of *B2* scenarios is also increasingly independent but more environmentally friendly (IPCC 2000). The most optimistic scenario is *B1*, which is of a more integrated world that is also more ecologically friendly. There still is rapid economic and population growth (similar to *A1*), but there are reductions in material intensity and the introduction of clean and resource efficient technologies with an emphasis on global solutions to economic, social, and environmental stability (IPCC 2000). With the "best case" emission scenario global surface mean temperature continues to rise up to and beyond the end of the century. With scenario A1B the rise is greater. However, even the most aggressive efforts to reduce greenhouse gas emissions will have little effect in reducing climate change for decades and sea-level rise will continue for much longer. The most optimistic of the IPCC SRES scenarios show that the global mean surface temperature will continue to rise to the end of the century and beyond as a result of the residence time of past and current anthropogenic emissions of greenhouse gases into the atmosphere. While mitigation measures can be implemented now, the residence time (atmospheric lifetime) of greenhouse gases, primarily carbon dioxide, varies from 50–200 years, forcing global mean temperature to continue to rise. Therefore, globally, we are locked minimally into a number of decades of climate change that cannot be diminished even by aggressive mitigation. Adaptation is an essential and urgent response (Dickinson 2007).

7.4 Paths for integration

Two paths for the integration of adaptation into development are being followed. These are through international official development assistance (ODA) and through the Climate Convention (UNFCCC). As part of the development assistance approach, the bilateral donor agencies are taking steps to incorporate assessment of climate risks and adaptation into their planning and projects. This trend is being supported at the OECD level (Gigli and Argawala 2007). In addition the World Bank and the regional development banks are also starting to factor climate risks and adaptation into project designs and loans. In the spring of 2007, 40 participants joined together in Geneva to discuss and share "screening tools" that were designed to identify projects at risk and to give an indication of the magnitude of the risks. Identified were information generation, databases, and platform tools, including PRECIS (Providing Regional Climates for Impacts Studies); computer-based decision tools, CRiSTAL (see Dickinson, 2007, for adaptation model summaries); and adaptation/risk management processes tools including Preparedness for Climate Change (IISD 2007).

While there is much practical sense in the integration (also referred to as "mainstreaming") of climate change into development in this manner, it has been criticized by developing countries as an evasion of responsibility by the developed (donor) countries. This critique is also linked to the social construction of climate change as an environmental pollution issue in the Climate Convention. Historically and up to the present day the developed countries have been responsible for the major part of carbon emissions into the atmosphere, but it is the developing countries, (especially the poorer ones) that are most vulnerable to climate change and will suffer most. Arguing on the basis of the internationally accepted "polluter pays principle" the developing countries have insisted that the developed countries should help meet costs of adaptation. At the time of the negotiation of the Climate Convention this was accepted in principle and incorporated into Article 4.4 which states:

The developed country Parties … shall also assist the developing country Parties that are particularly vulnerable to the adverse effects of climate change in meeting costs of adaptation to those adverse effects.

This was understood to mean that funds would be provided through the Convention process that would be *additional* to official development assistance in recognition of the sources of pollution. The concern is that mainstreaming (integration) of adaptation funding into regular development assistance would result in the loss of the important principle of *additionality*. It would no longer be possible to identify and measure the additional contribution. In response to this claim some funds have been provided through the Convention process. The funds include the Special Climate Change Fund (SCCF), the Least Developed Countries Fund (LDCF) and the GEF Trust Funds. Voluntary contributions to the funds are made by some of the donor countries and these are administered by the Global Environment Facility (GEF). The GEF was established by the donor countries for the special purpose of administering a new category of international funding for the support of multi-lateral environmental agreements including the Climate Convention as well as other conventions on biodiversity, desertification, the ozone layer, and persistent organic pollutants.

One result of this development is the "two pathway" approach to the integration of adaptation – through development assistance and through the Convention and its financial mechanism the GEF. A complicated "regime" for the integration of adaptation is thus emerging in which there is a need to coordinate and harmonize the allocation of funds from different sources each with its own set of rules and requirements and political drivers. A common complaint from developing countries is that it takes an inordinate amount of time to access the GEF funds and that transactions costs are too high.

There is a further complication. In both of the adaptation tracks towards integration – through development assistance and through the Climate Convention process – there is also a need to consider the relationship of adaptation to mitigation. In addition to the belief that talk of adaptation detracts from the urgency to mitigate, the misunderstanding and misinformation surrounding adaptation lead to criticisms by the public and media that it is a "do nothing" approach. Adaptation also presents more difficult measurement problems. Adaptation is mainly place-based and the benefits are realized locally. As such there is no credible way to cumulate them from local to national and international levels. This has been described as a problem of "upscaling" (Burton *et al.* 2007b). The success or otherwise of mitigation can be measured at the global scale by the level of greenhouse gas concentrations in the atmosphere. This is reflected in the proliferation of mitigation models compared to adaptation models (Dickinson 2007). The recent Stern Review did implement the use of models to analyze the economic impact of climate change; however the models (for example PAGE2002) were altered from their previous form specifically for the Stern Review in a questionable manner. The mitigation models do not adequately factor in the potential for adaptation. The state-of-the-art models overlook the fact that there can be both tradeoffs and co-benefits with the integration of adaptation and mitigation. Adapting to heat waves by the expansion of air conditioning can reduce the number of deaths while simultaneously increasing carbon emissions. Planting (or not cutting) trees in upland watersheds can help to reduce peak flood levels by slowing the rate of run-off and at the same time contribute to carbon storage. So far little attention has been given to the integration of adaptation and mitigation measures either in models or in practice. An approach to this is being developed under the AMSD (Adaptation and Mitigation in the Context of Sustainable Development) project based at the University of British Columbia (Burton *et al.* 2007b; Bizikova *et al.* 2009). The project is in response to a perceived need to promote adaptation at the local level in harmony with mitigation. AMSD highlights that mitigation and adaptation are a part of larger objectives all aiming towards increased sustainability. The participatory process suggested by AMSD includes the identification of local sustainable development scenarios that are linked with climate change impacts, then identifying the integration between mitigation and adaptation and finally determining an AMSD strategy (Bizikova *et al.* 2009). Current studies on integration of responses to climate change show that a focus on development provides a suitable context for a mixture of adaptation and mitigation. Much depends on the magnitude and rate of climate change in particular local contexts (Bhandari *et al.* 2007; Golkany 2007; Wilbanks and Sathaye 2007).

AMSD emphasizes that the linkages between adaptation and mitigation are highly context-specific and place-based and depend on the priorities guiding development. In this way, AMSD emphasizes that climate change adaptation and mitigation are a part of the wider development goals in transition towards sustainability. This is congruent with the views of Wilbanks and Sathaye (2007). The effectiveness of adaptation and mitigation measures on their own is limited, especially those that aim for behavioral changes without challenging the underlying development pathway. However, to advance interconnected adaptation and mitigation responses to climate change within the context of sustainable development, a number of methodological challenges need to be addressed including the development of tools to assess tradeoffs between the responses in respect of uncertainties within climate scenarios. This is being done in a way that accounts for a diversity of local values and development priorities and helps to create local involvement and support for actions and policies.

7.5 The post-2012 regime

The year 2012 marks the end of the first commitment period under the Kyoto Protocol. The Parties to the Convention face a formidable negotiating challenge in extending the Kyoto Protocol or replacing it with a new regime. Whatever form it takes the post-2012 agreement must now include more effective action on adaptation including financial support. It would clearly also be helpful if adaptation under the Convention were to be integrated into sustainable development and harmonized with ODA. This should also be done in such a way that adaptation and mitigation are mutually supportive. The negotiations were launched at the 13th Conference of the Parties to the Convention (COP 13) in Bali, Indonesia in December 2007 and are due to be completed at COP 15 in Copenhagen in December 2009. In addition to mitigation and adaptation the negotiations towards a post-2012 agreement will focus on technology development and sharing (transfer) and finance.

7.6 Research contributions from integrationists

As the post-2012 negotiations accelerate and intensify, they will necessarily be based on what knowledge is available and how well this is communicated to the negotiators. The IPCC Fourth Assessment (IPCC 2007) provides a positive and influential foundation and Chapter 18 of the Working Group II report delivers a useful summary of what little is known about the potential synergies and tradeoffs between adaptation and mitigation.

There are a number of opportunities to integrate the diversity of actors, scales, sectors, and governance structures in order to benefit from the synergies and avoid adverse tradeoffs. A local and regional focus is required to create concrete alternatives and directions within given development pathways. These individual small-scale actions could create a rich basis of examples to push the actions at national and global scales. In fact, effective actions at larger scales tend to be limited without demonstrations of effective bottom-up development (Wilbanks 2007). Larger-scale actions can be shaped and fine-tuned in association with smaller-scale stakeholders and, in fact, in large part global strategies will be implemented through local actions (Burton *et al.* 2007b).

Policy makers at the local level are in the difficult situation of trying to reconcile a wide diversity of local development visions with tradeoffs over limited resources, at a time when more actions, both in mitigation and adaptation, will be needed in order to tackle future climate change impacts, as well as to protect us from climate-related surprises. Therefore, viewing adaptation and mitigation as separate fields of action and policy without direct linkages may work against the implementation of opportunities that are perhaps not the most significant contribution to emission reduction, or avoided climate damage, but which can still offer tangible local benefits (Bizikova and Cohen 2006). Largely lacking however is an integrated policy analysis of the various options that might be considered for the integration of adaptation. What might be the best balance between integration into development assistance and the Convention process? How can the two tracks be harmonized and made mutually supportive and not duplicative? What institutional arrangements are needed under the Convention and how will adaptation be linked to mitigation? What additional funds will be required and supplied by the developed countries and should these continue to be on a voluntary basis or would mandatory contributions be more appropriate? How might credit for financial contributions be formally linked to a country's carbon emissions? Where might the additional funds be found and how would they be managed and by what institution? On what basis of estimated need or vulnerability could the funds be allocated?

There are many such questions and very few answers. For this and other reasons the "solutions" that emerge in the post-2012 agreement are likely to be temporary, flexible, and approximate and to allow for further learning by experience. The need for research and practical experiments on how to integrate adaptation will not end in 2012. It will only just be beginning.

References

Bhandari, P. M., S. Bhadwal, and U. Kelkar 2007. Examining adaptation and mitigation opportunities in the context of the integrated watershed management programme of the Government of India. *Mitigation and Adaptation Strategies for Global Change*, **12**, 919–933.

Bizikova L. and S. Cohen 2006. Climate change adaptation, mitigation and linkages with sustainable development. *International Workshop, Final Report*, April 19–21, Vancouver. Vancouver: Adaptation and Impacts Research Division, Environment Canada and Institute For Resources, Environment and Sustainability.

Bizikova, L., J. Robinson, and S. Cohen 2009. Linking climate change and sustainable development at the local level. *Climate Policy*, **7**, 271–277.

Burton, I., L. Bizikova, T. Dickinson, and Y. Howard 2007a. Integrating adaptation into policy: upscaling evidence from local to global. *Climate Policy*, **7**, 371–376.

Burton, I., T. Dickinson, and Y. Howard 2007b. Upscaling adaptation studies to inform policy at the global level. In C. van Bers *et al.* (eds.), *Global Assessments: Bridging Scales and Linking to Policy*. Bonn: Global Water System Project, pp. 68–74.

Dickinson, T. 2007. *The Compendium of Adaptation Models for Climate Change*, 1st edn. Downsview: Adaptation and Impacts Research Division, Environment Canada.

Gigli, S. and S. Argawala 2007. Stocktaking of progress on integrating adaptation to climate change into development co-operation activities. OECD. Retrieved from http://www.oecd.org/ dataoecd/ 33/62/39216288.pdf

Golkany, I. 2007. Integrating strategies to reduce vulnerability and advance adaptation, mitigation and sustainable development. *Mitigation and Adaptation Strategies for Global Change*, **12**, 755–786.

IISD 2007. *Sharing Climate Adaptation tools*. Retrieved from http://www.iisd.org/ pdf/2007/ sharing_climate_adaptation_tools.pdf

IPCC 2000. *IPCC Special Report: Emission Scenarios. Summary for Policy Makers*. Retrieved from http://www.ipcc.ch/ipccreports/sres/emission/index.htm

IPCC 2007. *IPCC Fourth Assessment Synthesis Report.* Retrieved from
    http://www.ipcc.ch/
UN 2007. *The Millennium Development Goals Report.* United Nations. Retrieved from
    http://www.un.org/millenniumgoals/pdf/mdg2007.pdf
Wilbanks, T. J. 2007. Scale and sustainability. *Climate Policy,* **7**, 25–42.
Wilbanks, T. J. and J. Sathaye 2007. Integrating mitigation and adaptation as responses to
    climate change: a synthesis. *Mitigation and Adaptation Strategies for Global
    Change,* **12**, 957–962.

8

Stakeholders in integrated regional assessment

ANN FISHER AND BERND KASEMIR

8.1 Introduction

Integrated regional assessment (IRA) processes provide insights when a complex issue has the potential for regional impacts and perhaps solutions, and when discipline-by-discipline analysis is inadequate for capturing the fullness of interactions and implications for the region. Yet scientists and policy makers complain that adverse public opinion often overturns recommendations based on "good science" and "good policy proposals" (e.g., National Research Council 1989; The Presidential/Congressional Commission 1997). This complaint gives an important motivation for a chapter on stakeholders in a book about integrated regional assessment: the prospect for closer matches between judgments by stakeholders and by scientists and policy makers (Kasemir *et al.* 2000a, b; Fisher 1991). Improving stakeholders' acceptance of the need for actions and their support of recommended actions is compelling from a practical perspective. Even more important, stakeholders can help identify key components of the issue and provide access to data and other special knowledge that otherwise would not be available to scientists and policy makers (e.g., Gregory 2000). Thus effective stakeholder participation has the potential for leading to a more comprehensive assessment that draws upon a wider range of information, as well as leading to more acceptance of the assessment process and resulting recommendations.

In principle, applied processes of stakeholder participation can draw upon a long tradition of research on participatory methods. Participatory methods that have been discussed in the literature include mediation or alternative dispute resolution, planning cells or citizen panels, consensus conferences, and focus groups (Dahinden 2000). In contrast to the other methods mentioned here, mediation or alternative dispute resolution aims at settling conflicts between organized interests, as a possible alternative to court settlement. Mediation processes have been conducted for a number of environmental and technology policy issues (Bingham

1986). Planning cells (Dienel 1992), citizen panels (Crosby 1987), and related procedures for cooperative discourse (Renn *et al.* 1995) are conducted to facilitate stakeholder participation on concrete policy options, for example siting of infrastructure projects or technical alternatives for waste treatment. They are similar to consensus conferences, where lay panels try to find a consensus on issues of technology or environmental policy. Developed in Denmark, consensus conferences have been conducted in the Netherlands and the UK. In the USA a consensus conference was organized on democratic control of the telecommunication industry (Sclove 1995); another explored global climate change issues (Citizens Jury 2002). Focus groups (Morgan and Krueger 1998) are small group discussions on a specific topic. They have been used extensively in marketing applications, but recently have received increasing attention as a method to generate qualitative data for academic research (Goss and Leinbach 1996).

However, the literature contains few models that would explain why one stakeholder involvement approach might be more effective than another in practice (Chess and Hance 1994; O'Connor and Bord 1994; Fisher *et al.* 1995; Chess and Purcell 1999). Instead, most of the literature fits into two categories. One category contains conceptual descriptions and occasional evaluations of stakeholder participation in assessment. The second category consists of guidelines based on what has seemed effective but without much conceptual basis to explain why (e.g., Lundgren 1994, US EPA 1997, Morgan *et al.* 2005, Moser 2005). In addition, participatory methods have most often been applied in cases where an issue is more or less local in scope, is highly salient for a number of stakeholders, and where action could yield relatively quick results. Examples include where to locate a waste facility or highway, minimizing water quality and odor impacts from concentrated animal feeding operations, or determining whether an apparent cancer cluster is real.

Engaging stakeholders is more challenging for issues beyond immediate local concerns that are complex, global, uncertain, and have lags before impacts will be evident. Climate change is a prime example. There is a widely published debate on the nature and magnitude of uncertainties concerning regional impacts of climate change (for the case of flooding probabilities, see e.g., Kleinen and Petschel-Held 2007). Also, the question whether delaying or accelerating mitigation measures is a more prudent policy response to uncertainties has been widely discussed (see e.g., Dessler and Parson 2006, for a theoretical discussion and Viscusi and Zeckhauser 2006, for an empirical study of attitudes on this issue). However, the literature has little to offer, either conceptually or descriptively, about methods for stakeholder participation in such assessments. For the first time, this chapter brings together discussions and lessons from two exceptions, as the basis for synthesizing what is known about effective stakeholder engagement in the IRA process. The first exception is the Mid-Atlantic Regional Assessment (MARA), which was one of several

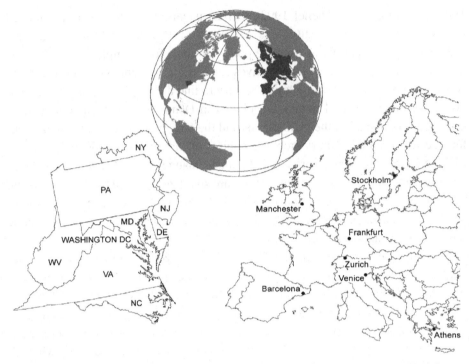

Figure 8.1. The Mid-Atlantic Region (left) and Europe with locations of ULYSSES groups (right).

components of the first US National Assessment[1] of potential consequences of climate variability and change. The second exception is represented by the Urban Lifestyles, Sustainability, and Integrated Environmental Assessment (ULYSSES) project in Europe, which used a series of structured focus groups to explore the issues of climate change and energy use (for detailed discussion see Kasemir *et al.* 2000a, b, 2003b). The aim of the ULYSSES project was to develop and test methods for public participation in integrated assessments of complex environmental problems like climate change.

This chapter describes stakeholder participation for these two case studies to extract insights for the role and value of stakeholder involvement in other integrated regional assessments. The following two sections discuss objectives and methods used for stakeholder participation in MARA and ULYSSES; the regions addressed by these two projects are illustrated in Figure 8.1.

The output from stakeholder involvement differs in these examples because it is conditioned by how the researchers and the stakeholders viewed the stakeholder opportunities, responsibilities, and tasks. The final section of this chapter then

[1] http://www.usgcrp.gov/usgcrp/nacc/default.htm; accessed February 24, 2006.

summarizes lessons learned about stakeholder involvement in IRA, and the outlook for the role of stakeholders in future IRAs.

8.2 Stakeholders and the MARA

8.2.1 Scope of the MARA

The Global Change Research Act of 1990 mandated that the US Global Change Research Program (USGCRP) assess and report on the potential impacts to the nation from global change (Public Law 101–606). The first US National Assessment focused on impacts from climate variability and climate change (rather than other global changes). Recognizing that some impacts will differ across regions, the USGCRP collaborated with federal agencies to sponsor 16 regional assessments (including the MARA) that span the nation and its territories. The collaboration also included nationwide assessments for five crosscutting sectors: coastal areas and marine resources, fresh water, agriculture, forests, and human health (National Assessment Synthesis Team 2000).

As defined for this assessment, the Mid-Atlantic Region (MAR) included parts or all of eight states (see Figure 8.1). Less than 4% of the MAR land area is urban, but more than half of its 35.2 million people live in its six largest urban areas: Philadelphia, Pittsburgh, Baltimore, Washington DC, Richmond, and Norfolk (US Bureau of the Census 1998).

In the context of the overall US National Assessment, the MARA goals were to identify:

- how people and their surroundings are affected now by climate variability and how they will be affected by climate change;
- how individuals and communities can take advantage of opportunities and reduce vulnerabilities resulting from climate variability and change; and
- what additional information and research are needed to improve decision making related to impacts from climate variability and change.

Note that these goals focus on impacts of climate change, not on mitigation of the emissions that cause global warming; other activities explore these issues, such as the international assessments of the Intergovernmental Panel on Climate Change.[2]

8.2.2 Goals for stakeholder participation

Active participation by stakeholders in the assessment process was an objective of the US regional and sectoral assessments, as well as for the overall synthesis.

---

[2] http://www.ipcc.ch

Stakeholders include anyone who might be affected by changes in climate and its variability. Thus for a regional assessment, stakeholders include the people who live in the region, people who do not live in the region but who own at-risk property or visit there, or have values for its special places even though they might not intend to visit them. For the MARA, stakeholders were defined as those who might be affected by climate change in the region or who might make decisions based on output from the assessment.

The MARA team identified six goals for stakeholder involvement (O'Connor *et al.* 2000):

Helping identify assessment questions that address stakeholder concerns. Stakeholder input can help set priorities for which impacts to assess and which impacts will receive more in-depth assessment. Interactive communication with stakeholders is the only way to identify potential impacts of particular concern to them.

Enhancing technical quality of the assessment. Stakeholders sometimes know about or have access to data otherwise not available to the core assessment team. Whether as citizens untrained in technical disciplines or as scientists, stakeholders sometimes can identify gaps or biases in the methods or data sets – early enough to make adjustments.

Providing a forum for sharing ideas among stakeholders with diverse constituencies. Meetings of the Advisory Committee provide a non-adversarial setting; focusing on the science and the process of the assessment can lead to discovering unexpected areas of common interest.

Making the process more legitimate to third parties. Broad stakeholder involvement can enhance the legitimacy of the assessment process and its findings. Engaging a broad group of stakeholders makes it more difficult for critics to assert that the results would have been different if others had been included in the process.

Improving stakeholder awareness of potential impacts as well as adaptation strategies. Participating in the MARA process should build stakeholder knowledge about both positive and negative potential impacts related to climate change and about options for taking advantage of opportunities or coping effectively with damages.

Facilitating dissemination of assessment findings. Stakeholders can help by informing their own constituencies and by advising the MARA team about what people want to know, how to present the information so it is useful and useable, and about dissemination strategies such as public information opportunities.

8.2.3 Stakeholder participation process

Stakeholder involvement began with the planning for a September 1997 "scoping" workshop (Fisher *et al.* 1999). A 17-member steering committee represented public interest groups, industry, state and federal government agencies, river basin commissions, and other universities. At a May 1997 meeting and through phone calls, faxes, e-mail, and regular mail, the Steering Committee helped to identify workshop goals, whom to invite as participants and as plenary speakers, and the workshop agenda. Workshop goals were to summarize scientific agreements as well as uncertainties about global climate change and its potential regional impacts, to explore the role of the news media in communicating about climate change, to begin developing a regional network of stakeholders who might provide additional input and feedback about assessing and communicating regional impacts, and to elicit stakeholder input on potential regional impacts they could envision or were concerned about.

More than half of the (approximately 90) participants responded to a questionnaire mailed before the workshop; results (summarized at the workshop) served as a baseline for a follow-up questionnaire to evaluate how perceptions had changed as a result of participation. Also before the workshop, participants received background papers summarizing the available literature on what the impacts might be for the MAR. This information was reinforced by plenary sessions.

Earlier research by the MARA team found that global warming is not salient for most people, but that they have a sense of unease about potential impacts (Lazo *et al.* 1999; O'Connor *et al.* 1999). These findings, along with USGCRP guidance, the background papers, and the Steering Committee's advice, led to choosing the following impact categories for working sessions at the 1997 workshop:

- economic growth, industry, and commerce (including energy and transportation);
- ecosystems;
- water resources (quantity and quality);
- natural hazards;
- human health;
- agriculture, forestry, and fisheries.

Each participant was assigned to one of these working groups; each group devoted nearly six hours to its topic during the two-and-a-half-day workshop.

Participants' deliberations were summarized in working group reports presented at the concluding session, identifying issues they judged to deserve special attention in an assessment of potential impacts from climate change in the Mid-Atlantic Region. Rapporteur Baruch Fischhoff synthesized the summaries, pointing out similarities and differences among the groups' conclusions. Stakeholders provided additional feedback after the workshop. (Workshop results are summarized below in the section on how stakeholders affected the MARA.)

While awaiting authorization to initiate the assessment, the MARA team kept in touch with the September 1997 workshop participants and others who had expressed interest. Upon receiving the go-ahead, a broad-based stakeholder Advisory Committee was established to enable interactive communication as a routine part of the MARA. The intent was to form an Advisory Committee small enough

Table 8.1. *MARA Advisory Committee members.*

Category	Number
Citizens groups	25
Business and industry	19
State and local governments, commissions	22
Federal government researchers	13
Academic researchers	13
Total	92

to focus constructively on a set of important issues yet large enough to represent the groups likely to experience substantial impacts in the region. The recruitment process was informal and inclusive, building on the 1997 workshop. Drawing upon individuals and organizations that express skepticism about global warming as well as those that support actions to reduce greenhouse gas emissions, representatives were recruited from business and industry along with experts from research organizations. To keep the size manageable, elected officials were not recruited. However, every individual who has sought to participate has been welcome to join the Advisory Committee. Table 8.1 reports the distribution of Advisory Committee members.

The Advisory Committee was not expected to be statistically representative of the MAR stakeholders, but to reflect a range of stakeholder perspectives related to the region's potential impacts from climate change. It was assumed that differing levels of involvement across stakeholders reflected their judgment about the importance of potential impacts for perspectives such as theirs.

Stakeholders had access to information about the MARA process and results through its web site.[3] The web site was updated frequently, provided contact information as well as a link for stakeholders to provide input to the MARA team, and links with related sites.

At a June 1998 workshop, nearly 40 university and government researchers exchanged information about state-of-the-science resources (e.g., data sets, studies, expertise) that the MARA team might use for assessing regional impacts from climate variability and climate change. The list of impact topics differed somewhat from that explored during the 1997 workshop:

- forests, farming, and land use (commercial, recreational, and aesthetic resources);
- coastlines and inland bays;
- fresh water resources;

---

[3] http://www.cara.psu.edu/mara

- health (hot summers, natural hazards, and infectious diseases);
- ecosystem vulnerability and resilience; and
- communication.

The participants discussed potential structures and processes for implementing the MARA. They emphasized the need to communicate assessment findings in a format that actually would be useful to potential users.

Because researchers are only one category of stakeholders, an October 1998 meeting (with 65 participants) elicited input from the full Advisory Committee about what research questions should be addressed. For small-group two-hour working sessions, each participant chose two of the following six impact topics:

- forests, farming, and land use: possible impacts for commerce and recreation;
- coastlines and inland bays: compounding the pressures from development;
- fresh water resources: changes in water quantity and quality;
- extreme weather events: what happens when they change;
- changing climate and changing health risks;
- issues for industry and commerce.

Participants summarized their working group deliberations in plenary sessions, and the summaries were synthesized in a final session. Those Advisory Committee members unable to attend could provide input and feedback by phone, fax, mail, and e-mail.

Each of the 63 participants at a May 1999 meeting of the Advisory Committee chose two of the 1.5-hour small working group sessions on these seven topics:

- coasts,
- forests,
- water,
- health,
- ecosystems,
- industry,
- agriculture.

They provided feedback on the write-up of these topics in the draft preliminary assessment report (which had been mailed to them two weeks earlier). The next day, they participated in two hour-and-fifteen-minute small-group sessions to provide advice about displaying findings, developing materials, and disseminating the assessment results to a wide audience. They also provided input for planning the next stages in the assessment.

Meetings among Advisory Committee members had the advantage of providing a forum for discovering common interests, understanding, and concerns related to the impacts of climate change in the Mid-Atlantic Region. Because meetings are

Table 8.2. *Contacts with Advisory Committee.*

Purpose/time period	Number of times contacted	Number of responses
On scenarios, outlines, draft chapters, April 1999 draft report, agenda and May 1999 workshop	29	47
On September 1999 and March 2000 *Overview* drafts	10	44
2000 *Foundations* drafts	5	27
Miscellaneous (other than Advisory Committee)		>100

time-consuming and expensive, more routine exchanges took place approximately bi-monthly using e-mail, fax, regular mail, or telephone calls. These exchanges included updates on the MARA team's progress, requests for feedback on drafts ranging from very preliminary outlines of research questions to formal comments on the supporting technical *Foundations* document (Fisher *et al.* 2000a), and occasional information about the US National Assessment or related science announcements.

Table 8.2 summarizes how often and why Advisory Committee members were contacted between the formal start of the MARA on May 15, 1998 and February 2001; by then, the MARA activities essentially had ended or been merged with its successor, the Consortium for Atlantic Regional Assessment (CARA). The table omits contacts related to the 1997 scoping workshop, which was before the Advisory Committee was established. It also omits contacts about meeting logistics, the multiple (very helpful) comments received from stakeholders during the meetings, and requests for copies of MARA reports. Many of the contacts included stakeholders beyond the formal Advisory Committee; Table 8.2 also summarizes their responsiveness.

Interactions with stakeholders reinforced our earlier findings about perceptions of global climate change and its potential impacts (O'Connor *et al.* 2000). Most people are aware of global warming, but it ranks below their concerns about issues that affect their daily lives and about other environmental issues (Bord *et al.* 1998; Roper Starch 2000). This lack of salience explains much of the challenge in achieving sustained interest and participation in the MARA process.

8.2.4 Results: how stakeholders affected the MARA

What to assess

Background papers prepared for the September 1997 workshop and the USGCRP guidance suggested that the MARA team focus on fresh water, forestry, and agriculture; these sectors are sensitive to current climate and we expected data

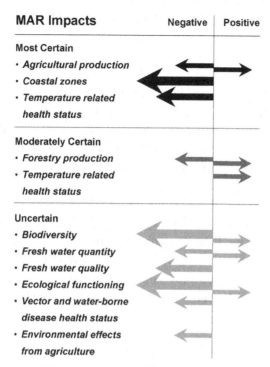

Figure 8.2. Potential Mid-Atlantic Region Impacts from Climate Change (Fisher *et al.* 2000b).

and methodology to be available for a timely assessment. However, two surprises emerged at the September 1997 workshop (Fisher *et al.* 1999). Participants expressed much more concern than anticipated about potential health impacts of climate change, and about coastal impacts along the Mid-Atlantic Ocean as well as for the Chesapeake and Delaware Bays.

Continuing contact with the Advisory Committee, especially at the October 1998 workshop, demonstrated constancy of concern about potential health and coastal impacts. Largely as a result of the stakeholder participation process, the team added health impacts (which we expected to be less sensitive to climate in the MAR) and coastal impacts (expected to be much more of an issue in other regions). Concerns about coastal impacts ranged from sea-level rise and accompanying storm damages to health impacts to ecological impacts (especially if marshes and wetlands might be flooded). Thus the set of impacts examined in the MARA was expanded to include: forests, agriculture, fresh water quantity and quality, coastal zones, human health, and ecosystems.

Figure 8.2 shows the projected damages and benefits in the MAR. The topics match those both experts and lay people expected to be important with respect to the region's impacts from climate change. The assessment showed some of these impacts are likely to be much larger than others (as indicated by the relative size of the arrows).

The uncertainty issue

The Advisory Committee also encouraged the MARA team to address issues of uncertainty as clearly and comprehensively as possible. This comment led to segmenting the *Overview* summary figure into sections showing the confidence in projections, as well as to additional attention to this issue throughout the document (Fisher *et al.* 2000b). Figure 8.2 shows the summary figure.

Making sure they are heard

As summarized in Table 8.2 the MARA Advisory Committee was invited to provide input and feedback for workshop agendas, draft outlines on what to assess and what methods to use, scenarios for analysis, and draft reports. Their responsiveness suggests that they recognized how important the MARA team considered their input to be. The MARA team's responses to comments received on drafts are documented in the section of the web site following each final report.[4]

Expertise and data

Advisory Committee members helped identify additional experts to provide input or to review MARA plans and interim products. A few Advisory Committee members became formal collaborators in the assessment process.

Communicating findings

At the 1997 workshop, a key session provided reporters' perspectives on the decline in news coverage of the environment (Fisher *et al.* 1999). Their frank statements about the requirement that news media make a profit suggested the need to develop a suite of strategies for disseminating assessment results. They also reinforced the need for presenting results in ways that would be useful to potential users – ranging from reporters to individuals deciding whether to renovate their beachfront home to government officials faced with decisions about ensuring future water supplies. One challenge is finding effective ways for getting people's attention to information that they currently do not view as salient even though it might be useful input to decisions (by individuals, organizations, and communities) affecting the long-term future.

On the other hand, other MARA research also demonstrated that people are aware that they lack knowledge about potential impacts from global warming, and are receptive to information about such impacts (Lazo *et al.* 2000, Kinnell *et al.* 2002,

---

[4] http://www.cara.psu.edu/mara

Ready *et al.* 2006). Many MARA Advisory Committee members demonstrated this receptivity by continuing to provide input about next steps and feedback on interim and final products. In response to solicitations about what information stakeholders want, and what formats would make the information most useful, they recommended access to lots of information about potential regional impacts of climate change, in many formats, all easy to understand and use (Fisher *et al.* 2000c).

The primary responses to this challenge were the MARA web site and the *Overview* (Fisher *et al.* 2000b). About 2000 copies of the *Overview* were distributed within its first year. Both the *Overview* and *Foundations* reports can be downloaded from the MARA web site, along with selected high-quality figures and overheads so that users can tailor their own presentations.

Relatively few journalists viewed regional impacts of climate change to be grist for a news story, but the MARA team responded promptly to their requests for information. For example, a central Pennsylvania public television station produced two 15-minute programs on the MARA process and results.

MARA team members accepted invitations to summarize the results for public groups. These ranged from university geology clubs to local chapters of the Audubon Society, to specialty conferences on topics such as coastal bays. For others in the US National Assessment, MARA methodological and procedural steps were described at meetings and in publications such as *Acclimations*.[5]

A communications challenge arose: although the MARA team had a wide range of disciplinary skills, their primary expertise was in research and assessment. Many target audiences were particularly vulnerable groups, often in sub-regions, and mostly interested in strategies and materials especially suited to their needs. Ultimately, the information should be made available to planners at the community and corporate level, in formats that make it clear why the information is useful for the decisions they are making. Limited resources made it infeasible to add formal outreach expertise to the initial MARA effort, which relied largely on distribution of the *Overview* report and web access to its back-up *Foundations* report. A two-page summary flyer was also printed and distributed. Many Advisory Committee members and other stakeholders indicated that they found these useful and assisted in disseminating them.

Summary

Stakeholder participation affected the MARA in five major ways: (1) the choice of impact categories to assess; (2) identification of data and methods otherwise not available to the assessment team; (3) addition of key expertise to the assessment

---

[5] Available at: http://www.usgcrp-gov/usgcrp/Library/nationalassessment/newsletter/default.htm

team; (4) additional attention to the uncertainties in the climate, socio-economic, and ecological projections and in how these would be affected by climate change; and (5) the emphasis on multiple methods and channels for displaying and presenting the MARA findings so that stakeholders find them useful and useable.

The lessons learned in MARA were instrumental in its successor project: The Consortium for Atlantic Regional Assessment (CARA). A key MARA finding was that many worrisome potential impacts from climate change are the same as impacts from poorly managed land use. This finding led to a combined CARA focus on climate and land use. Stakeholders again played key roles in determining what impacts to examine and how to present the results so that they would be useful to stakeholders ranging from regional development officials to individual citizens. Stakeholders participated in groups, as well as individually through an e-Delphi computer technique that elicited their evaluations of the CARA website. Stakeholders' needs to have information at small geographic scales led to the presentation of climate and land-cover data and projections for regions as small as counties – with tutorials about the models and cautions about the limits in projecting changes for small areas. Although CARA focuses on the mid- and upper-Atlantic region of the USA, its methods can be adapted to other regions. Its website is intended to be user-friendly; see http://www.cara.psu.edu/. Stakeholder engagement in developing, evaluating, and revising CARA is reviewed in Dempsey and Fisher (2005).

8.3 Stakeholders and ULYSSES

8.3.1 Scope of ULYSSES

The ULYSSES project was begun at a critical time in integrated assessment (IA) research. By 1995, integrated computer models on complex environmental problems like climate change had been constructed, and there had been some experience in stakeholder interactions with those models and with the research teams that created them. But these interactions were usually with high-level actors in public policy. Part of the IA community felt that views other than those of scientists and policy makers were being neglected in integrated assessments, and that there was a lack of tools for including broader public views.

Until then, integrated assessment research had relied heavily on computer modeling. Enabling participation of other stakeholders, like ordinary citizens, would require devising ways for lay people to interact meaningfully with computer tools. This was the motivation for the European ULYSSES project, which developed the IA focus group methodology to facilitate such interaction. This was felt to be important for making really *integrated* assessments feasible, where different views and different types of knowledge could be included. In order to make the methodology developed by the project available to other researchers interested in including citizen and stakeholder views in integrated assessments, the project team compiled a Handbook on Public Participation in Sustainability Science (Kasemir *et al.* 2003b) as one of the major outputs of the project.

Lay participants are expected to have the strongest views and be most knowledgeable about local circumstances. This makes it necessary to adapt assessment methods to local circumstances, including cultural contexts. For the ULYSSES project, which was set up to explore options for including citizen participation in integrated assessments, this meant that experiences with similar but slightly varying methods should account for local differences. Some parameters of the ULYSSES local studies were shared (all study areas were to be cities or highly urbanized areas). However, the studies were tailored to local contexts to allow conducting focus groups in regions from the South to the North of Europe. This was seen as important in order to avoid developing a method that would only work in specific parts of Europe, and in order to explore public views on issues related to climate change across as much of Europe as was feasible within the project. Focus group studies were conducted in Athens, Barcelona, Frankfurt, Manchester, Stockholm, Venice, and Zurich (see Figure 8.1).

8.3.2 Goals for stakeholder participation

The central goal of the ULYSSES project was to develop methods for facilitating stakeholder participation in integrated assessments, especially the participation of ordinary citizens. Why was this seen as essential? Consider the issue of climate change and its relation to regional lifestyles, which was the topic chosen as an example for developing participatory tools in ULYSSES.

Initially results from the natural sciences were essential in raising awareness of the problem of climate change. A second phase has now started because of initiatives such as the Kyoto Protocol to the UN Framework Convention on Climate Change. While we continue to need better studies on the changing climate system and its impacts on ecosystems and human society, there is enough evidence to point to the need for action. But concrete steps in climate policy, both for mitigation and adaptation, require combining actions at the level of international environmental diplomacy with actions involving diverse stakeholders. These range from peasants to forest managers, from tourist operators to inhabitants of coastal zones, and from financial investors to ordinary citizens (Brown 1999). Citizens must be involved because climate mitigation measures will "require consumer and worker cooperation as well as citizen consent" (Kempton 1991) for successful implementation (see also Löfstedt 1992). Without integrating citizens' points of view, environmental policy runs the risk of getting stalled in the implementation phase.

Since Kyoto, many initiatives have developed that attempt to link citizens more directly to climate policy development (for example the "Cities for Climate Protection Campaign" of ICLEI, the International Council for Local Environmental Initiatives). However, the long-term success of such attempts will depend on whether research on global environmental problems can be made more accessible to ordinary citizens, and better related to local action contexts (Burgess *et al.* 1995). In turn, accomplishing this requires integrating natural with social sciences, and combining academic research with stakeholder involvement. The IA Focus Group method developed in the ULYSSES project is one step toward making this feasible.

8.3.3 *Stakeholder participation process*

ULYSSES and the related Swiss project CLEAR (Climate and Environment in Alpine Regions; Jaeger and Pahl-Wostl 2000) used specially designed focus groups to explore the inclusion of ordinary citizens' views in integrated assessments of climate change and sustainability issues. Focus groups are widely used in marketing and public opinion research. In contrast to these rather simple applications, the research discussed here has explored a more complicated use of focus groups as parts of integrated assessment (IA) procedures. Striving to provide more useful information to decision makers than can be achieved with traditional disciplinary research, IAs aim at integrated pictures of complex decision situations – rather than highly detailed but not integrated pieces of knowledge. There has been growing interest in employing participatory techniques in integrated assessments, but much of this has been limited to panels of experts.

Conventional focus group techniques are not well suited to provide information for integrated assessments. Thus focus group methods were adapted in several ways in the ULYSSES project. These adaptations include a longer and more structured discussion process that allows exploration of spontaneous associations by the participants (e.g., in collage work). It also allows their interaction with current research findings (usually by the use of computer models in the focus groups), before the participants themselves summarize their views on the focal topic. Because of these methodological adaptations, the resulting focus groups were called IA focus groups to distinguish them from other types.

To obtain a broad sample of European opinion, ULYSSES and CLEAR conducted IA focus groups in seven metropolitan areas – Athens, Barcelona, Frankfurt, Manchester, Stockholm, Venice, and Zurich. Several focus groups were conducted in each area because the views of a single group may be biased and not be indicative of the views of the larger population. Including pilot groups, approximately 600

Table 8.3. *Number of citizens involved in ULYSSES and CLEAR Groups.*

Region	Main groups	Pilot groups	Sum
Frankfurt	72	36	108
Venice (approx. numbers)	54	40	94
Manchester[a]	24	16	40
Barcelona	38	16	54
Stockholm	36	23	59
Athens		20	20
Zurich (incl. groups in CLEAR project)	132	92	224
Total	356	243	599

[a] Manchester region, including St. Helens.

people from across Europe participated in these groups (see Table 8.3). In addition, special sessions were conducted with regional decision makers and representatives of the financial industry and the news media.

The purpose of ULYSSES was to study views that emerge in discussions among different types of participants, because this was thought to be more indicative of public opinion dynamics than a study of the views of isolated segments of the population. As in MARA, the groups were not expected to be statistically representative of the European population, but to reflect the range of typical citizens' perspectives on the topics under discussion. For this reason, it was seen as important that the focus groups be diverse with respect to age, gender, income, educational level, and attitudes toward the environment. Potential participants were screened to accomplish this, with quotas for manual workers and college graduates; those who felt that environmental problems were important and those who did not; and those who favored environmental regulation and those who did not. Recruitment was handled either by the research teams themselves or by survey research firms. Each IA focus group consisted of six to eight people. Moderate financial incentives were given, e.g., corresponding to the remuneration of local policy makers for committee work, to compensate participants for engaging in the intense and time-consuming process described below.

The exact procedure for conducting the focus groups varied somewhat among research regions, but the general format was the same. The members of each group met with moderators on five separate occasions, each session lasting about two and a half hours. The sessions were much more structured than those of conventional focus groups. The groups were designed so that participants first could express spontaneous feelings about climate change and energy use (e.g., by collage work), before expert knowledge was made available to them (usually in the form of output from computer models). The last step was for participants to synthesize their views. The three distinct phases to the ULYSSES focus groups are described in more detail below for one variation of the general format (more details are provided in chapters of the Handbook on Public Participation in Sustainability Science by Kasemir *et al.* (2003a) concerning the first phase, Dahinden *et al.* (2003) concerning the second phase, and Querol *et al.* (2003) concerning the third phase of the IA focus group procedures).

The *first phase*, which took one session, started by discussing which environmental problems in general the participants were concerned about, before focusing on climate change in particular. Then, the group was divided into two; each smaller group was asked to produce a collage by cutting out and arranging images from various magazines. One collage was to show how the area in question would look if energy use were to evolve for the next 30 years much as it had in the past; the other collage was to show how the region would look if energy use were reduced by 50%.

In the *second phase*, which lasted for three sessions, the group was exposed to experts' opinions about climate change and asked to focus more systematically on the issue. Interactive computer models, supported by the moderators, were used to facilitate these discussions. These sessions used both models with a global change scope and models with a regional scope. Participants were introduced to the most important variables in the models and they were shown examples of scenarios generated by the models. Then at their request, they were assisted in exploring specific variables (e.g., the range of sea-level rise) or specific interactions (e.g., the relation of economic growth to emissions development) in the models' different scenarios. Sometimes the research team suggested a set of questions that could be explored with the help of the models' scenarios. At the end of the interaction, the participants discussed their views on these scenarios, and on the models used.

In the *third phase*, which consisted of the final session of the process, the focus group participants were asked to produce written reports stating their view(s) on the issue of climate change and the policy options for addressing it. For this purpose, they were given a list of specific questions, such as:

- do you think there is a problem of climate change (in this region, or world-wide)?
- given this, how should we live in 30 years in this region?
- what should be done so that we can get there?

The moderators left the room while the group worked on its report for about one and a half hours. Participants were encouraged to note areas of disagreement as well as areas of agreement. At the end, the group presented its report to the moderators and commented on it. In addition to these written group reports, participants in some groups kept diaries throughout the whole focus group process, either individually or for the whole group.

The sessions were analyzed in a more structured way than is usual with conventional focus groups. For example, one procedure that has been found to be especially useful is the following. The group discussions were videotaped, selectively transcribed, and then scanned for discussion sequences relating to the topics of primary interest, with the help of software that supports qualitative content analysis. Exploring the patterns of such discussion sequences helps to identify the relevant "ideal types" of lay views on scientific and policy questions, not just on the basis of the number of times a view is expressed but also in terms of the meaningfulness of the vision reflected in it. An "ideal type analysis" simplifies qualitative data by focusing on possible futures of a given social system and relating such alternative futures to moral choices facing this system. This approach is thus especially suited for studies targeted at views of social actors and their potential impact on future development, which are central issues in integrated assessments that strive to combine expert with lay knowledge and global change issues with the regional dimension.

8.3.4 Results: European citizens' views as explored by ULYSSES

ULYSSES was targeted at exploring possibilities for an interface between expert and lay views for integrated environmental assessment. One central methodological result was that computer models are a very promising medium for supporting this interaction (see Dahinden *et al.* 2003). On the substantive side, ULYSSES examined citizens' views across Europe on energy use and climate change issues. Citizens who participated in the focus groups often saw the prospect of climate change as very frightening. They adopted an ethical rather than an economic approach to framing their discussions of climate impacts. Expectations of the deterioration of basic livelihoods around the world, ecological destruction, and danger to future generations made a wait-and-see policy unacceptable to them. Participants tended to be in favor of mitigation measures even in the face of scientific uncertainty – that is, they usually based their views of the climate issue on the precautionary principle. In these respects, their opinions can be said to be close to a "deep-ecology perspective." Their views parted company with that perspective, however, when it came to discussing climate change mitigation. Participants in these groups readily framed mitigation measures in economic terms and the cost dimension was important to them. In this respect, their views were closer to an "economic-management perspective." While most participants were supportive of small increases in energy prices, they usually rejected large ones. Although they often saw large reductions in energy use as more desirable than business as usual, participants thought that this reduction should and could be achieved without greatly raising energy prices. Developing the products necessary to make this possible was often seen as the task of research and industry. From these results one may infer that climate policy will only be acceptable to the European public if it focuses on developing low-cost options for substantially reducing energy use.

It is interesting to compare the ULYSSES results to those of other major studies of citizens' views. The international Health of the Planet survey on environmental attitudes aimed "to give voice to individual citizens around the world concerning the environment and economic development issues affecting their lives" (Dunlap *et al.* 1993a, b; Dunlap and Mertig 1997). The Health of the Planet survey was carried out in 24 nations both in developed and in developing regions around the world; citizens in six nations (Canada, the USA, Mexico, Brazil, Portugal, and Russia) were asked specifically about global warming (Dunlap 1998). Other studies have targeted US residents' views on global climate change. For example, Bord *et al.* (1998) mailed a questionnaire to a representative sample, while Kempton (1991, 1997) relied more on semistructured interviews of a small but diverse sample of citizens. In contrast, the European participants of ULYSSES IA Focus Groups had the opportunity to debate expert opinions on climate change and energy use in in-depth group debates, and had access to relevant computer models in this process. What are major similarities and differences between the findings of these other studies and of ULYSSES?

First, the other studies found a high level of citizen concern about climate change, but their findings differ on the impacts of highest concern. In the six-country study conducted in the context of the Health of the Planet survey, respondents saw global warming effects on natural systems (either by creating various ecological problems or by changing weather patterns) as more worrying than effects on social systems. "Choices that people have about where to live and economic well-being are the least likely realms to be seen as being harmed – suggesting that economists' emphasis on economic impacts (see, for example, Nordhaus 1994) ignores the public's crucial concerns . . . " (Dunlap 1998: 485). However, Kempton's findings suggest that species preservation is not highly valued in the abstract by the US public. Instead rising food costs, for example, were perceived as more worrying than species extinction, and most of his informants justified species preservation in anthropocentric terms. Our findings are more in line with Dunlap's results about climate change impacts: Most European citizens participating in the IA focus groups expressed concern about geophysical, ecological, and human health impacts, but not about the economic impacts of climate change.

Second, concerning mitigation of environmental degradation in general, the "Health of the Planet" survey overall found some willingness to pay higher prices if this supports environmental protection. The role of technology was perceived ambiguously, seen both as a cause of environmental problems and as a part of the solution. Bord *et al.* (1998) found that US citizens are willing to make modest sacrifices (e.g., installing insulation, replacing older appliances, purchasing a fuel-efficient car), but they are much less willing to reduce their use of fuels for heating and air conditioning, and especially for driving. The ULYSSES research on views of European citizens can give more depth to these results. While the IA focus group participants supported some rises in energy prices, they rejected hard-hitting price increases. They expected technology to achieve decreased energy use without high costs or loss of comfort. Under these conditions, a strong reduction of energy use was often seen as more positive than business as usual. This is in contrast to Kempton's findings that US respondents saw reductions in energy use much less positively: "When analysts talk about reducing energy use, lay people [in the United States] tend to interpret it as decreasing energy services." (Kempton 1991, p. 203). Given the much more positive views of ordinary Europeans found by the ULYSSES project on this issue, European policy makers may have more opportunity to address climate change by focusing on significant reductions in energy use than do their US counterparts. However, there is at least one match between Kempton's results of US citizens' views and the views of ordinary European citizens participating in ULYSSES IA focus groups: he found that average Americans "are aware that measurable environmental disturbance is now global in scale, . . . and they want decisions to be based not only on cost and benefits but also on our responsibility to leave a healthy planet for our descendants." (Kempton 1991, p. 208). Our IA focus group results indicate that European citizens strongly tend to agree with that.

8.4 Differences and common lessons in MARA and ULYSSES

As discussed above, substantial evidence (e.g., Bord *et al.* 1998; Dunlap 1998; Kempton 1997; Löfstedt 1992; O'Connor *et al.* 1999b) indicates striking similarities as well as important differences among citizens of different nations regarding their perceptions of global climate change, its potential impacts, and the desirability of alternatives to reduce risks related to climate change. This evidence about differences in perceptions suggests the need for tailoring both the assessment and its stakeholder component to the region. Part of the tailoring will include specification of goals for the overall assessment scope as well as for stakeholder involvement. For instance, the MARA goals were to explore how climate change might affect the region, and what might be done to ameliorate negative impacts and take advantage of positive impacts (rather than explicitly how to reduce climate change). This scope was different and in a sense complementary to that of ULYSSES. The ULYSSES project aimed to develop and demonstrate a methodology for stakeholder participation in order to facilitate the better inclusion of citizens' views in integrated assessments of complex environmental problems.

A key difference between the MARA and ULYSSES is thus that the latter was designed at the outset primarily as a research and demonstration activity about stakeholder participation. In contrast, the MARA was oriented more toward developing projections of potential regional impacts from climate change, with stakeholders envisioned as having a role in defining the scope of impacts to consider. The role of MARA stakeholders initially was fuzzy, because the first National Assessment was a new endeavor, with evolving guidelines for its characteristics and components (Fisher *et al.* 2000c). The National Assessment activities also had to cope with uncertainties and delays in funding, and uncertainty about how long the assessment process (and thus the expected stakeholder involvement) might continue (Morgan *et al.* 2005). These factors made it more difficult to start with a coherent design for stakeholder participation. Specific expectations about stakeholder opportunities and responsibilities – including the scope and depth of input and influence, and the frequency and timing of contact and meetings – evolved as the National Assessment and its individual component studies received funding and learned from one-another's experience and from feedback provided by stakeholders. The uncertainties about the role of stakeholders made it difficult to decide what early information should be collected from or about MARA stakeholders that could serve as a baseline for evaluation later. For example, documentation about stakeholders who decline to participate, or who drop out of the process, could provide insights about the representativeness of the stakeholder process, and about groups that do not perceive themselves to be stakeholders. In contrast to MARA, the focus of ULYSSES on developing and testing participatory procedures made it feasible to work on a coherent design for stakeholder participation directly from the outset of the study.

Another difference between MARA and ULYSSES is how stakeholders interacted with one another and how they saw their role in the assessment. For the MARA, the primary stakeholder contingent was the Advisory Committee, a group of approximately 90 individuals with diverse backgrounds and interests and who lived in dispersed parts of the MAR. They were charged with considering the entire region, even though they often wanted to know more about impacts in their own locality. Meetings were in a single location, with all Advisory Committee members invited to attend (with travel aid for those who needed it). Most of the interactions took place by mail, e-mail, phone, and fax, because of the time and expense of large group meetings. For ULYSSES, overall approximately 600 citizens met in a number of smaller groups (six to eight members per group, with each group based in one of seven cities and meeting five times). The number of meetings, their timing, and specific expectations of what would be produced during the meetings

Table 8.4. *Steps for stakeholder participation, and how they were accomplished.*

Step	MARA	ULYSSES
Identify stakeholders	Snowball technique, starting with MARA team members and climate change experts	Careful and costly screening led to group compositions that reflected a range of typical perspectives
Identify goals for stakeholder participation	Steering Committee for scoping workshop	Determined by ULYSSES team (to develop methods for stakeholder participation in IA)
Identify goals for IRA	MARA team, plus workshops and comments elicited on mailed materials	Local ULYSSES research teams in the seven study areas targeted by the project
Participate in IRA itself	Experts workshop; 1-on-1 interaction between stakeholders and MARA team; collaboration on work includes co-authorship	In-depth series of 5 IA meetings with each focus group; 3 of these involved the use of computer models
Review of IRA process and findings	Elicited comments on several stages of mailed drafts and at workshops	IA Focus Group members synthesized their views in citizens' reports, analyzed by researchers
Assist in disseminating findings	Workshops, presentations, web-based reports, graphs, overheads	Handbook on Public Participation in Sustainability Science

were specified at the outset. Each group of ULYSSES stakeholders was charged with considering the global climate change issue specifically in the context of their own urban regions – each of which is much smaller than the Mid-Atlantic Region.

A specific feature of ULYSSES was the development of methods to facilitate stakeholder use (assisted as needed) of computer models, which allowed in-depth discovery and discussion of how changes in key parameters of lifestyles and climate policy could affect the future, as well as about the role of uncertainty. While interactive computer use was not part of the MARA methodology, stakeholders suggested parameters they thought might be important, as well as ranges for those parameters. This information was combined with guidance from the National Assessment process, and then used with scenarios for global climate change to simulate impacts within the MAR. Ranges were used to demonstrate how the results vary with different parameter values. For an overview of these and other similarities and differences between MARA and ULYSSES, see Table 8.4 on major steps in the two processes.

An important lesson of both projects is that effective stakeholder participation is not something that can be done as an aside. It requires substantial resources over an extended period of time. The longer the time period over which stakeholders are expected to participate, and the less tangible or more distant the outcome of their participation, the more difficult it will be to build and maintain an effective stakeholder network. This implies the need for continuing interaction to maintain their long-term commitment to participating in the IRA.

Future integrated assessments might consider combining the idea of using Advisory Committees made up of stakeholders as explored in MARA with the in-depth IA focus group interaction as explored by ULYSSES. Also, they might further explore the possibilities of combining stakeholder dialogues with interactive computer use in concrete regional assessment processes. In this context, more detailed studies of the interface between participatory integrated assessment (PIA) of global change and participation in local and regional planning on responses will be important (for a discussion on combining PIA and participatory regional planning in the implementation of the European Water Framework Directive, see Ridder and Pahl-Wostl, 2005). It is highly probable that future stakeholder involvement will rely increasingly on internet exchanges, which will reduce the distance barriers for large regions. On the other hand, it will remain essential to bring the stakeholders together periodically for face-to-face interactions as a way to reinforce the importance of their task and keep them committed to participating. It is expensive to bring people together from long distances, but the long-term benefits of in-person meetings appear to be worth the costs, based on the experiences of the two projects discussed here.

Acknowledgements

This chapter draws upon the research conducted by the MARA and ULYSSES teams, and necessarily upon publications that describe these activities. We especially rely upon the Fisher *et al.* (2000c) paper, and upon the Kasemir *et al.* (2003b) Handbook. Partial support for the MARA has been provided by the US Environmental Protection Agency's Cooperative Agreement No. 826554, with many contributions from colleagues in the Center for Integrated Regional Assessment at the Pennsylvania State University (supported by the US National Science Foundation Grant SBR-9521952). Partial support for CARA came from the US Environmental Protection Agency's Cooperative Agreement No. 830533. ULYSSES was supported by DG Research of the European Commission (Fourth Framework Programme/RTD Programme Environment and Climate) and by the Swiss Federal Office for Education and Science. There were considerable synergies between ULYSSES and the Swiss CLEAR project. CLEAR was financed by the Swiss National Science Foundation through the Priority Program "Environment."

References

Bingham, G. 1986. *Resolving Environmental Disputes. A Decade of Experience.* Washington, DC: The Conservation Foundation.

Bord, R., A. Fisher, and R. O'Connor 1998. Public perceptions of global warming: United States and international perspectives. *Climate Research*, **11**, 75–84.

Brown, K. S. 1999. Taking global warming to the people. *Science*, **283**(5 March), 1440–1441.

Burgess, J., C. Harrison, and P. Filius 1995. *Making the Abstract Real: a Cross-Cultural Study of Public Understanding of Global Environmental Change.* London: University College London, Department of Geography.

Chess, C. and B. J. Hance 1994. *Communicating with the Public: Ten Questions Environmental Managers Should Ask.* New Brunswick, NJ: Center for Environmental Communication, State University of New Jersey-Rutgers.

Chess, C. and K. Purcell 1999. Public participation and the environment: do we know what works? *Environmental Science & Technology*, **33**(16), 2685–2692.

Citizens Jury: *Global Climate Change*: March, 2002: http://www.jefferson-center-org/vertical/Sites/%7BC73573A1-16DF-4030-99A5-8FCCA2FOBFED%7D/uploads/%7BF29315AB-1654-4CBB-B29B-CBA9FEFC2C15%7D.PDF

Crosby, N. 1987. Citizen panels: a new approach to citizen participation. *Public Administration Review*, **46**, 170–178.

Dahinden, U. 2000. *Demokratisierung der Umweltpolitik: Oekologische Steuern im Urteil von Bürgerinnen und Bürgern.* Baden-Baden: Nomos Verlagsgesellschaft.

Dahinden, U., C. Querol, J. Jäger, and M. Nilsson 2003. Citizen interaction with computer models. In B. Kasemir, J. Jäger, C. C. Jaeger, and M. T. Gardner (eds.). *Public Participation in Sustainability Science. A Handbook.* Cambridge: Cambridge University Press.

Dempsey, R. and A. Fisher 2005. Consortium for Atlantic Regional Assessment: information tools for community adaptation to changes in climate or land use. *Risk Analysis*, **25**(6), 1495–1509.

Dessler, A. E. and E. A. Parson 2006. *The Science and Politics of Global Climate Change. A Guide to the Debate.* Cambridge: Cambridge University Press.

Dienel, P. C. 1992. *Die Planungszelle: Eine Alternative zur Establishment-Demokratie.* Opladen: Westdeutscher Verlag.

Dunlap, R. E. 1998. Lay perceptions of global risk. Public views of global warming in a cross-national context. *International Sociology*, **13**(4), 473–498.

Dunlap, R. E. and A. G. Mertig 1997. Global environmental concern: an anomaly for postmaterialism. *Social Science quarterly*, **78**(1), 24–29.

Dunlap, R. E., G. H. Gallup, and A. M. Gallup 1993a. *Health of the Planet: Results of a 1992 International Environmental Opinion Survey of Citizens in 24 Nations.* Princeton, NJ: The George H. Gallup International Institute.

Dunlap, R. E., G. H. Gallup, and A. M. Gallup 1993b. Of global concern. *Environment*, **35**(9), 6–15 and 33–39.

Fisher, A. 1991. Risk communication challenges. *Risk Analysis*, **11**(2), 173–179.

Fisher, A., S. Emani, and M. Zint 1995. *Risk Communication for Industry Practitioners: an Annotated Bibliography*, plus Appendix. McLean, VA: Society for Risk Analysis.

Fisher, A., E. Barron, B. Yarnal, C. G. Knight, and J. Shortle 1999. *Climate Change Impacts in the Mid-Atlantic Region: a Workshop Report.* Sponsored by US Environmental Protection Agency Cooperative Agreements No. CR 826554 and CR 824369. University Park, PA: Pennsylvania State University.

Fisher, A., D. Abler, E. Barron, *et al.* 2000a. *Preparing for a Changing Climate: the Potential Consequences of Climate Variability and Change: Mid-Atlantic Foundations.* Prepared for USGCRP First National Assessment, sponsored by US Environmental Protection Agency Cooperative Agreement No. CR 826554. University Park, PA: Pennsylvania State University. http://www.cara.psu.edu/mara

Fisher, A. *et al.* 2000b. *Preparing for a Changing Climate: the Potential Consequences of Climate Variability and Change: Mid-Atlantic Overview.* Prepared for USGCRP First National Assessment, sponsored by US Environmental Protection Agency Cooperative Agreement No. CR 826554. University Park, PA: Pennsylvania State University. http://www.cara.psu.edu/mara

Fisher, A, J. Shortle, R. O'Connor, and B. Ward 2000c. Engaging stakeholder participation in assessing regional impacts from climate change: the Mid-Atlantic Regional Assessment. In D. Scott, B. Jones, J. Andrey, *et al.* (eds.), *Climate Change Communication: Proceedings from an International Conference.* Warriner: University of Waterloo, pp. A4.1–A4.9.

Goss, J. D. and T. R. Leinbach 1996. Focus groups as alternative research practice: experience with transmigrants in Indonesia. *Area*, **28**(2), 115–123.

Gregory, R. S. 2000. Valuing environmental policy options: a case study comparison of multiattribute and contingent valuation survey methods. *Land Economics*, **76**(2), 151–173.

Jaeger, C. C. and C. Pahl-Wostl (eds.) 2000. CLEAR. *Integrated Assessment* (special issue), **1**(4), 339–349.

Kasemir, B., U. Dahinden, A. Gerger, *et al.* 2000a. Citizens' perspectives on climate change and energy use. *Global Environmental Change*, **10**(3), 169–184.

Kasemir, B., D. Schibli, S. Stoll, and C. C. Jaeger 2000b. Involving the public in climate and energy decisions. *Environment*, **42**(3), 32–42.

Kasemir, B., U. Dahinden, A. Gerger Swartling *et al.* 2003a. Collage processes and citizens' visions for the future. In B. Kasemir, J. Jäger, C. C. Jaeger, and M. T. Gardner (eds.). *Public Participation in Sustainability Science. A Handbook.* Cambridge: Cambridge University Press.

Kasemir, B., J. Jäger, C. C. Jaeger, and M. T. Gardner (eds.) 2003b. *Public Participation in Sustainability Science. A Handbook.* Cambridge: Cambridge University Press.

Kempton, W. 1991. Lay perspectives on global climate change. *Global Environmental Change*, **1**(3), 183–208.

Kempton, W. 1997. How the public views climate change. *Environment*, **39**(9), 12–21.

Kinnell, J., J. K. Lazo, D. J. Epp, A. Fisher, and J. S. Shortle 2002. Perceptions and values for preventing ecosystem change: Pennsylvania duck hunters and the prairie pothole region. *Land Economics*, **78**(2), 228–244.

Kleinen, T. and G. Petschel-Held 2007. Integrated assessment of changes in flooding probabilities due to climate change. *Climatic Change*, **81**, 283–312.

Lazo, J. K., J. Kinnell, T. Bussa, A. Fisher, and N. Collamer 1999. Expert and lay mental models of ecosystems: inferences for risk communication. *RISK: Health, Safety & Environment*, **10**(1), 45–64.

Lazo, J. K., J. Kinnell, and A. Fisher 2000. Expert and layperson perceptions of ecosystem risk. *Risk Analysis*, **20**(2), 179–193.

Löfstedt, R. E. 1992. Lay perspectives concerning global climate change in Sweden. *Energy and Environment*, **3**(2), 161–175.

Lundgren, R. 1994. *Risk Communication: A Handbook for Communicating Environmental, Safety, and Health Risks.* Columbus, OH: Battelle Press.

Morgan, D. L. and R. A. Krueger 1998. *The Focus Group Kit*, vols. 1–6. London: Sage.

Morgan, M. G., R. Cantor, W. C. Clark, *et al.* 2005. Learning from the U.S. National Assessment of Climate Change Impacts. *Environmental Science & Technology*, **39**(23), 9023–9032.

Moser, S. C. 2005. Stakeholder involvement in the first U.S. national assessment of the potential consequences of climate variability and change: an evaluation, finally. *Report prepared for National Research Council, Committee on Human Dimensions of Global Change*. Washington, DC: National Research Council.

National Assessment Synthesis Team 2000. *Climate Change Impacts on the United States – the Potential Consequences of Climate Variability and Change*. New York: Cambridge University Press (also available at http://www.nacc.usgcrp.gov/, accessed February 24, 2006).

National Research Council 1989. *Improving Risk Communication*. Washington, DC: National Academy Press.

Nordhaus, W. D. 1994. Expert opinion on climatic change. *American Scientist*, **82**(January–February), 45–51.

O'Connor, R. E. and R. J. Bord 1994. The two faces of environmentalism: environmental protection and development on Cape May. *Coastal Management*, **22**(2), 183–194.

O'Connor, R. E., R. J. Bord, and A. Fisher 1999a. Risk perceptions, general environmental beliefs, and willingness to address climate change. *Risk Analysis*, **19**, 461–471.

O'Connor, R. E., R. J. Bord, A. Fisher, *et al.* 1999b. Determinants of support for climate change polices in Bulgaria and the U.S. *Risk Decision and Policy*, **4**, 255–269.

O'Connor, R. E., P. J. Anderson, A. Fisher, and R. J. Bord 2000. Stakeholder involvement in climate assessment: bridging the gap between scientific research and the public. *Climate Research*, **14**(3), 255–260.

Querol, C., A. Gerger Swartling, B. Kasemir, and D. Tàbara 2003. Citizens' reports on climate strategies. In B. Kasemir, J. Jäger, C. C. Jaeger, and M. T. Gardner (eds.), *Public Participation in Sustainability Science. A Handbook*. Cambridge: Cambridge University Press.

Ready, R., A. Fisher, D. Guignet, R. Stedman, and J. Wang 2006. A pilot test of a new stated preference valuation method: continuous attribute-based stated choice. *Ecological Economics*, **59**(3), 247–255.

Renn, O., T. Webler, and P. Wiedemann (eds.) 1995. *Fairness and Competence in Citizen Participation – Evaluating Models for Environmental Discourse*. London: Kluwer.

Ridder, D. and C. Pahl-Wostl 2005. Participatory integrated assessment in local level planning. *Regional Environmental Change*, **5**, 188–196.

Roper Starch 2000. *Roper Green Gauge 2000: Rising Concerns*, Roper Starch Worldwide, November.

Sclove, R. E. 1995. *Democracy and Technology*. New York: Guilford Press.

The Presidential/Congressional Commission on Risk Assessment and Risk Management 1997. *Framework for Environmental Health Risk Management*, vol. 1. http://www.riskworld.com, accessed February 24, 2006.

US Bureau of the Census 1998. *Historical National Population Estimates*. http://www.census.gov/popest/archives/1990s/#state, accessed February 24, 2006.

US Environmental Protection Agency 1997. *Community-Based Environmental Protection: A Resource Book for Protecting Ecosystems and Communities*, 230-B-96-003. Washington, DC: EPA.

Viscusi, W. K. and R. J. Zeckhauser 2006. The perception and valuation of the risks of climate change: a rational and behavioral blend. *Climatic Change*, **77**, 151–177.

9

A framework for integrated regional assessment

C. GREGORY KNIGHT, ANN FISHER, AND THE CIRA TEAM

9.1 Introduction

The complexity of global climate change demands a comprehensive integrated approach to understand what is happening and to develop policies that address causal factors and ameliorate undesirable impacts. Most human interactions with the environment and social dynamics occur locally and regionally. Global climate change begins and accumulates from local places and regions; impacts are transmitted from global to regional and local scale. Thus the focal role of the region in understanding climate change is virtually self-evident.

In this chapter, we adopt the common understanding of the book that *integrated regional assessment* (IRA) means the application of interdisciplinary scientific knowledge to address the causes and consequences of global environmental change in a way that provides useful knowledge to stakeholders and decision makers at regional scales. Yarnal (Chapter 2, this volume) discusses in detail the nature of linear, end-to-end frameworks and models for IRA and suggests the importance of more systematic approaches that explicitly incorporate feedback mechanisms. Indeed, the idea of conceptual frameworks inspired formal quantitative models of the human–environment system at global and regional levels. All of these developments parallel in a broad way the evolution of schematic frameworks for the analysis of human responses to extreme geophysical events, popularly known as natural hazards, and the modifications of these frameworks for climate change (Kates 1985; Burton *et al.* 1993).

The purpose of this chapter is to illustrate one framework for examining the causes and consequences of global climate change at the regional level. This framework evolved at the Center for Integrated Regional Assessment (CIRA) to guide a team of researchers addressing regional contributions to global change and regional impacts and response choices to those changes as they might occur. The framework was also intended to illustrate to the stakeholder and policy maker communities the logic through which IRA was being addressed, without necessarily being a wiring diagram for the actual processes of change. In other words, the framework was developed as an heuristic to guide IRA as a process for developing knowledge in a researcher–stakeholder–policy maker context without being methodologically or quantitatively prescriptive. This chapter addresses the development of the CIRA framework from an evolutionary viewpoint, and illustrates how the framework is used in practice, using, among others, questions relevant to the Mid-Atlantic Region of the United States (Fisher *et al.* 2000; Fisher and Kasemir, Chapter 8, this volume).[1]

9.2 Development of the framework

In the late 1980s, a group of social and environmental scientists interested in the theme of human dimensions of global change (HDGC) had been informally organized in a new college-based research unit inspired by the Earth System Science concept noted in Chapter 1.[2] Seeking an opportunity for more formal interaction and research activity, the HDGC group developed the idea of a center for integrated regional assessment (CIRA). A simple heuristic guided the proposal-writing team when the concept for CIRA itself was being developed in 1994–1995.[3] This heuristic (Figure 9.1) suggested a feedback loop from the causes of climate modification to the changes themselves, thence to the consequences for the environment and society, and to the range of responses – spontaneous and purposeful – that could follow. The simple heuristic also recognized that responses could have important feedback on the causes of change in the first instance, thus completing the feedback loop. Important was the recognition that all of these process cascade through various scales, from local through regional to potentially global, in both causal and impact directions. In Figure 9.1, the four components of this simple heuristic were nested in concentric circles to connote this important scale dimension.

As CIRA moved from an aspiration to an institutional reality in 1996,[4] the team faced some daunting challenges. Its major support called for the identification of

[1] The framework helped to guide the Mid-Atlantic Regional Assessment (MARA) component of the US Global Change Research Program's National Assessment of the Consequences of Climate Change during 1997–2002, under direction of Ann Fisher with support of the Global Change Research Program of the US Environmental Protection Agency.
[2] The HDGC group (and later CIRA) worked under the auspices of the Earth Systems Science Center in the College of Earth and Mineral Sciences (later, the EMS Environment Institute) at the Pennsylvania State University, USA. Disciplinary colleges are a typical organization unit within large universities in North America.
[3] The Penn State team at that time was led by Diana Liverman (now at Oxford University) with considerable support by Robert Merideth (now at the University of Arizona).
[4] Liverman and Merideth left Penn State at the end of 1995; C. Gregory Knight became the coordinator of a revised and eventually successful research proposal to the Human Dimensions of Global Change Program at the US National Science Foundation in 1996.

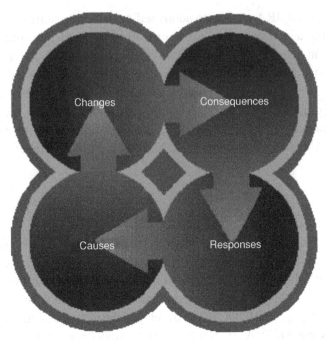

Figure 9.1. A simple heuristic for integrated regional assessment. The concentric circles connote differences in scale (e.g., global, nation, regional).

new methodologies for integrated regional assessment, and it would only be a year later that the team would begin doing IRA as well: the Mid-Atlantic Regional Assessment (MARA; subsequently the Consortium for Atlantic Regional Assessment, CARA). How would a disparate interdisciplinary group of scholars articulate their investigations? Could this group evolve from a loosely organized group of scholars doing related research to an articulated team? How could the team work effectively across disciplinary and intra-institutional boundaries? How could the team identify important linkages and relationships within the cascading impacts and feedbacks that the simple heuristic connoted? How could the CIRA research on climate change be located in a wider set of human–environment concerns? How could the team make explicit to stakeholders and policy makers the conceptual underpinnings of its responses to their concerns? During CIRA's first year, sub-groups had begun to explore linkages and methodological challenges, but a hoped for holism of the team had yet to evolve.

Thus in late 1997, the CIRA team organized a seminar for the senior researchers and graduate students to address these institution-building questions and explore the specific ways in which specific research competencies could be linked and elaborated, hopefully with a resulting synergism among the research team. The choice was made to think graphically as well as verbally, and to build upon the

Integrated regional assessment of global climate change

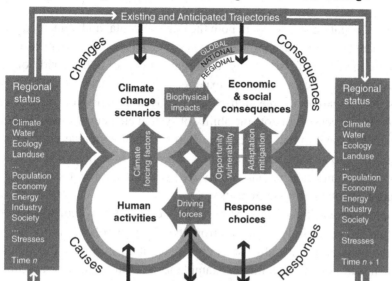

Figure 9.2. The CIRA framework for integrated regional assessment. The first formal publication of this framework was in CIRA annual reports (www.essc.psu.edu/cira) and subsequently in the Mid-Atlantic Regional Assessment (Fisher *et al.* 2000; Knight 2001).

initial heuristic that had guided the team's thinking at the proposal stage. A simple diagram seemed to be a method with which all contributors could identify, one based on conceptual issues rather than the names of specific disciplines. The challenges were addressed by common readings, discussions, and multiple approximations of an emerging framework graphic. The resulting framework is shown in Figure 9.2. A diagram similar to the core of the CIRA framework was subsequently included in the 2001 IPCC Third Assessment Report (Watson *et al.* 2002).[5]

The CIRA framework relies on long-established concepts of feedback, feedforward, and cascading systems (Bennett and Chorley 1978). *Feedback* refers to the way in which causal events are in turn affected by the results of their action. *Positive feedback* refers to deviation-amplifying processes – the outcomes multiply in subsequent cycles; examples include population growth and sound feedback at concerts. *Negative feedback* is a deviation-dampening process in which the outcome progressively decreases the effects of the causal event; examples include

---

[5] Neither the principal author of the 2001 Synthesis Report nor the IPCC secretariat has been able to document the origin of the IPCC graphic nor explain its up–down arrows between impacts and development paths.

a thermostat-controlled air conditioning system or computer-controlled signals responsive to traffic changes. In the realm of climate impacts, one can expect both positive and negative feedbacks in both the environmental and human domains. In climate change, global warming might experience positive feedbacks as industrial nations use increasing fossil energy to mitigate effects of hotter conditions; negative feedback could occur as nations experience early climate change impacts and address the mitigation of greenhouse gas (GHG) emissions with greater resolve. *Feedforward* processes connote the ability of system elements to anticipate change and act with respect to that expectation. Cases in nature include plant photoperiodic responses to seasonal change such as flowering in anticipation of a coming rainy season; examples in human society are saving for a rainy day or acting in the expectation of future adverse impacts.

Complex feedback–feedforward systems such as global human–environment interaction can be usefully addressed from the viewpoint of *cascading systems*, in which the outputs of one sub-system constitute, at least in part, the inputs for a subsequent sub-system. A common example is the hydrological system, in which water cascades from atmosphere to surface to soil water, rivers, and ground water, thence to lakes and ocean and back to the atmosphere via plant evapotranspiration and evaporation from water bodies. The cascading system conceptualization is not to ignore feedback or feedforward mechanisms, but to recognize that it is heuristically useful to think of climate change having biophysical effects that in turn have human implications, the regional response to which will have cascading feedbacks on the driving forces of climate change and thus on future climate, as well as on the future of the region. Such systems thinking is, of course, *one way* of conceptualizing the impacts of climate change, with no claim to exclusivity in perspective.

The same concepts of feedback, feedforward, and cascades apply to regions as systems of human–environment interaction. In some respects, working at the regional level would appear to simplify the enormous challenges of understanding global climate change, bringing the scale and resolution of analysis within manageable bounds. However, similar kinds of system dynamics also occur across scales, as regions contribute to global change, and global change impacts cascade downward in scale to the region.

9.3 Walking through the framework

The framework should first be viewed from its outside components: the current status of the region, including environmental and social parameters with specific attention to existing stresses; existing and anticipated trajectories in the absence of climate change; and the resulting future status of the region. This kind of scenario development could consist of iterations over discrete periods of time. The kinds of questions asked at this stage include:

(1) What is the current status of the region and what are the anticipated trajectories of change in the absence of climate change?
(2) What are the current dynamics of land use and other environmental change, population, economy, and development?
(3) What stresses already occur in the environment and human systems?
(4) What other global processes will drive the region's future, beyond that of climate change?
(5) What is the current vision for the future and what threatens achievement of that vision?

These are the kinds of questions that are typically addressed in the environmental, social, policy, and planning sciences, individually or together. The remainder of the framework is devoted to answering how climate change will affect the answers to these questions. What can the region do about causes and impacts for undesirable consequences? Clearly, concerns about climate need to be viewed in the context of what matters to individuals, businesses, governments, countries.

The subsequent elements of the framework involve, then, the contributions of the region to climate change (lower left of the diagram), the climate changes themselves that may affect the region, the biophysical changes that result from an altered climate, the economic and social consequences of environmental changes, the opportunities and vulnerabilities that result from these consequences, the opportunities to mitigate the consequences or adapt to them, the response choices among these alternatives, how the responses change the driving forces of human activities, and in turn how the altered human activities change climate forcing factors. Taking these elements in turn, each links a cascade of linked questions.

Although it is not obligatory, it is useful to begin at the lower left, asking what human activities in the region contribute to the global warming potential, such as energy production, other greenhouse gas emissions, and land-use changes. This question at the regional level is similarly raised at the national level in reports by individual countries to the United Nations Framework Convention on Climate Change (UNFCCC) in addition to National Communications to UNFCCC. Earlier, the US Department of Energy's Country Studies Program offered national methodologies for GHG documentation in developing and transitional countries (US Country Studies Management Team 1999). In the USA, the Global Change and Local Places project (AAG 2003) developed methods for documenting regional contributions to global GHG at the sub-national level.

Typically, climate change scenarios at the regional level would be developed from General Circulation Model (GCM) output over the region, accompanied by some relevant method of downscaling, from surface fitting to get regional trends of change among the GCM grids to statistical downscaling from the larger grids to regional resolution, use of neural net models for a similar purpose (Hewitson and Crane 1994), nested meso-scale climate models within GCMs and regional climate modeling. Climate analogs could also be useful, asking for the consequences if actual extreme climate events reoccur or become common (Rosenberg *et al.* 1993). From these activities, climate scenarios can be developed. Within this the relevant questions become:

(1) How might the global climate change? What plausible picture of the future can be developed under relevant development and emissions scenarios? So the question is: how would global climate change given certain development trajectories?
(2) What climatic scenarios can be developed by downscaling from GCMs to the regional level?
(3) To what extent do these models agree in both the direction and magnitude of climate change in the region?
(4) How much uncertainty is there in climate scenarios, and what are their major differences?

Clearly, the temporal and spatial scale of the climate scenarios will be driven by the size and nature of the region and by the dominant questions raised by the stakeholder and policy-maker communities. In many cases there may be greatest interest in weather extremes; in others it may be in broader issues of monthly and seasonal climate. Scenarios could also include critical climate variables for natural ecosystems, forests, crops, or disease vectors.

The next component, moving clockwise, poses questions of how climate change could initiate cascading impacts through the environmental system. Such impacts may be related more to climate extremes than means, although both will be important. Important questions include:

(1) What are the ecological and physical impacts of the climate scenarios that would have significant economic and social consequences?
(2) How would climate change affect forests, agriculture, ecosystems, and water resources?
(3) What is the geographic and temporal distribution of these changes?
(4) What would be the impact of warmer climate on the winter snow melt and on runoff regimes? Would rivers have greater or lesser seasonal change in flow? Would flood regimes change? What would extended periods of low flow mean in terms of water pollution?
(5) How would natural, unmanaged ecosystems respond? Are there sufficient migration corridors for flora and fauna in the face of climate change?
(6) How would the phenology and productivity of managed ecosystems – forests and farms – be affected? Would changes in growing seasons occur? Would early winter warm periods encourage early blossoming and subsequent loss to frost of fruits?

Would severe storms and hail threaten grain crops? Would longer growing seasons permit growth of warmer-season crops or greater yields?

(7) What would climate change mean in terms of the population dynamics of the vectors and hosts of human, livestock, and crop diseases?

Complex interactions within the environmental realm are considerably obscured by a simple arrow. For example, changing patterns of vegetative cover resulting from climate change may have an effect on regional hydrology equal to, if not greater than, those of climate change itself. Such interactions and feedbacks are important and should be part of the investigation.

Once there are some reasonable scenarios of possible biophysical impacts, the assessment then turns to the question of the human consequences of these changes. Some of the consequences may be direct and palpable. Focal questions concerning the economic and social consequences include:

(1) What is the impact on the regional economy and on social dynamics such as employment and migration?
(2) Would there be increasing or decreasing flood threat?
(3) Would seasonal water shortages create an economic bottleneck to development?
(4) How much greater investment would be needed for water pollution control?
(5) Would there be impacts on outdoor recreation? Would tourism seasons be shortened (winter sports) or lengthened (summer holidays)?
(6) How would forest dynamics affect forest harvesting?
(7) How would farming operations be affected?
(8) Is the building stock of the community suitable for different scenarios of climate change?
(9) How would transportation networks be affected?

Stakeholder and policy-maker communities play an important role in directing analysis to issues that are particularly salient for the region, given the multiplicity of questions that could be raised. It is important, however, for the researchers to present a sufficiently comprehensive list of potential types of impacts that stakeholders and policy makers guiding the research are motivated to suggest specific concerns that might otherwise be overlooked.

Some of the consequences may be identified as particularly threatening, raising questions of how society might mitigate these consequences or adapt to them; some consequences may offer new opportunities, suggesting questions about how society might exploit these opportunities, perhaps to some degree off-setting negative consequences that cannot be totally buffered. In recognition of the ability of society to develop responses to the impacts, the team added the reverse arrow between impacts and responses in recognition of feedforward (expectation and planning) and feedback (impact dampening or enhancing) phenomena. Example questions to be addressed include:

(1) How would these consequences help to identify opportunities that may result as well as vulnerabilities that are created?
(2) Would there be new opportunities for agricultural production? For outdoor recreation?
(3) What is the geographical and social distribution of new opportunities?
(4) What mechanisms – structural or social – exist to adapt to changes or mitigate them?
(5) What are the economic consequences and how would those consequences cascade through the regional spatial economy?
(6) Are there differences in the intensity and direction of impacts across places, social groups, income levels, etc.?

Clearly, the answers to these reciprocal questions depend on society's responses to the identified impacts. There are many categories of potential responses; policies, laws, investments, and subsidies, spatially-relevant locational decisions, and technology are among them. Stakeholders and policy makers are essential in identifying specific kinds of responses. Questions to be addressed by researchers with assistance from these advisors will include:

(1) What are the responses that the regional government and society might take?
(2) In what time frame would critical decisions need to be made?
(3) What socio-economic and environmental monitoring is necessary to maintain vigilance?
(4) Can mechanisms be found to deal with social and spatial inequities in negative impacts?
(5) How could cooperation be established with regional neighbors on issues of mutual concern, such as common management in shared river basins?
(6) How do response choices expand or limit adaptations to inevitable climate change?
(7) How would social and regional equity be maintained?

Human responses, in turn, have a further effect on driving forces of human activities that contribute to climate change. Some of the responses will constitute intentional or inadvertent no regrets strategies, actions that will be justified on grounds other than climate change (such as energy conservation) but could help to diminish the human driving forces and actions that create climate change. Such actions are in effect negative feedbacks in the system. Other actions can be positive feedbacks on global warming, such as using greater amounts of energy to air condition living and working space or expanding rice paddies to feed growing populations. Taking together the driving forces and actions affecting human impacts on the environment, the kinds of questions raised at this stage include:

(1) What is the effect on driving forces of human activity – such as energy demand and conservation, economic development, population dynamics, and urbanization?
(2) How would human activities change in the face of alterations in these driving forces?

Table 9.1. *The importance of scale in sample domains.*

Scale	Hydrology	Society	Economy	Policy
Global	*Global climate and hydrology*	*Global population*	*Dynamics of the global economy*	*International treaties and agreements*
↑↓ National	↑↓ *Regional and seasonal hydrology*	↑↓ *Employment and life quality*	↑↓ *National growth and income*	↑↓ *National policy, law, and regulations*
↑↓ Regional	↑↓ *River basin seasonal and annual dynamics*	↑↓ *Differential migration*	↑↓ *Regional role in national economy*	↑↓ *Regional reaction & implementation*
↑↓ Local	↑↓ *Water resources & hazards*	↑↓ *Population & land use*	↑↓ *Economic growth or stagnation*	↑↓ *Local implications and impacts*

(3) Would greater wealth → bigger is better → greater energy demand even if energy use per unit of wealth decreases?

(4) Would responses provide positive or negative feedback on GHG emissions?

(5) Could energy conservation decrease GHG emissions?

(6) Could reforestation as a strategy for sequestering carbon help to mitigate threats of flooding or decrease available runoff?

(7) What would be the region's contribution in terms of greenhouse warming potential in the future?

Clearly, as the cycle of causes–changes–consequences–responses proceeds at the regional level, the region is embedded in an environmental and social matrix at wider levels, and the interactions are complex. One kind of interaction of particular importance is the policy realm. Here, we address the questions of how the region contributes to and is affected by policy at the national and global level:

(1) How would responses, driving forces, and activities be encouraged or constrained by policy changes at various spatial scales?

(2) How would the region react to carbon taxes or other energy policy?

(3) Would policies not directly related to climate change encourage or impede the ability of the region to address global warming issues and impacts?

Scale is an issue that pervades all of the analysis (Table 9.1). Indeed, many of the analytical tools brought to integrated regional assessment involve isolating regional and local outcomes from within models operating at wider scales; downscaling from larger-scale models and processes; and providing upward information from the regional scale to national and global dimensions. There are a number of research frontiers in the area of multi-scale analysis and modeling (Polsky and Munroe, Chapter 4, this volume); IRA should take advantage of accomplishments in this area as they evolve. Among the questions that need to be addressed that involve interactions across scales are:

(1) How would impacts be transmitted across scales and among regions? Through natural systems (e.g. water flow; plant and animal dispersion)? Via socio-economic processes (e.g. transportation, migration, flow of money)?
(2) How important are these trans-scale and trans-boundary effects compared to impacts within the region?
(3) Do impacts or responses in the focus region assist with or exacerbate challenges in other regions or across scales?
(4) Which transmitted impacts create threats, and which offer opportunities?
(5) Do salient scale interactions also involve political and institutional boundaries?
(6) How do political and institutional factors impede or assist in adaptation to or mitigation of cross-scalar negative impacts? In exploiting new multi-scale opportunities?

Finally, the framework takes us back to the original questions, reframed in the presence of climate change impacts.

(1) What is the current status of the region and how would anticipated trajectories of change be altered by climate change?
(2) How would climate change affect current dynamics of land use and other environmental change, population, economy, and development?
(3) How would climate change exacerbate or ameliorate stresses already occurring in the environmental and human systems?
(4) What other global processes would drive the region's future, beyond that of climate change, and how would these processes be altered in the presence of climate change?
(5) What is the current vision for the future and how would climate change threaten achievement of that vision?

The role of the stakeholder and policy-maker advisors is again salient: they must help to define the issues of importance and raise questions that the researchers might otherwise have overlooked. In the end of any stage of analysis, it is intended that a picture will emerge of the most important potential climate change impacts and how society might deal with them. The results of the first stage of the Mid-Atlantic Regional Assessment are illustrated in Chapter 8 (Fisher and Kasemir, this volume).

9.4 Assessing the framework

The CIRA authors make no claim for any universality or correctness of a framework that we have found useful in guiding our own work. We recognize that the framework diagram does not include some important elements of the IRA process:

(1) how one works with communities, stakeholders, policy makers, and the public;
(2) how one chooses the spatial, temporal, and conceptual limits to a specific activity;
(3) how one chooses the most appropriate methodology at each stage in the conceptual cascade;
(4) how one assesses and communicates uncertainty; and
(5) how one can move from a framework to a model.

Nevertheless, we have found the framework diagram and its underlying conceptual linkages useful both within and outside the research community. Within the team, it has helped us build vital conceptual linkages between framework elements and the disciplinary skills we bring to the assessment process. It would be an illusion to think the framework totally overcame tensions between disciplines and individual ways of addressing research questions. But it helped us move appreciably along the road from individuals in disciplines to a team working in an interdisciplinary manner. To the outside world, we have found the framework helpful in communicating the cascading processes underlying global change processes playing out in the regional context. It has also been helpful in showing the non-academic world how our disciplines link together in a holistic way.

9.5 Future directions for IRA frameworks

As a result of the CIRA experience with its framework, we encourage others to develop parallel approaches, particularly frameworks that more explicitly incorporate IRA as a process not simply a product. Certainly the CIRA framework as it now exists will be revised as our hands-on experience with integrated regional assessment grows. In some settings, a relatively simple, informal IRA process will be enough to make the issues clear and identify desired actions. In other situations (e.g., for longer time spans and larger regions), it will be helpful to refine the modeling approach to handle the multiple complex interactions and goals. We also believe the framework can inspire a modeling approach to climate change impacts at regional scales, taking the experience in numerical modeling in cascading end-to-end assessment described elsewhere in this book into a modeling system complete with feedback elements and the ability to experiment with alternative futures.

Acknowledgements

The CIRA Team worked with primary support from the United States National Science Foundation (Grant SBR-9521952). At the time of the reported framework development, C. Gregory Knight and Ann Fisher were director and co-director of CIRA; the team developing and refining the framework also included at various times senior members David G. Abler, Eric J. Barron, Jeffrey Carmichael, Robert G. Crane, Mary M. Easterling, William E. Easterling, Amy K. Glasmeier, Jeffrey Lazo, Diana Liverman, Stephen Matthews, Robert Merideth, Robert E. O'Connor, Adam Rose, James S. Shortle, Marieta P. Staneva, Liem Tran, Zili Yang, Brent Yarnal, and along with (then) graduate students Heejun Chang, Stephen Lachman, and Colin Polsky. This chapter was drafted on behalf of the team by Knight with assistance from Fisher.

References

Association of American Geographers Global Change and Local Places Research Team 2003. *Global Change and Local Places: Estimating, Understanding, and Reducing Greenhouse Gases.* Cambridge: Cambridge University Press.

Bennett, R. J. and R. J. Chorley 1978. *Environmental Systems: Philosophy, Analysis and Control.* Princeton, NJ: Princeton University Press.

Burton, I., G. F. White, and R. W. Kates 1993. *The Environment as Hazard.* 2nd edn. New York: Guilford Press.

Fisher, A., D. Abler, E. Barron, *et al.* 2000. *Preparing for a Changing Climate: the Potential Consequences of Climate Variability and Change, Mid-Atlantic Overview, March 2000.* University Park, PA.

Hewitson, B. and R. G. Crane (eds.) 1994. *Neural Nets: Applications in Geography.* Dordrecht: Kluwer.

Kates, R. W. 1985. The interaction of climate and society. In R. W. Kates, J. H. Ausubel, and M. Berberian (eds.), *Climate Impact Assessment*, SCOPE 27. Chichester: Wiley.

Knight, C. G. 2001. Regional Assessment. In A. S. Goudie (ed.), *Encyclopedia of Global Change* vol. 2. New York: Oxford University Press, pp. 304–308.

Rosenberg, N. (ed.) 1993. *Towards an Integrated Impact Assessment of Climate Change: the MINK Study.* Dordrecht: Kluwer.

US Country Studies Management Team 1999. *Climate Change: Mitigation, Vulnerability, and Adaptation in Developing and Transition Countries.* Washington, DC: US Country Studies Program.

Watson, R., D. L. Albritton, T. Barker, *et al.* 2002. *Climate Change 2001, Synthesis Report.* Cambridge: Cambridge University Press.

10

The global context of integrated regional assessment

JILL JÄGER

10.1 Introduction

Research on global change is an international and interdisciplinary endeavor that has made huge advances in recent years. In a book that integrated much of the work carried out under the auspices of the International Geosphere–Biosphere Programme (IGBP),[1] Steffen *et al.* (2004) pointed out that global change is much more than climate change and its full extent and complexity has only been realized very recently. There is a broad range of biophysical and socio-economic changes taking place. These changes are interlinked. For example, consider the nitrogen cycle (Figure 10.1).

Human activities add 1.5 times more nitrogen to the atmosphere than natural terrestrial processes.[2] In particular nitrogen is added to the atmosphere through fossil fuel use and fertilizer use. Nitrogen released to the atmosphere can, in sequence, increase tropospheric ozone concentration, decrease atmospheric visibility, and increase precipitation acidity. Following deposition it can increase soil acidity, decrease biodiversity, pollute groundwater, and cause coastal eutrophication. Once emitted back to the atmosphere, it can contribute to climate change and decreased stratospheric ozone. This illustrates the multiple drivers of global change (e.g. the demand for energy and food) as well as the multiple impacts (on the atmosphere, soil, vegetation, and oceans) and the chains of interactions that take place. As such it is one example of why an *integrated* approach to the understanding of global change is required.

Steffen *et al.* (2004) pointed out further, however, that global change is taking place in varying ways at different locations, each with its own set of characteristics being impacted by a location-specific mix of interacting changes. The global environment and the phenomenon of global change are both heterogeneous, and

---

[1] www.igbp.kva.se   [2] SCOPE (2007).

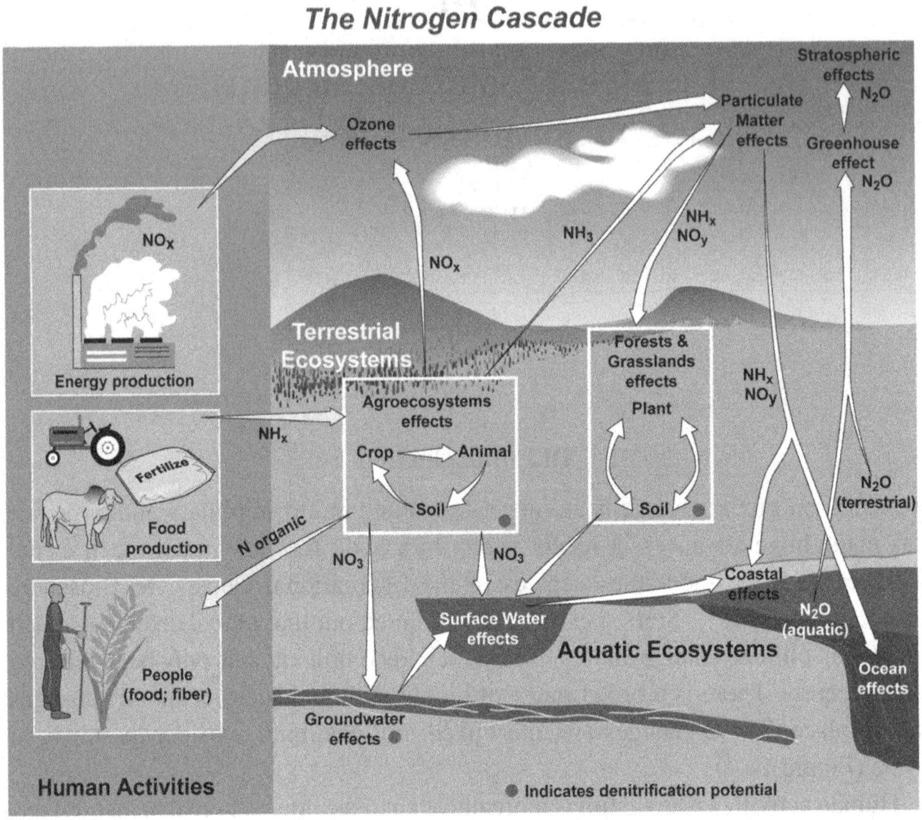

The Nitrogen Cascade

Figure 10.1. The nitrogen cycle, human impacts, and interconnections (used with permission of SCOPE from UNESCO–SCOPE (2007)).

the variety of human–environment relationships is vast. Therefore, Steffen *et al.*, p. 7, conclude:

To cope with or adapt to global change requires analyses that couple the particular characteristics of a location or region with the nature of the systemic, globally connected changes to the Earth's environment as they interact with other factors affecting the location or region.

Given the need for integration of disciplinary approaches and a regional perspective in order to improve understanding of global change, Steffen *et al.* explored the nature of "integrated regional studies." They identify approaches that are appropriate for this kind of analysis: vulnerability analyses; identification of hotspots of risk; and identification and simulation of syndromes of environmental degradation. Further, they find that in order to be effective as a tool for Earth Systems analysis integrated regional studies must have a number of characteristics, including:

- the ability to show how the region functions as an entity and how that functioning might change;
- work at disciplinary boundaries across natural and social sciences and address all relevant aspects of marine, terrestrial, atmospheric, social, economic, and cultural components and processes within and across the region;
- consideration of the two-way linkage between the region and the global system.

One example of an integrated regional study is the Large-scale Biosphere–Atmosphere Experiment in Amazonia (LBA). This is described in Section 10.2 of this chapter. A similar effort that started more recently is the Monsoon Asia Integrated Regional Study (MAIRS) described in Section 10.3. Then the chapter moves to consider integrated regional assessments carried out within the auspices of major international assessment processes – the Intergovernmental Panel on Climate Change (IPCC) and the Millennium Ecosystem Assessment – in Section 10.4. In Section 10.5, the syndrome and vulnerability approaches are described and the chapter concludes in Section 10.6 with a short discussion of the approaches presented in the chapter as an introduction to the following chapters.

10.2 A regional assessment in Amazonia (LBA)

The Large-scale Biosphere–Atmosphere Experiment in Amazonia (LBA) was an international initiative led by Brazil (Nobre *et al.* 2002) to study the Amazon region in an integrated way. The LBA began in the early 1990s. The aim was to understand the climatological, ecological, biogeochemical, and hydrological functioning of the region, the impact of land use and future climate change on those functions, and the interactions between Amazonia and the Earth System. In total there were more than 100 closely linked and integrated studies involving about 600 scientists and students from South and North America, Europe, and Japan.

Two key questions integrated the various interdisciplinary approaches (Steffen *et al.* 2004: 278). The first question was concerned with how Amazonia functions as a regional entity with respect to the cycles of water, energy, carbon, trace gases, and nutrients. The second question asked how changes in land use and climate will affect the biological, chemical, and physical functions of Amazonia, including sustainable development in the region and the influence of Amazonia on global climate.

Nobre (in Steffen *et al.* 2004: 278) suggests that the success of the initiative depended on a number of important factors:

- being able to build from the beginning on a framework that allowed the integration of the individual parts;
- taking a modular approach;

- integrating national and regional development concerns into the global changes questions;
- attracting scientists from within Brazil, from the Amazon Basin and from the international scientific community.

The program used various techniques to scale up the individual studies to the basin level including remote sensing, modeling, and the use of transects based on eco-climatic and land-use intensity gradients. It also contributed to the enhancement of research capacities and networks within and between the Amazonian countries.

10.3 A regional assessment in monsoon Asia (MAIRS)

Together with the Earth System Science Partnership (ESSP),[3] START (global change SysTem for Analysis, Research and Training) and its regional networks in East Asia, South Asia, and Southeast Asia are undertaking integrated regional studies of global change in monsoon Asia (Monsoon Asia Integrated Regional Study – MAIRS). The long-term objectives of the integrated regional studies that will ultimately combine field experiments, process studies, and modeling components are:

(1) to better understand how human activities in regions are interacting with and altering natural regional variability of the atmospheric, terrestrial, and marine components of the environment;
(2) to contribute to the provision of a sound scientific basis for sustainable regional development; and
(3) to develop a predictive capability for estimating changes in global–regional linkages in the Earth system and to recognize on a sound scientific basis the future consequences of such changes.

The key issues for such integrated regional studies include considerations of what the region will be like in another two to five decades; what the consequences of global changes will be for the region; and what the consequences of regional changes will be for the global Earth system. The studies will consider:

- major demographic, socio-economic, and institutional drivers for change, including scenarios of change related to urbanization and industrialization, energy production and biomass burning, land use/cover change, and water resources harvesting, including dam construction;

---

[3] The Earth System Science Partnership (http://www.essp.org) is a collaboration between the international global change research programs (International Geosphere–Biosphere Programme, IGBP, World Climate Research Programme, WCRP, the International Human Dimensions Programme on Global Environmental Change, IHDP, and DIVERSITAS, an international program on biodiversity science). The objective of the partnership is the integrated study of the Earth System, the ways that it is changing, and the implications for global and regional sustainability.

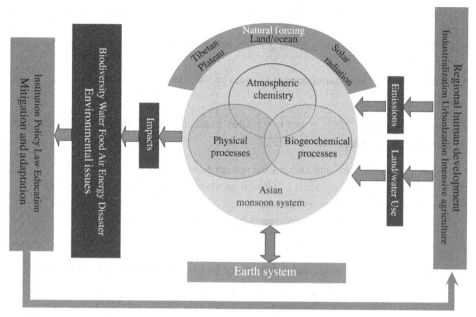

Figure 10.2. The MAIRS conceptual framework (Fu and de Vries 2006; used with permission of the MAIRS International Project Office).

- effects on regional and atmospheric composition/pollution, regional water cycle and coastal systems, and local ecosystem structure and function;
- impacts on biogeochemical cycles and the physical climate system, including its variability at different scales;
- potential impacts of global and other feedback effects on the regional biospheric life support system, including food systems, water resources, and health.

The integrated regional studies are being preceded by a first phase of rapid assessment, undertaken jointly by START, its regional programs, and the Scientific Committee on Problems of the Environment (SCOPE). Three sub-regional rapid assessment projects for China/East Asia, South Asia, and Southeast Asia are systematically reviewing current knowledge regarding regional aspects of global change in monsoon Asia, in order to highlight gaps in knowledge and uncertainties, and define research priorities for the integrated regional studies.

The concept of an integrated Asian monsoon system was originally proposed by Asian scientists (e.g. Fu 1997). From the Earth system science point of view, the Asian monsoon system is a coupled physical/biological/chemical/social system (Fu *et al.* 2002).

The conceptual framework for the MAIRS program is shown in Figure 10.2.

Key research questions in MAIRS are:

- is the Asian monsoon system resilient to this human transformation of land, water, and air?
- are societies in the region becoming more, or less, vulnerable to changes in the Asian monsoon?
- what are the likely consequences of changes in the monsoon Asia region on the global climate system?

To answer these questions about key issues in an integrated manner, four research themes have been proposed:

- rapid transformation of land and marine resources in coastal zones;
- multiple stresses on ecosystems and biophysical resources in high mountain zones;
- vulnerability of ecosystems in semiarid zones due to changing climate and land use;
- changes in resource use and emissions due to rapid urbanization.

10.4 Integrated regional assessment in global assessment processes

10.4.1 Intergovernmental Panel on Climate Change

The Intergovernmental Panel on Climate Change (IPCC) was established by the World Meteorological Organization (WMO) and the United Nations Environment Programme (UNEP) in 1988 in response to increasing concern about anthropogenic climate change.[4] The mandate of the IPCC is to produce a state-of-the-art assessment of climate change, including the scientific basis of concern, possible impacts, and possible response options. To that end, the IPCC has to date produced four assessments (1990, 1995, 2001, 2007). In addition to the main reports, technical reports and special reports have also been produced. The work of IPCC has always been divided between three working groups. Working group I has always been on the science of climate change. Working group II has always considered impacts of climate change. Working group III now considers the options for limiting greenhouse gas emissions and otherwise mitigating climate change.

With regard to integrated regional assessment, the work of all three working groups is of relevance but most of the regional focus of IPCC is found in working group II. In the Second Assessment Report (1995), the charge to working group II was to review the state of knowledge concerning the impacts of climate change on physical and ecological systems, human health and socio-economic sectors. In addition, working group II was charged with reviewing available information on the technical and economic feasibility of a range of potential adaptation and mitigation strategies. Thus, the 1995 report took a sectoral cut in reviewing the impacts of climate change. This immediately spurred interest in the potential regional impacts.

---

[4] All reports of the IPCC can be found at www.ipcc.ch

The Subsidiary Body for Scientific and Technological Advice (SBSTA) of the UN Framework Convention on Climate Change (UNFCCC) therefore requested IPCC to produce a report addressing the degree to which human conditions and the natural environment are vulnerable to the potential effects of climate change. The report (Watson *et al.* 1997) consists of vulnerability assessments for 10 regions: Africa, Arid Western Asia, Australasia, Europe, Latin America, North America, the Polar regions, Small Island States, Temperate Asia, and Tropical Asia. The regional approach identified wide variation in the vulnerability of different populations and environmental systems.

The Third Assessment Report (2001) moved away from the "single sector" approach taken in the Second Assessment Report, noting that climate change is likely to have multiple impacts across sectors and synergistic effects with other socio-economic and environmental stresses, such as desertification, water scarcity, and economic restructuring. The report pointed out that relatively few studies had attempted to integrate regionally or even identified segments of the population that are most at risk from climate change. It suggested however that the assessment of vulnerability provided an integrative approach at the regional level that could deal with multiple environmental and socio-economic pressures.

The Third Assessment Report made further methodological advances over the previous report. Framing questions regarding issues such as vulnerable populations, policy questions, and scale of the assessment were raised at the outset of the assessment, often in conjunction with representative stakeholders (see Carter *et al.* 1994; Downing 2000). The assessment also recognized that the choice of temporal scales, regional extent, and resolution should be related to the focus of the assessment. It was noted that often, more than one scale is required and linkage to global assessments may be necessary to understand the policy and economic context (e.g., Darwin *et al.* 1995). The most common set of methods and tools used in the studies that formed the basis of the assessment were various forms of dynamic simulation modeling, such as crop–climate models or global vegetation dynamic models, some of which are used also in studies described in later chapters of this book. The Third Assessment Report noted many of the data and methodological challenges still faced in estimating regional impacts of climate change, including the particular challenge of taking adaptation into account. Adaptation can significantly reduce people's vulnerability to climate change, as shown by Dickinson and colleagues (Chapter 7, this volume). However, adaptation can take many different forms and is correspondingly difficult to model. The 2001 report concluded that there were no integrated assessment models available that could adequately represent or guide the full range of adaptation decisions. In summary, the assessment noted that despite the growing number of country-level case studies, knowledge about climate change and climate change impacts at the regional level remained limited.

By the time that the Fourth Assessment Report was written, additional studies, particularly in regions that had previously been little researched, enabled a more systematic understanding of the regional impacts of climate change. As in the Third Assessment, attention focused on "key vulnerabilities" and research results that had become available since the previous report were used to update the findings. An important advance since the 2001 report was the completion of impacts studies for a range of different development pathways taking into account not only projected climate change but also projected social and economic changes. That is, the assessment was able to integrate more pressures on the regions. It was possible to show that large differences in regional population, income, and technological development under alternative scenarios can be a strong determinant of the level of vulnerability to climate change.

In summary, the IPCC has moved from a consideration of single sector impacts of climate change to a more integrated view of regional vulnerability to multiple pressures that include climate change. The regional resolution (at the level of "Africa" or "the Polar regions") is still extremely broad, which limits the use of the results for policy making. However, the tools and methods that form the basis of the studies assessed by the IPCC are also being used for studies at a resolution that is more policy-relevant, as shown in later chapters of this book.

10.4.2 The Millennium Ecosystem Assessment

The Millennium Ecosystem Assessment (MEA)[5] was launched by UN Secretary-General Kofi Annan in June 2001 and was completed in March 2005. It was designed to meet assessment needs of the Convention on Biological Diversity, the Convention to Combat Desertification, the Ramsar Convention on Wetlands, and the Convention on Migratory Species, as well as needs of other users in the private sector and civil society. The MEA focused on how changes in ecosystem "goods and services" (food, timber, water purification, flood protection, biodiversity, etc.) have affected human well-being (health security, livelihood security, cultural security, etc.), how ecosystem changes may affect people in future decades, and what types of responses can be adopted at local, national, or global scales to improve human well-being and contribute to poverty alleviation.

In addition to its distinct focus on ecosystems and human well-being, the MEA included another pioneering aspect that distinguished it from past "global" assessments. It was conducted as a "multi-scale" assessment with integral assessment components being undertaken at local community, watershed, national, and regional scales, as well as at the global scale. The MEA sub-global assessments were designed to meet needs of decision makers at the scale at which they are undertaken,

---

[5] For detailed information on the MEA reports see www.millenniumassessment.org

strengthen the global findings with on-the-ground reality, and strengthen the local findings with global perspectives, data, and models. There are 18 MEA-approved sub-global assessments, and an additional fifteen with an associated status (see Box 10.2 below). Assessments at sub-global scales are needed because ecosystems are highly differentiated in space and time, and because sound management requires careful local planning and action. Local assessments alone are insufficient, however, because some processes are global, and because local goods, services, matter, and energy are often transferred across regions. Box 10.1 gives an example of the aims of one of the sub-global assessments.

Box 10.1
Assessment of the Colombian coffee-growing regions

Goals

To support the development of a more representative, effective, and viable Andean Protected Area System; to identify conservation opportunities in rural landscapes while developing and promoting sustainable use and management tools for biodiversity conservation; to expand, organize, and disseminate the Andean biodiversity knowledge-base to stakeholders and policy makers; and to promote intersectoral coordination to address root causes of biodiversity loss in the Colombian Andes.

Ecosystem services addressed

Provisioning: coffee, other food, water quality, cultural;
Supporting and regulating: biodiversity, maintenance of soil structure, and chemical properties.

Key features of the assessment

Due to the wide variety of environmental conditions present in the coffee growing region, 86 ecotopes with homogeneous topographic, climate, and soil characteristics were defined. These were the local units of analysis for the assessment. The ecosystem assessment focused on the following issues: use values of ecosystem functions in coffee growing regions and their implications for policy making and planning; assessment of biodiversity in rural landscapes; and in situ conservation.

The lessons learned and the initial results of the sub-global assessments working group were published in 2005[6] and the individual assessments are being published separately.[7] The sub-global assessments provided useful insights on the influence

[6] Ecosystems and human well-being: multi-scale assessments. Findings of the Sub-Global Assessments Working Group, Island Press.

[7] Reports on the Southern African Millennium Ecosystem Assessment (SAfMA) were released in September 2004. Reports on the Assessment of Colombian Coffee-Growing Regions were released in March 2005.

Box 10.2

**Approved and associated assessments within the MEA process
(Source: Millennium Ecosystem Assessment,
http://www.millenniumassessment.org/en/Multiscale.aspx)**

Approved assessments

- British Columbia, Canada
- Integrated Assessment of the Salar de Atacama, Chile
- Ecosystem Management and Social-Ecological Resilience in Kristianstads Vattenrike and River Helgeå catchment
- Stockholm Urban Assessment
- Coastal, Small Island, and Coral Reef Ecosystems in Papua New Guinea
- India Local Ecosystem Assessments: Karnataka and Maharashtra of Western Ghats
- Ecosystems and People: the Philippines MA
- Alternatives to Slash-and-Burn (ASB) Crosscutting Sub-Global Assessment
- Norwegian Millennium Ecosystem Assessment, Pilot Study
- Integrated Ecosystem Assessment of Western China
- Southern African Sub Global Assessment (SAfMA)
- Downstream Mekong River Wetlands Ecosystem Assessment, Vietnam
- Portugal Millennium Assessment
- Assessment of the Northern Range of Trinidad, Trinidad & Tobago
- Assessment of the Caribbean Sea (CARSEA)
- Altai-Sayan Ecoregion
- Vilcanota Sub-Region, Peru
- Local Ecosystem Assessment of the Higher and Middle Chirripo River Sub-Basin, Cabecar Indigenous Territory of Chirripo, Costa Rica

Associated assessments

- Arab Region Millennium Ecosystem Assessment: Supporting Decision Making for the Sustainable Use of Ecosystems
- Biodiversity, Local Knowledge, and Poverty Alleviation: Sinai Sub-Global Assessment
- Arafura and Timor Seas Sub-Global Assessment
- Indonesia Sub-Global Assessment
- São Paulo City Green Belt Biosphere Reserve Assessment
- Ecological Function Assessment of Biodiversity in the Colombian Andean Coffee-Growing Region
- Assessment of the Central Asian Mountain Ecosystems
- The Great Asian Mountains Assessment (GAMA)
- The Upstream Region MA of the Great Rivers, Northwest Yunnan, China
- Fiji Sub-Global Assessment
- Environmental Service Assessment in Hindu-Kush Himalayas Region – Tradeoffs and Incentives
- Indian Urban Assessment with focus on Western Ghats

of scale and knowledge systems on the relationship between ecosystems and human well-being. Local assessments showed the importance of key relationships between ecosystem services and drivers of ecosystem change that were often not seen at global scales, especially those related to cultural services. The conceptual framework developed for the MEA was useful for the sub-global assessments, although it required modifications to take particular local circumstances into account. However, assessments that were led primarily by indigenous communities developed alternate frameworks, which, they felt, reflected better the communities' views on the relationship between people and ecosystems. In particular, these alternate frameworks placed more emphasis on cultural services and the spiritual aspects of human well-being.

Because the MEA focused on the linkages between ecosystems and human well-being, its conclusions had a somewhat different emphasis than other assessment approaches that focus on "single issues," such as climate change. For example, while the MEA does confirm that major problems exist with tropical forests and coral reefs, from the standpoint of linkages between *ecosystems and people*, the most significant challenges involve dryland ecosystems. These ecosystems are particularly fragile, but they are also the places where human population is growing most rapidly, biological productivity is least, and poverty is highest.

The MEA also documented areas where more information is required. It was noted that at a local and national scale, relatively limited information exists about the status of many ecosystem services and even less information is available about the economic value of non-marketed services. Moreover, the costs of the depletion of these services are rarely tracked in national economic accounts. Basic global data on the extent and trends in different types of ecosystems and land use are surprisingly scarce. Furthermore, the MEA concluded that models used to project future environmental and economic conditions have limited capability for incorporating ecological "feedbacks," including non-linear changes in ecosystems, or behavioral feedbacks such as learning that may take place through adaptive management of ecosystems.

10.5 The vulnerability and syndrome approaches

As discussed in the previous section, one advance in recent years has been the move from single-sector analyses of the impact of climate change to the analysis of vulnerability to global change.

Vulnerability assessment is described in detail by Leary and Beresford (Chapter 6, this volume). In studies of vulnerability over the last few decades at least two main strands of research can be distinguished. The first concentrated on the field of natural hazards research, looking at human vulnerability related to physical threats and disaster reduction (e.g. Cutter 1996 or World Bank 2005). It has focused on vulnerability in relation to environmental threats, such as flooding, droughts, or earthquakes. Vulnerability to these extreme events depends on their likelihood and the place where they occur. In the face of global environmental change it is not only the occurrence that matters but also change in frequency and magnitude (e.g. changes in flood frequency and magnitude), which can change drastically as a consequence. This field also examines the environmental threat posed by the slower, long-term process of climate change. Most research has resulted in analyzing the dynamics in hazardous areas and impacts that occurred.

The second strand of research looked at socio-economic factors in relation to human vulnerability (e.g. Adger and Kelly 1999 or Watts and Bohle 1993). It has shown that in the face of (non-)environmental threats, socio-economic factors are equally important. Exposure to the threats is to a large extent determined by socio-economic factors, as is the ability to cope with those threats. This has been shown in many cases, where exposed to the same threats, impacts varied enormously for different communities and people. Poverty, conflict and lack of entitlements are some of the principle determinants.

In recent years, these two strands of research have been combined in a number of studies, in recognition of the fact that both aspects, namely, natural hazards and environmental changes and socio-economic factors, together determine human vulnerability to environmental change. This emerging, more comprehensive approach looks at multiple stresses from different domains and in this way comes closer to the concept of sustainable development, which requires integrating the economic, environmental, and social dimensions within one framework.

Another approach to looking at interactions between humans and the environment in a particular place is the "syndrome approach," which was introduced in the 1990s to obtain a global overview of current non-sustainable dynamics and mechanisms of global change (WBGU 1996; Schellnhuber *et al.* 1997).

Syndromes focus on the better understanding of *non-sustainability* while taking into account that there is a close interaction between global environmental change and rapid developments in the socio-economic sphere. Thus, the syndrome approach deals with the following problems:

- a multitude of non-sustainable cases of human–environment interaction has to be covered to cope with the high degree of interconnectedness of global change;
- the approach should not rely on the paradigm of only one of the contributing sciences;
- different kinds of knowledge collected in different disciplines have to be combined, in particular more qualitative-oriented knowledge from social and political sciences with more quantitative knowledge from the natural sciences.

The approach is based on the hypothesis that it is possible to identify a limited number of *typical* dynamic cause–effect patterns (syndromes). These should be general enough that each relevant observed case of problematic (i.e., non-sustainable) human–environment interaction could be subsumed under one syndrome. So the multitude of observed problematic cases (from Canadian cod overfishing to deforestation in Kalimantan to groundwater pollution with nitrite in Europe) is reduced to *several* typical cause–effect patterns without striving for a disciplinary first principle explanation. Two complementary methods were developed to identify and verify these typical cause–effect patterns:

- a systematic procedure for the inductive syndrome identification in multi-disciplinary expert groups (WBGU 1995);
- indicator based identification of the spatial distribution of the hypothesized syndromes, opening falsification possibilities (Lüdeke *et al.* 2004).

The vulnerability and syndrome approaches have been combined in UNEP's Fourth Global Environmental Outlook (Kok *et al.* 2006). In GEO-4 "archetypes of vulnerability" are defined as specific representative patterns of the interactions between environmental change and human well-being. That is, similar to the syndromes approach, typical dynamic patterns are identified. For example, a "dryland archetype" is identified – a typical pattern of human–environment interaction found in many arid and semiarid regions of the world.

The archetypes of vulnerability (seven patterns are described in GEO-4) illustrate the basic processes whereby vulnerability is produced. The archetypes are simplifications of real cases, in order to show the basic processes whereby vulnerability is produced within a context of multiple stressors. This may allow policy makers to recognize their particular situations within a broader context – providing regional perspectives and important connections between regions and the global context and insights into possible solutions. The archetypes are also not mutually exclusive – in some ecosystems, countries, sub-regions, regions, and globally, a mosaic of these selected (and other) patterns of vulnerability may exist. This makes policy responses a complex challenge. By analyzing the vulnerability of human–environment systems to multiple stresses (drivers and pressures), challenges and opportunities within and beyond the environmental policy domain are identified.

10.6 Discussion

The scientific need for integrated regional assessment has been recognized by the global change research community and by the international assessment programs that integrate knowledge to make it useful for the policy process. As a result, integrated assessments for regions such as monsoon Asia have started. Other integrated assessments at the regional level, such as the Arctic Climate Impact Assessment,[8] show the value of taking a regional approach to discussing interactions between humans and the environment. The international processes have moved from a focus on single-sector impacts of environmental change to a consideration of vulnerability to multiple stresses and of ecosystem services and human well-being. These foci necessitate a regional approach, because vulnerability unfolds within the differing contexts of different places.

The following chapters illustrate advances made in the methods and tools available for integrated regional assessments. The results of these tools are also used in international assessment processes, such as the IPCC, MA, and GEO.

One further issue should be mentioned, however, in a chapter on the global context of integrated regional assessment and that is "capacity building." Within the global change research community START is responsible for capacity building. In addition to its role in MAIRS, described in Section 10.3 of this chapter, START has supported capacity building in integrated regional assessment through the project on Assessment of Impacts and Adaptation to Climate Change[9] and through a training institute on vulnerability.[10] Within the assessment community, significant efforts are made by UNEP to provide capacity building related to the Global Environmental Outlook process (GEO). To this end the *GEO Resource Book*[11] was completed in 2007. It builds on advances in the science and practice of integrated environmental assessment and reporting (IEA) based on UNEP's GEO-4, and the rich experience with IEA capacity building using the earlier GEO training manual. The modules in the resource book provide a comprehensive guide to the process of carrying out an integrated assessment at sub-global level.

References

Adger, W. N. and P. M. Kelly 1999. Social vulnerability to climate change and the architecture of entitlement. *Mitigation and Adaptation Strategies for Global Change*, **4**, 253–266.

Brazil, Ministry of Science and Technology 2005. *The Large Scale Biosphere–Atmosphere Experiment in Amazonia.* http://lba.cptec.inpe.br/lba/index.php?p=1&lg=eng#

Carter, T. R., M. L. Parry, H. Harasawa, and S. Nishioka 1994. *IPCC Technical Guidelines for Assessing Climate Change Impacts and Adaptations.* London: Department of Geography, University College London and Tokyo: Center for Global Environmental Research, National Institute for Environmental Studies, p. 59.

---

[8] http://www.acia.uaf.edu/
[9] http://www.start.org/Program/AIACC.html
[10] http://www.start.org/Program/advanced_institutes_3.html
[11] http://www.iisd.org/measure/learning/assessment/training.asp

Cutter, S. L. 1996. Vulnerability to environmental hazards. *Progress in Human Geography*, **20**(4), 529–539.

Darwin, R., M. Tsigas, J. Lewandrowski, and A. Raneses 1995. *World Agriculture and Climate Change: Economic Adaptations. Agricultural Economic Report* No. 703. Washington, U.S. Department of Agriculture, Washington, DC, USA, 86 pp.

Downing, T. 2000. *Climate Change Vulnerability: Linking impacts and adaptation.* United Nations Environment Programme and Oxford: Environmental Change Institute.

Fu, C. B. 1997. Concept of "General Monsoon System," an Earth system science view on Asian Monsoon. *Proceedings of the International Workshop on Regional Climate Modeling of the General Monsoon System in Asia.* Beijing, China, pp. 1–6.

Fu, C. B. and F. P. de Vries 2006. *Initial Science Plan of the Monsoon Asia Integrated Regional Study.* Beijing: MAIRS International Project Office.

Fu, C. B., H. Harasawa, V. Kasyanov, *et al.* 2002. Regional–global interactions in east Asia. In P. Tyson (ed.), *Global–Regional Linkages in the Earth System.* Berlin: Springer-Verlag.

Kok, M. T. J., V. Narain, S. Wonink, and J. Jäger 2006. Human vulnerability to environmental change: an approach for UNEP's Global Environmental Outlook (GEO). In J. Birkmann (ed.), *Measuring Vulnerability to Natural Hazards: Towards Disaster Resilient Societies.* Tokyo: United Nations University Press.

Lüdeke, M. K. B., G. Petschel-Held, and H. J. Schellnhuber 2004. Syndromes of global change: the first panoramic view. *GAIA*, **13**(1), 42–49.

Nobre, C. A., P. Artaxo, M. A. F. Silva Diaz, *et al.* 2002. The Amazon Basin and land-cover change: a future in the balance? In W. Steffen, J. Jäger, D. Carson, and C. Bradshaw (eds.), *Challenges of a Changing Earth: Proceedings of the Global Change Open Science Conference.* Berlin: Springer-Verlag, pp. 137–141.

Schellnhuber, H.-J., A. Block, M. Cassel-Gintz, *et al.* 1997. Syndromes of Global Change. *GAIA*, **6**(1), 19–34.

Steffen, W., A. Sanderson, P. D. Tyson, *et al.* 2004. *Global Change and the Earth System.* Berlin: Springer-Verlag.

UNESCO-SCOPE 2007. *Human Alteration of the Nitrogen Cycle: Threats, Benefits and Opportunities, Unesco Policy Briefs* No. 4. Paris: UNESCO-SCOPE.

Watson, R. T., M. C. Zinyowera, R. H. Moss, and D. J. Dokken 1997. *The Regional Impacts of Climate Change: an Assessment of Vulnerability.* Cambridge: Cambridge University Press.

Watts, M. J. and H. G. Bohle 1993. The space of vulnerability: the causal structure of hunger and famine. *Progress in Human Geography*, **17**(1) 43–67.

WGBU 1996. *World in Transition: Ways Towards Global Environmental Solutions. Annual Report 1995.* German Advisory Council on Global Change. Berlin: Springer-Verlag.

World Bank 2005. *Natural Disaster Hotspots: a Global Risk Analysis.* Washington, DC: The World Bank.

11

The Asia–Pacific integrated model

MIKIKO KAINUMA, YUZURU MATSUOKA, TSUNEYUKI MORITA,*

AND KIYOSHI TAKAHASHI

11.1 Introduction

It is predicted that climate change will have a significant impact on society and economy in the Asia–Pacific region, and that adoption of measures to tackle climate change will force the region to carry a very large economic burden. Also, if the Asia–Pacific region fails to adopt such countermeasures, it has been estimated that its greenhouse gas (GHG) emissions will increase to become half of all global emissions by 2100.

The climate change issue has been recognized as one of the most important policy issues for the region's development. To tackle this issue Japan's role in the region has increased through greater contributions in official development assistance (ODA), technology transfer, research, and implementation of the clean development mechanism (Masui and Kobayashi 2000).

In order to promote the adoption of countermeasures, it is necessary to project as precisely as possible the greenhouse gas emissions in the region and also the impacts of climate change. The effects of countermeasures on emission reductions and impact abatement also need to be estimated, taking into account international cooperative efforts. Such projections and analyses require an integrated simulation model for the region (Morita 1997).

The Asia–Pacific Integrated Model (AIM) was developed by an Asian collaborative project team composed of the National Institute for Environmental Studies (NIES) and Kyoto University in Japan, and a number of research institutes in China, India, Korea, Thailand, Indonesia, and Malaysia. It estimates the emission and absorption of greenhouse gases in the Asia–Pacific region and the impact that they have on the natural environment, society, and the economy. The model is expected to evaluate climate change and contribute to policy making.

---

* Unfortunately, Professor Tsuneyuki Morita (1950–2003) did not live to see the publication of this chapter.

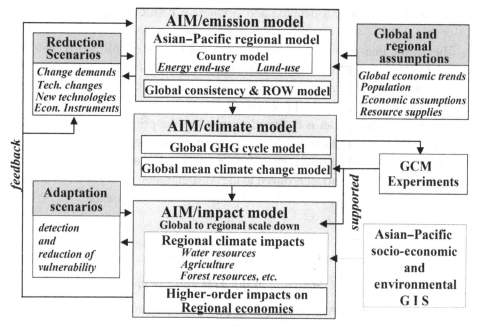

Figure 11.1. Structure of the AIM model.

11.2 Outline of AIM

AIM comprises three models: the AIM/emission model for projecting greenhouse gas emissions, the AIM/climate model for estimating global and regional climate change, and the AIM/impact model for estimating the impacts of climate change. Figure 11.1 shows the relationships between these models (Kainuma *et al.* 2002).

The AIM/emission model comprises models of social and economic activities that drive greenhouse gas emissions through energy consumption, changing land use, and agricultural and industrial production. At its heart lies the energy model, which is made up of a world model as well as country-specific models for the Asia–Pacific region. The world model is a top-down model that uses economic indices based on prices and elasticity to express the connection between energy consumption and production. The country-specific model is a bottom-up end-use model that focuses on the activities of the people who deal with industrial production and the consumption of energy as well as the change in technologies used in these countries. The model provides forecasts of the total energy consumption and production. For a long-term forecast, a top-down world model based on market equilibrium that forecasts world economic activity and change is indispensable. Likewise, to explain the exact direction of policy and its effects to policy makers and politicians, a detailed and persuasive bottom-up end-use model is essential.

The AIM/climate model is designed to link to other established models for calculating global and regional climate change. It comprises a carbon cycle module, a greenhouse gases concentration module, a radiative forcing module, a temperature change module, and a sea-level rise module.

The AIM/impact model is designed to calculate the primary impacts on water supply, agricultural production, forest vegetation, human health, and other aspects of society, and to make projections of higher-order impacts on the regional economy. The AIM/impact model is linked to the AIM/emission model through outputs of general circulation model (GCM) experiments and AIM/climate.

11.3 The AIM emission model

11.3.1 Structure of the AIM/end-use model

To reduce CO_2 emissions in each country, it is very important to identify what types of energy conservation technologies will be used to which extent, necessitating the development of an end-use energy-technology model. The AIM/end-use model has been developed for this purpose and can calculate the changes in energy consumption from technological substitution caused by changes in energy prices, in a bottom-up manner. Thus, it is possible to evaluate not only the efficiency of each individual policy, but also the effect when various policies are combined. By linking the technology selection model with the energy demand model, it is possible to estimate energy efficiency improvement based on the actual situation for each technology (Masui *et al.* 2006).

As shown in Figure 11.2, the AIM/end-use model first estimates energy service demands based on socio-economic factors, such as population, economic growth, industrial structure, and lifestyle, and then calculates what kind of technology will be used and to what extent (Kainuma *et al.* 1997). To compare and consider energy technologies, detailed technological data and energy data are prepared. Once the kind of energy technology to be used is known, the model calculates the energy necessary to provide the energy services and the amount of CO_2 emissions produced under each type of energy technology.

The key factor to mitigating CO_2 emissions is the extent to which energy-saving technology can be introduced at the end-use point. AIM focuses on the fact that substitution technology will be available according to energy price and estimates energy efficiency and energy consumption on the basis of each technology. Therefore, it is possible to evaluate the effectiveness of each policy or combine various policies. More than 400 technologies are evaluated in the Japanese model.

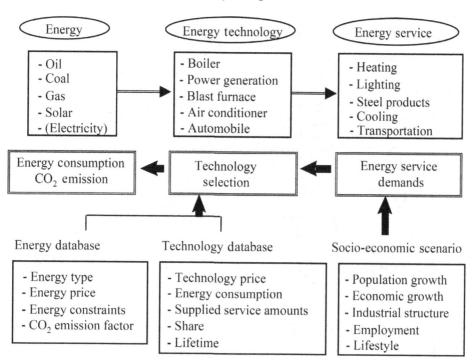

Figure 11.2. Structure of the AIM/end-use model.

An example of the end-use model is shown in Figure 11.3. This figure shows the structure of the Japanese steel industry model. This model considers three types of steel making processes: the blast furnace process, the electric arc furnace process, and the smelting reduction process. The blast furnace and the electric arc furnace are currently in operation. The smelting reduction method is an innovative iron manufacturing process. In the steel industry, the end-uses are the steel products. These products are produced with several types of technologies using external energy. Numerous types of internal services/energy are also produced in the steel-making process. There are several alternative technologies in each process. Which technologies should be used – conventional technology or energy-saving – is determined based on costs. This model has been applied to estimate CO_2 emissions and mitigation potentials in several countries in Asia such as Japan, China, India, and Korea (Kainuma *et al.* 2002).

11.3.2 *Case studies of CO_2 emission reduction possibilities in Japan*

According to the Kyoto Protocol, Japan needs to reduce its emissions by 6% from the 1990 level. The AIM/end-use model was applied to analyze the possibilities for achieving this target (Kainuma *et al.* 2000).

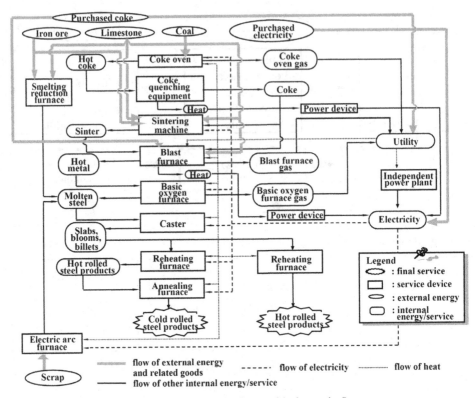

Figure 11.3. Structure in the steel industry in Japan.

Several assumptions were used for the projection of Japanese CO_2 emissions. Assumptions on socio-economic scenarios are listed in Table 11.1. Table 11.2 shows scenarios on value added used in this model and Table 11.3 shows scenarios on industrial products. Four cases were studied for the estimation of CO_2 emissions with a timeframe for simulation from 1990–2010.

Frozen case. The technology share is frozen at 1998 levels; only energy service demand is varied.

Market case. New technologies are introduced based on cost through a market mechanism. The payback period for new technologies was assumed to be less than 3 years.

Carbon tax case. A tax of 30 000 Yen/ton of carbon is introduced from the year 2000.

Carbon tax plus subsidy case. A tax of 3000 Yen/ton of carbon is introduced and the revenue from this is used for subsidizing new technologies.

In the frozen case, CO_2 emissions in 2010 are 18% higher than the 1990 level on the whole, though emissions in different sectors show variance. Emissions from the industry sector rise by 4%, while those from the residential sector rise by 31%.

Table 11.1. *Socio-economic assumptions in Japan.*

GDP growth rate	1999, 0.6%; 2000, 1.0%; 2001–2010, 2%
Oil price	2010, 30$/barrel (real)
Exchange rate	120 Yen/$
Nuclear power generation	56.2 MkW

Table 11.2. *Scenarios on value added.*

	(billion Yen)			
	1990	1995	2000	2010
Total	429 860	465 037	483 360	589 213
Total (excl. imputed interests, import tax, etc.)	451 938	486 724	505 901	616 691
Agriculture, forestry, fishing	10 921	9 653	10 249	10 293
Mining	1 122	861	909	791
Construction	43 428	44 781	47 731	58 721
Food	12 322	12 844	11 130	11 191
Textiles	2 514	2 148	2 214	1 998
Pulp and paper	3 366	3 122	3 287	3 530
Chemicals	9 375	11 036	15 418	19 293
Oil and coal products	4 143	3 735	4 584	4 181
Ceramic	4 382	4 303	4 596	5 311
Steel	7 082	7 373	7 360	7 775
Non-ferrous metals	2 384	2 300	2 210	2 437
Metal and machinery	56 470	62 937	62 347	82 211
Other manufacturing	19 181	16 657	18 901	21 317
Others (service)	275 249	304 875	314 973	387 640

Table 11.3. *Scenarios on industrial products.*

	(thousand tons)			
	1990	1995	2000	2010
Steel products	111 710	100 023	90 979	82 280
Cement products	86 849	91 499	89 319	85 810
Ethylene products	5 810	6 944	7 076	7 005
Paper products	28 086	29 659	30 631	35 339

Similarly, the commercial sector emissions increase by 30%, the transportation sector emissions increase by 38%, and emissions in the energy conversion sector are 16% higher. The market case has a 10% increase in CO_2 emissions in 2010. In this case, the industry sector decreases emissions by 3%, the residential sector increases by 17%, the commercial rises by 18%, while the transportation sector is 37% higher, and the energy conversion sector is 9% higher than the respective 1990 levels.

Since emissions increase in both of the above cases, policies must be introduced in order to meet the Kyoto target. In the carbon tax case, the industry and residential sectors experience a decrease in emissions of 11% and 13%, respectively, while there is no change in emissions in the commercial sector. On the other hand, the transportation sector emissions increase by 22% and the energy conversion sector emissions decrease by 5%. The overall effect of the carbon tax case is a 3% decrease in emissions by 2010. In the last case, that is the carbon tax plus subsidy case, emissions decrease by 10% in the industrial sector, and by 12% in the residential sector, increase by 1% in the commercial sector, and by 22% in the transportation sector and decrease by 5% in the energy conversion sector. The overall emission reduction achieved in this case is 2%. The effect in this case is similar to that of the pure tax case although the tax level is 10 times lower.

As it can be seen, the above policy simulations show a maximum decrease of emissions by 3%, which means that in order to achieve the 6% reduction required by the Kyoto Protocol, some drastic policy initiatives are needed.

11.3.3 Analysis of the CO_2 emission reduction potentials in China

In the Kyoto Protocol, there are no emission targets for developing countries. However, it is recognized that emissions in the developing countries are growing rapidly. The model was used to estimate greenhouse gas emissions in Asian countries, especially in China, India, and Korea, and to analyze various policy scenarios to reduce emissions. The following is a brief description of some of the analyses in China (Jiang et al. 1998, 1999; Hu et al. 2002; Jiang and Hu 2006).

CO_2 emissions projections were made for China for three scenarios: frozen, market, and policy cases. The CO_2 emissions in China grow by more than 2.5 GtC in the frozen case and by 2 GtC in the policy case by 2030. About a 13% reduction is calculated in the policy case compared to the market case. The reduction potential is high in the steel making industry, the rural residential sector, and the urban residential sector.

11.3.4 AIM/top-down model

A general equilibrium model has also been developed to estimate costs of reducing CO_2 emissions as well as secondary effects on production through

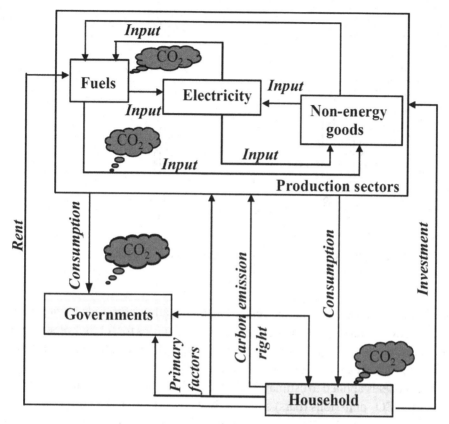

Figure 11.4. Structure of the world economic model.

international markets (Kainuma *et al.* 1999). Figure 11.4 shows the structure of the AIM/top-down model. The model has three sectors – production, household, and government – in each region. CO_2 and other greenhouse gases are emitted by each of these sectors. The production of electricity and of non-energy goods uses fossil fuels in the production sector, and the use of automobiles and other direct uses of fossil fuels emit CO_2 in the household and government sectors. It is assumed that the household sector has carbon emission rights and distributes them to the other sectors and within the household sector itself. Fossil fuels cannot be used without carbon rights. The price of carbon rights depends on several factors such as emission targets and the method of emission trading. The household sector also supplies primary factors to the production and government sectors. An agent in the household sector determines consumption and saving. The marginal propensity to save is a calibrated function of a weighted aggregate of regional and global rates of return on fixed capital. A regional investment is calculated with the gross domestic product (GDP) growth rate and regional and global rates of return. Investment is

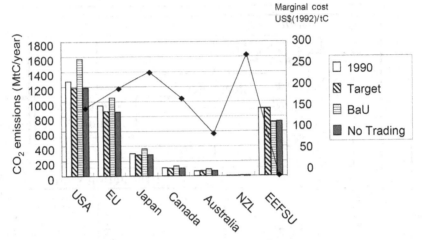

Figure 11.5. CO_2 emissions and marginal costs in 2010 (no-trading case).

balanced with saving on a global scale. The model allows for trade in intermediate goods. AIM assumes identical preferences in all countries for foreign versus domestic goods, i.e., the elasticity of substitution is the same for all regions. Domestic and export goods are not perfect substitutes.

The model is applied to estimate the effects of post-Kyoto scenarios. Figure 11.5 shows the CO_2 emissions (bars) and marginal costs (line) in the no-trading case in 2010. The left-hand bar for each region shows the 1990 emission level, the second bar shows the target emission level, the third bar shows the business-as-usual (BaU) emission level, and the right-hand bar shows the emission level in the no-trading case. The BaU emission of the former Soviet Union is less than the target, reflecting the economic deterioration of the region. The difference is the so-called "hot-air."

The line graph shows the marginal costs to achieve the emission targets. The emissions of the former Soviet Union are below the 1990 level until 2030 in the BaU case, so no policy intervention is necessary in 2010. On the other hand, CO_2 emissions in New Zealand in the reference case are 11 MtC in 2010 and New Zealand has to reduce CO_2 emissions by a large amount compared to the 1990 level, so the marginal cost is the highest for New Zealand followed by Japan, the EU, the USA, and Australia.

The changes in energy trade versus GDP caused by the Kyoto Protocol in 2010 were also calculated. The results showed that the reduction in imports was the highest in Europe followed by the USA. The change in exports was the highest in the Middle East. The economic impact on the USA was the highest followed by Europe.

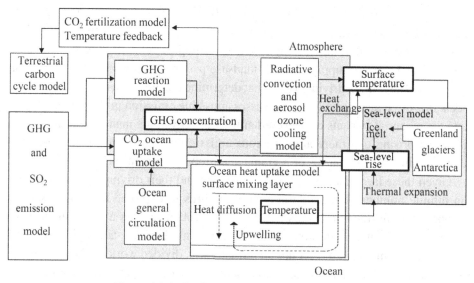

Figure 11.6. Outline of the AIM/climate model.

11.4 AIM/climate model

Several types of climate change model were developed to represent the CO_2 and heat absorption processes of the ocean, and the resulting sea-level rise. The basic structure of the model is shown in Figure 11.6.

The model begins with greenhouse gas emissions, which are input into a GHG concentration model. The GHG concentration model comprises several modules that represent the hemispherical and altitudinal characteristics of the gases in the atmosphere. Carbon dioxide is assumed not to decay, but is absorbed by the ocean and terrestrial ecosystems. A simple upwelling diffusion model, an advective-diffusion model or convolution approximations of ocean general circulation model (OGCM) experiments estimate ocean absorption. As for other GHGs, the decaying processes are modeled with the first-order reactions. The kinetic coefficients in the equations were calibrated by comparison with the outputs of more complex, but realistic models. Also the effect of moisture in the stratosphere is calculated using methane concentration. The cooling effect of decreased lower stratospheric ozone is calculated with GHG concentrations, and the cooling effect of aerosol sulfates is calculated with sulfur dioxide emissions. Sea-level rise is calculated from the expansion of seawater due to the temperature increase, the melting of the continental glaciers, and changes in the ice sheets of Greenland and Antarctica.

The model is applied to estimate climate change based on 259 published emissions scenarios including the Intergovernmental Panel on Climate Change (IPCC) emissions scenarios (Morita 1998; IPCC 2000a, b). Emissions scenarios provide an important input for the assessment of future climate change. Future GHG emissions depend on numerous driving forces including population growth, economic development, energy supply and use, land-use patterns, and a host of other human activities. These main driving forces that determine the emissions trajectories in the scenarios often also provide input to assess possible emission mitigation strategies and the potential impacts of unabated emissions. Given the many different uses, it is not surprising that there are numerous emission scenarios in the literature and that the number of scenarios of regional and global emissions is growing.

Given these large ranges of future emissions and their driving forces, there are an infinite number of possible alternative futures to explore. The IPCC Special Report on Emissions Scenarios (SRES; IPCC 2000a) covers a finite, albeit very wide, range of future emissions. Four scenario "families" are adopted to describe future developments: rapid economic growth (A1), divided world (A2), sustainable development (B1), and regional stewardship (B2). The temperature is estimated to increase between 0.8 and 2.1 °C by 2050 and between 1.3 and 4.7 °C by 2100. The corresponding sea-level rise is estimated to be between 6 and 45 cm by 2050 and 13–97 cm by 2100.

11.5 AIM/impact model

The AIM/impact model estimates the impacts of climate change in the Asia–Pacific region. AIM/impact comprises a water balance model, a vegetation change model, a health impact model, an agriculture production model, and an economic evaluation model for climate change damage.

The output of the AIM/emission model is entered into the AIM/climate model in order to project future changes in global mean temperature. The projected global mean temperature, the observed climate data with high spatial resolution, and the spatial pattern of climate change provided by GCM experiments are combined to produce climate change scenarios, which are used as the basic assumptions for the AIM/impact model. The structure of the impact model is shown in Figure 11.7. It calculates the primary impacts on water supply, agricultural production, forest vegetation, human health, and other aspects of society, and makes predictions of higher-order impacts on the regional economy.

11.5.1 Impact on water resources

Hydrological impacts are one of the most important aspects of climate change. Changes in the magnitude, frequency, and duration of hydrological factors

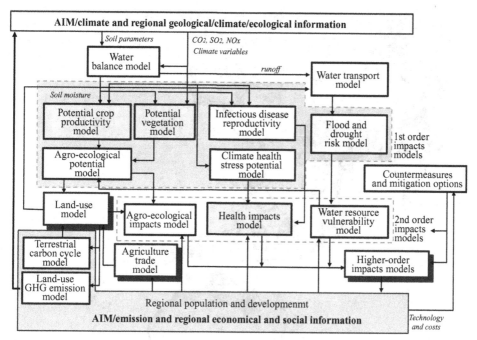

Figure 11.7. Structure of the AIM/impact model.

influence the availability of water resources, flooding intensity, as well as agricultural and natural terrestrial ecosystems. A rainfall–runoff process sub-module was developed as one of the basic sub-modules of the AIM/impact model. This sub-module consists of water balance and water transport components, and is intended to provide basic hydrological information to the impact models of other sectors. Specifically, it creates gridded high-resolution datasets of surface runoff, soil moisture, evapotranspiration, and river discharge.

The major parameters of the hydrological model are elevation, soils, and vegetation, as well as precipitation, temperature, and potential evapotranspiration. Soil and vegetation characteristics were set, in addition to elevation conditions at their current value.

The water balance component of the model is based primarily on the models of Thornthwaite and Mather (1955) and their successors. The water balance among precipitation, snowmelt, evapotranspiration, and surface runoff is calculated for each grid cell in the simulation region. A number of climatological and geographical data sets were prepared from various sources. Soil moisture capacities were estimated using current vegetation classes and soil textures (Vorosmarty *et al.* 1989; Webb and Rosenzweig 1993).

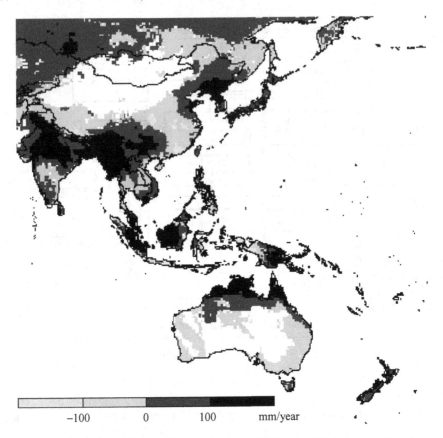

Figure 11.8. Changes in runoff calculated by AIM/impact based on the transient experiments of the CCSR/NIES model.

In the water transport component, the network topology of streams was determined from digital elevation data and modified with various hydrological maps of the analyzed regions. Modeling of surface water retention time in each cell followed Vorosmarty *et al.* (1989).

Figure 11.8 shows the changes in runoff calculated by the AIM/impact model based on the results of the transient experiments of the Center for Climate System Research (CCSR), Tokyo University/NIES climate models. Atmospheric CO_2 concentrations were assumed to increase at an annual rate of 1%. The difference between the mean runoff for the 10 years from 2050 to 2059 and that from 1980 to 1989 were calculated. Both increases and decreases in runoff were found, depending on the region. For example, increases in runoff in southwestern China and the Indian sub-continent were observed. Decreases were observed in central China and the Tibetan plateau.

11.5.2 Impact on agriculture

The productivity of crop land is strongly controlled by environmental changes. For example, future climate change should have a profound impact on the potential yields of crops, and as a result, influence the distribution of cropping patterns in the Asia–Pacific area.

In order to evaluate the impact of climate change on agriculture, a study was made based on a framework that can evaluate related direct and indirect effects. The basic assumptions are: (1) climate change will directly affect land and water resources (primary effect), and (2) changes in land and water resources will affect economic activities (secondary effect).

For a preliminary assessment of impacts on agriculture, a potential crop production model coupled with regional climate and soil environment data was developed. The method of potential agricultural productivity estimation is based on that used by the Food and Agriculture Organization (FAO 1978–1981). Days suitable for crop cultivation (growing period) are counted using climate data, and the crop growth during the growing period is simulated biophysically according to the growth characteristic parameters of each crop. The direct impact of CO_2 concentration on crop growth (CO_2 fertilization) is not considered in this study.

Rice and wheat are important staple crops throughout much of Asia so changes in their future productivity were calculated for the medium socio-economic scenario (IS92a) and medium climatic sensitivity (2.5 °C). For current climate, data were taken from the monthly average data sets of Legates and Willmott (1989) with 0.5° grid cells. For the future (2100) climate pattern, the equilibrium GCM output of the Canadian Climate Center was interpolated into 0.5° grid cells, adjusted to have the globally averaged temperature estimated under the assumed scenario and overlaid with the data from Legates and Willmott (1989). Soil characteristics data were taken from the *Soil Map of the World* (FAO/UNESCO 1994) with 5 minute grid cells. For rice, the northern portion of Asia can expect an increase in production, while the reverse is expected in central China and northern India.

11.5.3 Impacts on malaria

Air and water pollution, as well as solid and hazardous wastes, affect human health directly. Global climate changes will also affect human health in the future in many ways. For example, climate change will result in increasing temperatures and changes in vegetation close to the ground. This will allow the habitat of the Anopheles mosquito, which is the malaria vector, to expand. In addition, the development period of the malaria protozoan will shorten and its reproductive potential will increase. As a result, it is predicted that the global malaria risk will increase. In order to estimate the risk quantitatively a sub-model has been developed within AIM/impact (Matsuoka *et al.* 1995).

The major components of the malaria sub-model are the relationship between sporogony and temperature, and the eco-climatic index model which shows the climatic response of vectors. These components are complemented by the soil moisture sub-module and outputs from the GCM experiments. The primary climatic variables of this framework are surface temperature and precipitation distributed spatially and temporally for both the current situation and those expected under $2\times CO_2$ climate conditions.

The malaria potential was calculated for current and expected future climate conditions. It was concluded that the population expected to be at risk of malaria infection will increase by 30% in the Asia–Pacific region.

11.6 Concluding remarks

The AIM model has been applied to the assessments of emissions scenarios, climate change scenarios, climate impact scenarios, mitigation costs, and new policy integration in order to respond to the research needs of international organizations, national governments, and non-governmental organizations. These assessments include the IPCC emissions scenarios, mitigation scenarios for global climate stabilization, land-use related mitigation scenarios, implications and economic impacts of the Kyoto Protocol, the effect of the United Nations Framework Convention on Climate Change (UNFCCC) clean development mechanism in competition with emission trading, the potential of GHG reductions in Asian developing countries, and policy design for Asia–Pacific collaboration (Masui *et al.* 2000).

Further studies are needed to evaluate policy options to mitigate global climate change within the context of increasing economic activities and reducing local environmental problems (Matsuoka *et al.* 1999). If efforts are made to reduce CO_2 emissions even further than mandated by the Kyoto Protocol, the necessary costs will certainly become larger. These are the costs that industry and households will have to bear. On the other hand, there will be an increase in effective demand for producers of energy-saving technologies. There is a strong possibility that the indirect cost will become smaller through the activities of environmental industries. The business opportunities for environmental industries will also increase as projects to reduce CO_2 emissions jointly with developing countries are implemented. Earlier investment in environmental industries will reduce the total cost in the long term (Masui 2005).

Acknowledgements

This research is based on the results of studies entitled "Integrated Analysis of Mitigation and Adaptation Measures to Global Warming with Asia–Pacific Integrated

Model (B-52)" and "Comprehensive Assessment of Climate Change Impacts to Determine the Dangerous Level of Global Warming and to Determine the Appropriate Stabilization Target for Atmospheric GHG Concentration (S-4)," which are research topics of the Global Environmental Research Fund (GERF) of the Ministry of the Environment, Japan.

References

FAO 1978–1981. *Report on the Agro-Ecological Zones Project*, vols. 1–4, *World Soil Resource Report* 48. Rome: Food and Agriculture Organization of the United Nations.

FAO/UNESCO 1994. *Digital Soil Map of the World and Derived Soil Properties*. Rome: Food and Agriculture Organization of the United Nations.

Hu, X., K. Jiang, and H. Yang 2002. *AIM/Local Model System and Emission Inventory in 2000 in China, 7th AIM International Workshop*, Tsukuba, March 2002.

IPCC 2000a. *Special Report on Emissions Scenarios*. Cambridge: Cambridge University Press.

IPCC 2000b. *The IPCC Data Distribution Centre – Access to the GCM Archive*. http://ipcc-ddc.cru.uea.ac.uk/dkrz/dkrz_index.html

Jiang, K. and X. Hu. 2006. Energy demand and emissions in 2030 in China: scenarios and policy options. *Environmental Economics and Policy Studies*, **7**(3), 233–250.

Jiang, K., X. Hu, Y. Matsuoka, and T. Morita 1998. Energy technology changes and CO_2 emission scenarios in China. *Environmental Economics and Policy Studies*, **1**(2), 141–160.

Jiang, K., T. Masui, T. Morita, and Y. Matsuoka 1999. Long-term emission scenarios for China. *Environmental Economics and Policy Studies*, **2**(4), 267–287.

Kainuma, M., Y. Matsuoka, and T. Morita 1997. The AIM model and simulation, *Key Technology Policies to Reduce CO_2 emissions in Japan*. Tokyo: WWF Japan, pp. 39–57.

Kainuma, M., Y. Matsuoka, and T. Morita 1999. Analysis of post-Kyoto scenarios: the Asian–Pacific Integrated Model. *Special Issue of the Energy Journal*, **20**, 207–220.

Kainuma, M., Y. Matsuoka, and T. Morita 2000. The AIM/end-use model and its application to forecasting of Japanese carbon dioxide emissions. *Feature Issue of EJOR on Advances in Modeling: Paradigms, Methods and Applications*, **122**, 416–425.

Kainuma, M., Y. Matsuoka, and T. Morita (eds.) 2002. *Climate Policy Assessment*. Tokyo: Springer-Verlag.

Legates, D. R. and C. J. Willmott 1989. *Monthly average surface air temperature and precipitation: digital raster data on a 0.5° geographic 361 × 721 grid*. Boulder, CO: NCAR.

Masui, T. 2005. Policy evaluations under environmental constraints using a computable general equilibrium model. *European Journal of Operational Research*, **166**, 843–855.

Masui, T. and Y. Kobayashi 2000. Policy effects of the Clean Development Mechanism on global warming and air pollution abatement in China. *Proceedings of International Workshop on the Clean Development Mechanism*, Shonan, Japan.

Masui, T., G. Hibino, J. Fujino, Y. Matsuoka, and M. Kainuma 2006. Carbon dioxide reduction potential and economic impacts in Japan: application of AIM. *Environmental Economics and Policy Studies*, **7**(3), 271–284.

Matsuoka, Y., M. Kainuma, and T. Morita 1995. Scenario analysis of global warming using the Asian–Pacific Integrated Model (AIM). *Energy Policy*, **23**(4/5), 357–371.

Matsuoka, Y., M. Kainuma, and T. Morita 1999. Energy policy and reduction of air pollutant in Asia and Pacific region. *Report on 1999 Open Meeting of Human Dimensions of Global Environmental Change*, Shonan, Japan.

Morita, T. 1997. Several gaps between IAMs and developing countries. *Proceedings of the IPCC Asia–Pacific Workshop on Integrated Assessment Models*, CGER-I029-'97, pp. 125–138.

Morita, T. 1998. IPCC emission scenarios database. *Mitigation and Adaptation Strategies for Global Change*, **3**(2–4), 121–131.

Thornthwaite, C. W. and J. R. Mather 1955. The water balance. *Publications of the Climatology Laboratory of the Drexel Institute of Technology*, **8**(1), 1–104.

Vorosmarty C. J., B. Moore III, A. L. Grace, and M. P. Gildea 1989. Continental scale model of water and fluvial transport: an application to South America. *Global Biogeochemical Cycle*, **3**, 241–256.

Webb, R. S. and C. E. Rosenzweig 1993. Specifying land surface characteristics in general circulation models: soil profile data set and derived water-holding capacities. *Global Biogeochemical Cycles* **7**(1), 97–108.

12

Integrated regional assessment for South Asia: a case study

P. R. SHUKLA, AMIT GARG, AND SUBASH DHAR

12.1 Introduction

The South Asian (SA) region, comprising Afghanistan, Bangladesh, Bhutan, India, Maldives, Nepal, Pakistan, and Sri Lanka, is one of the most densely populated regions in the world. Comprising 3.5% of the total world area, the region holds about a quarter of the world's population. The region is characterized by diversity in geography, political and economic structures, and energy sources. The two northern arms of the Indian Ocean and vast coastlines, the Himalayan regions, the Thar Desert, and rich biodiversity provide the region with interesting environmental characteristics. The geographical diversity in the region has not only given rise to diverse climatic conditions but has also resulted in the availability of various energy resources, leading to a diverse energy mix in the region. In 2005, South Asia's commercial energy mix comprised of 44% coal, 34% petroleum, 13% natural gas, 8% hydroelectricity, and 1% nuclear (EIA 2006).

One of the major issues facing South Asian nations today is the rising energy demand essential to facilitate economic growth in the region. Dependence on fossil fuels to meet energy needs has led to rising emissions of greenhouse gases (GHGs). Also the exploitation of natural resources associated with rapid urbanization, industrialization, and economic development has led to increasing pollution, declining water quality, land degradation, and other environmental problems. The South Asian region also has several highly polluted mega-cities. Projected climate change impacts in the region include strengthening of the monsoon circulation, increases in surface temperature, sea-level rise, and increases in the magnitude and frequency of extreme rainfall events (IPCC 2007). These could result in major adverse impacts on the region's ecosystem and biodiversity; hydrology and water resources; agriculture, forestry and fisheries; mountains and coastal land; and human settlement and human health.

In this chapter, we review potential climate change impacts in South Asia.
Then we describe energy needs and cooperation in the region, showing through an
integrated modeling approach how the achievement of greenhouse gas mitigation
goals could be considerably enhanced through energy cooperation.

12.2 South Asia climate and climatic changes

South Asia's unique geography produces a spectrum of climates over the sub-
continent affording it a wealth of biological and cultural diversity. The diversity is
perhaps greater than any other area of similar size in the world. The Himalayan
region provides a northern-wall-like effect to the region with the northern plains
having a continental climate with fierce summer heat that alternates with cold
winters when temperatures plunge to freezing point. Western-central regions have
vast deserts, followed by highlands in the south. In contrast are the coastal regions
where the warmth is unvarying and the rains are frequent. The most important
feature of the regional climate is the season known as the monsoon, or the rainy
season. So significant is the monsoon season to the sub-continental climate that the
rest of the seasons are quite often referred to relative to the monsoon season. South
Asia is influenced by two seasons of rains accompanied by a seasonal reversal
of winds from January to July. During winters, dry and cold air blowing from a
northeasterly direction prevails over the Indian region. Owing to the intense heat
of the summer months the northern Indian landmass is heated up and draws moist
winds over the oceans causing a reversal of the winds over the region. This is called
the summer or the southwest monsoon. Changing global climate is projected to
alter the seasonal monsoon patterns (IPCC 2007).

Studies using a suite of state-of-the-art coupled atmosphere–ocean general circulation models (AOGCMs) and a regional climate model (RCM; HadRM2) for
projecting regional characteristics of rainfall and temperature over South Asia under
scenarios of increased greenhouse gas concentrations indicate marked increases in
both rainfall and temperature into the twenty-first century, particularly after the
2040s (Rupa Kumar *et al.* 2003; IPCC 2007). There is considerable intermodel
dispersion in the case of monsoon rainfall projections, while the models show a
better consensus in their temperature projections. Spatial patterns of rainfall change
projections indicate maximum increase over the northwest region, but the warming
is generally widespread over the whole region. The analyses carried out show that
there is a general tendency for the number of rainy days to show some decline
while the mean intensity of rainfall on a rainy day increases in the simulations
with increased GHG concentrations. The extremes in both rainfall and temperature
(maximum and minimum) generally show an increase over the region in the GHG

simulations. While the scenarios are indicative of the expected range of rainfall and temperature changes, the quantitative estimates still have large uncertainties associated with them. These climatic projections would have profound impacts on regional population, economy, and resource use.

12.3 Impacts of climate change in South Asia

The IPCC report (2007) identifies (with a greater than 90% confidence level) that food and fiber, biodiversity, water resources, coastal ecosystems, and land cover in South Asia are highly vulnerable to climate change impacts. The confidence level is slightly lower for human health. In the region vulnerable places include small island states (Maldives and Sri Lanka), densely populated poor coastal populations affected by increased flooding from the sea and rivers coupled with enhanced risk of extreme weather events (Bangladesh, India, and Pakistan), deserts (Thar Desert in India and Pakistan), and the Himalayan glaciers. The vagaries of rain-fed agriculture coupled with a projected 30% decrease in agricultural output due to climate change, and an endemic morbidity and mortality due to diarrhoeal diseases erupting from a disturbed hydrological cycle due to a changing climate (IPCC 2007) are further projected impacts. Lower incomes and lower capacity to deal with these impacts accentuate the misery further. Climate change is projected to impinge on sustainable development of most developing countries of South Asia, as it compounds the pressures on natural resources and the environment associated with rapid urbanization, industrialization, and economic development.

12.3.1 Hydrology and water resources

The impacts of climate change on water resources in South Asia will be more negative than positive (Cruz *et al.* 2007). Himalayan rivers will be adversely affected due to increased recession of glaciers, while ground water resources are contaminated and depleting fast due to excessive utilization. Himalayan glaciers cover about 3 million hectares or 17% of the mountain area as compared to 2.2% in the Swiss Alps (Cruz *et al.* 2007). They form the largest body of ice outside the polar caps and are the source of water for the innumerable rivers that support over a billion people in Afghanistan, Pakistan, Nepal, Bhutan, India, and Bangladesh. Glaciers in the Himalaya are receding faster than in any other part of the world (with one of the largest, the Gangotri glacier, at around 25 m/year) and there is a very high likelihood of them disappearing by the year 2035 and perhaps sooner if the Earth keeps warming at the current rate (WWF 2005). This could release large volumes of water in the initial years but once the glaciers melt away, there may be dry periods with very little water flowing in the rivers. This will cause havoc for infrastructure built on and around these rivers such as cities, dams, hydroelectric projects, bridges, and transportation networks. It could well be catastrophic for the vast human population and living organisms dependent on these river systems in South Asia.

Substantial elevation shifts of ecosystems in the mountains and uplands of Tropical Asia are also projected. At high elevation, weedy species can be expected to displace tree species, though the rates of vegetation change could be slow compared to the rate of climate change and constrained by increased erosion in the Greater Himalayas.

The projected decrease in the winter precipitation over the Indian sub-continent implies less storage and greater water stress during the lean monsoon period. Intense rain occurring over fewer days, which implies increased frequency of floods during the monsoon, will also result in loss of the rainwater as direct runoff, resulting in reduced groundwater recharging potential. It is projected that the gross per capita water availability in India will decline from $\sim$1820 m^3/yr in 2001 to as low as $\sim$1140 m^3/yr in 2050, with the poor and needy bearing the brunt of scarcity (Gupta and Deshpande 2004).

12.3.2 *Agriculture and food security*

The main direct agricultural effects will be through changes in factors such as temperature, precipitation, length of growing season, and timing of extreme or critical threshold events relative to crop development, as well as through changes in atmospheric CO_2 concentration (which may have a beneficial effect on the growth of many crop types). Indirect effects include potentially detrimental changes in diseases, pests, and weeds. Results of crop yield projections using HadCM2 indicate that crop yields could probably decrease up to 30% in South Asia even if the direct positive physiological effects of CO_2 are taken into account (Cruz *et al.* 2007). Taken together and considering the influence of rapid population growth and urbanization, the risk of hunger is projected to remain very high.

The livelihood of subsistence farmers, who make up a large portion of rural populations in South Asia, would be adversely affected. In regions where there is a likelihood of decreased rainfall, agriculture could be significantly affected. Inadequate irrigation coverage, heavy dependence on monsoons, and paucity of complementary inputs and/or institutional support systems severely constrains the adaptability of farmers. Swaminathan (2002) suggests a proactive three-pronged strategy of monsoon management for agriculture: (1) train people at different levels,

(2) conserve and manage water, and (3) develop contingency plans to suit different rainfall patterns and work out a compensatory production program.

12.3.3 Infrastructure

Huge investments are being made in new infrastructure projects all over South Asia. Infrastructures are long-life assets and are designed to withstand normal variability in the climate regime. However, with a changing climate, their system resilience is being subjected to a stress for which the infrastructure may not have been originally designed. The floods of 2007 in east and west India, increased cyclonic activities, and landslides caused by heavy rainfall in coastal ghat regions have caused heavy damage to existing infrastructure and indicate that the long-life assets are highly vulnerable. The latest IPCC report on Asia (Cruz *et al.* 2007), however, does not consider these threats to the existing and future infrastructure in South Asia.

12.3.4 Forests

Although there are uncertainties with respect to projections of the impacts of climate change on forest ecosystems, evidence is growing to show that climate change coupled with socio-economic and land-use pressures is likely to have adverse impacts on forest biodiversity, biomass productivity, and carbon sink and/or carbon uptake rates. Quantitative estimates of projected changes in forest biome types can be obtained on the basis of the number of regional circulation model (RCM) grids (out of a total of about 1500) that change from one biome type into another (Ravindranath *et al.* 2003). At 1% increase in CO_2 concentration about 70% of the existing forests are likely to experience a change (64% using GCM projections). In general, increased CO_2 is expected to lead to an increase in the net primary productivity. This has the effect of converting grassland into woodlands and woodlands into forests. The tropical seasonal forest, especially in the northeast, is likely to change into tropical rainforest due to a large increase in rainfall expected in that region. The changes expected in the colder regions are also similar, with tundra likely to change to boreal evergreens, and boreal evergreens into temperate conifers.

12.3.5 Coastal, low-lying areas

Changes in climate will affect coastal systems through sea-level rise and an increase in storm-surge hazards and possible changes in the frequency and/or intensity of extreme events. Even under the most conservative scenario, sea level will be about 40 cm higher than today by the end of twenty-first century, increasing the annual number of people in coastal populations flooded from 13 million to 94 million (IPCC 2007). Almost 60% of this increase will occur in South Asia (along coasts from Pakistan through India, Sri Lanka, Maldives, and Bangladesh to Burma). Using a coarse digital terrain model and global population distribution data, it is estimated that more than 1 million people will be directly affected by sea-level rise in 2050 in the Ganges–Brahmaputra–Meghna delta alone (Ericson *et al.* 2005). The Asian Development Bank (ADB 1994) study indicates that for a 1 m sea-level rise, about 50% of the total Indian population residing along the coast will be impacted. Socio-economic impacts from migration would be felt in major cities and ports, tourist resorts, commercial fishing, coastal agriculture, and infrastructure. Losses of mangroves and damage to corals are two examples of the impacts that changes in the coastal zone would have on natural ecosystems.

12.3.6 *Energy and electricity*

Almost half of population without electricity access in the world lives in South Asia (IEA 2006). To achieve the Millennium Development Goals, the number of people without access to electricity would need to fall to under a billion by 2015 and this implies lighting around 20 million households in South Asia alone. The South Asian story on energy access is equally dismal. While urban centers are enhancing their per capita commercial energy consumption by over 5% per annum, in the rural areas over three-quarters of the population are dependent on solid fuels for their energy needs, predominantly biomass. This creates excessive indoor air pollution, causing adverse health impacts while putting high stress on depleting forest and tree cover.

An increase in atmospheric temperature results in a higher rate of water evaporation from reservoirs, thus reducing available reserves for power generation. The limited availability of water in the reservoirs will require the addressing of complex water management issues in the areas of power generation and irrigation in multipurpose reservoirs. The efficiency of thermal generation plants, other industrial thermal installations, and engines used in vehicular transport are directly related to the atmospheric temperature. Therefore, any increase in atmospheric temperature results in lower plant efficiencies. Also, with increased atmospheric temperature, the efficiencies of cooling equipment drop, which affects all forms of industrial, commercial, and domestic sector installations involving a cooling component. Furthermore, there will be increased demand for air-conditioning and ventilation due to high temperatures. This in turn increases GHG emissions per unit of energy output.

Table 12.1. *Natural gas in South Asia: a snapshot in 2005.*

	Reserves (tcf)	Production (bcf)	R/P Ratio	Consumption (bcf)	Imports (bcf)	Share in energy
Bangladesh	10.6	494.4	21	494.4		74%
India	32.5	1055.9	31	1269.2	213.3	8%
Pakistan	26.8	967.2	28	967.2		48%

Note: tcf is trillion cubic feet, bcf is billion cubic feet.
Source: adapted from Energy Information Administration "Country Energy Profiles." Downloaded on June 11, 2008 from http://tonto.eia.doe.gov/country/index.cfm

12.4 Regional cooperation in energy

Regional cooperation[1] is among the key principles of sustainable development, exhorted in the Rio Declaration on Environment and Development as well as subsequent international declarations. The countries in South Asia have diverse energy resources, which provides the most compelling argument for cooperation – coal in India, gas in Bangladesh, hydroelectric power in Himalayan nations: Bhutan and Nepal. Afghanistan and Pakistan have strategic locations on transit routes connecting the region with the vast gas and oil resources of Central Asia and the Middle East, while Bangladesh is a transit route for Myanmar gas. The region is very deficient in crude oil and is therefore a net importer. Regional cooperation makes eminent sense to improve energy availability and energy security.

12.4.1 Regional cooperation in the gas market

There is an emergent gas market in South Asia involving three large countries: India, Pakistan, and Bangladesh. Despite significant exploration and infrastructure developments in recent times, gas is still below market potential in terms of fulfilling energy needs in India. Bangladesh and Pakistan have greater comparative advantages in using gas (Table 12.1) due to significant domestic gas endowments in Bangladesh and Pakistan's nearness to gas resources in the Middle East and Central Asia.

India already imports gas with most of the supplies coming from Qatar. Pakistan has announced a liquid natural gas (LNG) import policy and plans to commission an LNG terminal as early as 2011 (PIP 2007). Bangladesh has proven plus probable

---

[1] Principle 9 of the Rio Declaration on Environment and Development 1992 exhorts that the "States should cooperate to strengthen endogenous capacity-building for sustainable development by improving scientific understanding through exchanges of scientific and technological knowledge, and by enhancing the development, adaptation, diffusion and transfer of technologies, including new and innovative technologies."

Table 12.2. *Demand projections for natural gas.*

Country	Total primary energy (Mtoe)	Natural gas (Mtoe)	Share of primary energy
Year 2010			
Bangladesh	27	19	68%
India	467	41	9%
Pakistan	79	39	49%
Year 2020			
Bangladesh	62	42	68%
India	850	89	10%
Pakistan	177	78	44%
Year 2030			
Bangladesh	114	78	68%
India	1550	181	12%
Pakistan	361	163	45%

Note: demand projections for the three countries are the official projections. For Bangladesh these are based on a report prepared by an expert committee appointed by the government of Bangladesh (Ahmed *et al.* 2002), for India from the Integrated Energy Policy (GoI 2006), and for Pakistan based on Vision 2030 (GoP 2007).

reserves of 0.44 tcm (Table 12.1); however, the cumulative demand by 2020 is 0.47 tcm (Ahmed *et al.* 2002) and therefore they will have to start importing before 2020 if there are no significant new finds.

In India, several new pipeline projects are underway to link the rising gas demand in western and northern regions with large gas fields in the south. There are also proposals for expanding the gas transmission network so that all major cities are covered; this will create new markets and further increase the demand for gas. India, however, has abundant availability of coal which is a competitor for gas in the power sector and this limits the share of gas in the total energy mix (Table 12.2). The projections of commercial energy demand for Bangladesh and Pakistan show a continued reliance on gas (Table 12.2). In general, the gas demand for all three countries is beyond their domestic resources availability.

Global LNG trade has increased substantially over the last decade owing to a substantial reduction in liquefaction and transport costs (Jensen 2003), though LNG is expensive compared to piped natural gas at transport distances below 4000 km (Cornot-Gandolphe *et al.* 2003). Thus, natural gas pipelines from Iran and Turkmenistan can be economical to supply the large gas markets in Northern India and Pakistan, which are far from coasts where LNG imports can land. This has prompted policy makers in India and Pakistan to consider cooperative pipeline projects for importing gas from the surrounding gas-rich nations in the Middle East and Central Asia. Similarly, the economics of natural gas imports from Myanmar and South East Asia has prompted cooperation between Bangladesh and India.

The gas pipeline proposals have been discussed for a long time, e.g. the onshore pipeline projects involving Iran–Pakistan–India (IPI) (Pandian 2005; Tongia 2005; Verma 2007), Turkmenistan Afghanistan–Pakistan–India (Jung 2003; Pandian 2005) and Myanmar–Bangladesh–India (MBI) (Pandian 2005; Kumara-swamy and Datta 2006). In addition, two under sea pipelines – one from Qatar to Pakistan and the other from Iran to India (Pandian 2005) – have also been discussed. Historically, the implementation of trans-national gas projects has depended on cooperative agreements among nations (Hayes and Victor 2006) that reduce invest-ment risks. Studies have shown that in the gas pipeline projects involving transit countries (see Box 12.1), instances of supply interruptions by the transit countries are few, although a lot of attention is given to risks arising from supply interrup-tions (Hayes and Victor 2006). This means that Indian policy makers can reduce the emphasis they give to risks of supply interruptions by Pakistan through which gas from Middle East and Central Asia can come to India.

Box 12.1
Trans-national cooperation for gas pipelines

An Iran–India pipeline was initially proposed in 1989 and a memorandum of understanding was signed between India and Iran in 1993 (Pandian 2005). The proposed pipeline was slated to originate from the massive South Pars Gas fields in the Persian Gulf (Ali 2005). There were three options for transporting gas from Iran to India – overland, shallow sea, and deep sea. The deep sea option was not commercially viable. The proposal for an overland route was held up on account of geopolitical security concerns (Tongia and Arunachalam 1999). An alternative proposal for a pipeline passing through shallow territorial waters of Pakistan met with strategic resistance in both countries (Chaturvedi 2005).

Besides India's prevailing dependence on gas imports, the growing demand for gas in Pakistan would require gas imports in the future (GoP 2007). Simultaneously, improved relations between India and Pakistan have created fresh opportunities for cooperation to realize some pending gas projects. The Iran–India and Iran–Pakistan bipartite negotiations have now graduated into tripartite negotiations. This proposal is a land pipeline traversing around 1115 km through Iran, 705 km through Pakistan, and 850 km in India to connect to the existing pipeline network in India (PTI 2005a). The pipeline will have a capacity to transport 90 mscmd, out of which Pakistan's share will be 30 mscmd and India will receive 60 mscmd (PTI 2005b). Even at the higher end of the project cost estimates, which varies from US $4 billion (Tongia 2005) to US $7 billion (TNN 2006), the project would be economical. The realization of cooperative trans-national gas projects would bring economical gas supply to the region, thus promoting cleaner regional development and also delivering global benefits of reduced greenhouse gas emissions from the region.

Table 12.3. *Hydroelectric potential in South Asia.*

Country	Potential (GW)	Developed (GW)	Balance (GW)
Bhutan	30	1.4	18.2
India	149	30.9	103.0
Nepal	83	0.5	62.6
Pakistan	42	7.0	25.8
	304	40	210

The long-term future of the gas market in South Asia would require connecting with the huge gas resources in the adjoining regions. Regional cooperation within South Asia and with the neighboring regions would deliver co-benefits of greater energy security and a cleaner environment to the region (Shukla 2006). The argument for cooperation for clean energy gains greater strength as we look at benefits of cooperation in electricity in South Asia in the next section.

12.4.2 Electricity cooperation in South Asia

The electricity generation mix within South Asian countries shows significant variation, depending on their domestic energy resources. Bangladesh relies on gas (Faruque *et al.* 2002; EIA 2004), Bhutan, Nepal, and Sri Lanka on hydroelectricity (EIA 2004; Gautam & Karki 2005), India on coal (GoI 2006), and Pakistan on gas and hydroelectricity (Ahmed 2007). The energy security paradigm that underlies the electricity generation mix has made Bangladesh, India, and Pakistan, which constituted nearly 99% of electricity demand in 2005, highly fossil fuel dependent, economically inefficient, environmentally unfriendly, and unsustainable. Regional cooperation in South Asia can alter this path by: (1) increasing electricity trade which is now bilateral and negligible and (2) promoting development of hydroelectric potential.

Electricity trade requires a regional grid and a corresponding regulatory mechanism. This will rationalize regional electricity generation and also the use of primary energy resources besides improving the regional energy security. There is immense underdeveloped hydroelectric potential in South Asia (Table 12.3), especially on the Himalayan rivers which can be deployed with enhanced cooperation over sharing of river waters.

The per capita electricity consumption in 2003 was 91 kWh in Nepal and 218 kWh in Bhutan (UNDP 2006), as a large percentage of the population does not have access to electricity (EIA 2004). Increasing economic growth in the South Asia region will probably enhance the demand, though given the mountainous

Table 12.4. *Proposed cooperative hydroelectric projects on Himalayan rivers.*

Project	Generation (MW)	Flood control	Irrigation
Nepal and India			
Pancheshwar	5 600	Y	Y
Karnali	10 800	Y	Y
Sapta Kosi	3 300	Y	Y
Upper Karnali	300	N	N
Naumere	207	Y	Y
Bhutan and India			
Punatsangchhu I	1 095	Y	N
Punatsangchhu II	992	N	N
Mangdechu	360	N	N
Manas	2 800	Y	Y
Wangchu	900	Y	N
Bunakha	180	Y	N
Sankosh	4 060	Y	Y

Source: table prepared based on information from CEA (2007).

terrain and dispersed population, meeting such demand may require decentralized responses such as small hydroelectric projects. Exploiting most of the hydroelectric potential of Himalayan rivers would need large hydroelectric projects jointly developed by cooperation among several countries. Bhutan is already actively engaged with India (see Box 12.2) on such projects. A number of projects are earmarked for joint development by Bhutan and Nepal with India (Table 12.4). Nepal's Water Resource Strategy envisages exporting 15 GW of hydropower capacity by 2027 (Srivastava and Misra 2007). India (CEA 2007) and Bangladesh (Srivastava and

---

Box 12.2
Bhutan–India cooperation

India imported 1790 GWh of electricity in 2005 (CMIE 2007), 110 GWh from Nepal (Srivastav and Misra 2007) and the rest from Bhutan. Electricity export from Bhutan comes from three Himalayan hydroelectric projects that are integrated into India's eastern region grid. These projects were jointly developed by India and Bhutan, with India providing technical and financial inputs (CEA 2007). The electricity tariffs are agreed under long-term contracts. In 2003, electricity exports accounted for 72% of Bhutan's export revenues (Srivastav and Misra 2007). Enhanced cooperation would enable the development of Bhutan's as yet untapped but immense hydroelectric potential.

Box 12.3
Modeling experiences in South Asia

Research activities continue on commercial, traditional, and renewable energy resources in the South Asian countries. Over the past few years, researchers from India have acquired experience in using integrated assessment models (IAMs) and other component models. These models include the Asia–Pacific Integrated Model (AIM) developed by Japan's National Institute for Environment Studies (NIES). AIM (see Chapter 11, this volume) has been used in several Asian developing countries, e.g. China, India, Indonesia, and Korea. Different economic and energy system models such as the AIM/ENDUSE Model, MARKAL, EFOM, Edmonds–Reilly–Barns (ERB) Model and Second Generation Model (SGM) are used for integrating economic and energy policies with greenhouse gas mitigation.

Mathematical models of climate are the primary tools used to project the human influences on climate. There is growing experience in using general circulation models (GCMs). However, there is a large uncertainty associated with GCM projections on a regional scale, since GCMs are yet to realistically reproduce the observed features at this scale, particularly over the monsoon region. Several downscaling approaches are being used to derive the regional-scale features from large-scale model simulations (Rupa Kumar *et al.* 2002).

IAMs also provide an approach to impact assessment. Impacts are first assessed under a reference case scenario. The analysis is then repeated with a constraint on the future. The change in impacts represents the climate-related benefits of the policy. There is a need to share experiences with IAMs to enhance capacity for climate change policy assessment in South Asia.

Misra 2007) can import electricity, besides gaining from co-benefits of irrigation and flood control. Hydropower projects between India and Pakistan are governed by the Indus water treaty; however, development of the potential on the Indus river system will require going beyond the treaty by having joint development of water resources (Verghese 2005).

Cooperation in electricity needs a long-term horizon. Dams and transmission networks involve huge investments with long payback periods and there are lock-ins which, in a developing country context, create risks for investors (Shukla *et al.* 2004). Cooperation has happened mainly between India and Bhutan, which have very cordial relations at the governmental level. To take forward the cooperative regime a legal framework is needed that binds the member countries and that would instill confidence in investors. At the second level it is imperative to create institutions that decide electricity tariffs and water sharing arrangements.

12.5 Future energy cooperation in South Asia

A detailed analysis has been carried out on the economic, social, and environmental impacts of South Asian energy market cooperation. This analysis assumes that countries in South Asia reduce their individual energy consumption while maintaining their respective economic growth rates. This becomes possible due to exploiting the additional and cheaper power generation capacities in individual countries (such as hydroelectricity in Nepal and Bhutan) and selling them to neighboring countries; sharing cleaner energy resources of individual countries (such as natural gas in Bangladesh) with neighboring countries that use more coal (such as India); and providing safe passage to import natural gas (such as through Pakistan, Afghanistan, and Bangladesh) for high-demand regions (such as in India and Pakistan).

This analysis has been conducted using an integrated modeling approach (Figure 12.1). This framework (see also Kainuma *et al.*, Chapter 11, this volume) comprises three modules: top-down models, bottom-up models, and local models. These three modules are soft-linked through various parameters. The top-down models provide GDP and energy price projection outputs that are used as exogenous inputs to the bottom-up models. The bottom-up models provide a future energy balance output that is used for tuning the inputs of the top-down models. Similarly the bottom-up models provide detailed technology and sector-level emission projections that provide inputs to the geographical information system based energy and emissions mapping for the country. The framework depicted in Figure 12.1 is an update of previous work (see Garg *et al.* 2001) and is described in detail by Shukla *et al.* (2001). The important exogenous model specifications include the electricity demand trajectory, investment constraints, energy supply limitations, energy prices, technology costs, and technology performance parameters.

The analysis of regional cooperation (Heller and Shukla 2006; Shukla 2006) uses frameworks (Nair *et al.* 2003; Shukla *et al.* 2003) that integrate energy and electricity markets. The results shows significant direct, indirect, and spill-over benefits via economic efficiency, energy security, water security, and environment (Table 12.5). The indirect benefit of carbon saving (1.4 billion-ton of carbon or 5.1 billion-ton of CO_2) from this cooperation (over 20 years from 2010 to 2030) is 70% of what countries would have to mitigate under the Kyoto Protocol, including the USA, during the Kyoto period 2008–2012.

The regional electricity capacity profile under a cooperative regime has more renewable capacity (including large hydropower) than the combined capacities of individual country futures under moderate cooperation akin to current energy cooperation regimes. Owing to regional energy cooperation, coal consumption in 2030 is projected to decrease 9–13% compared to coal consumption in 2030

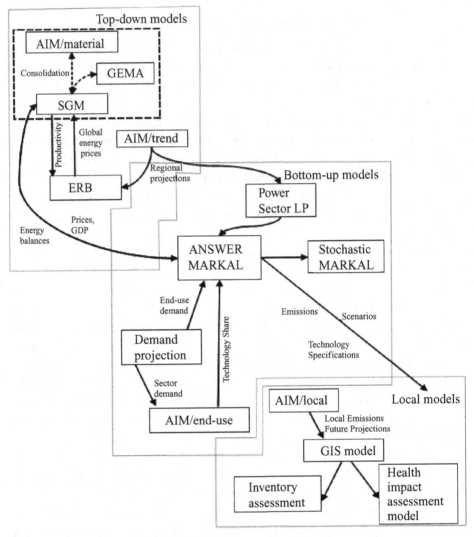

Figure 12.1. Soft-linked integrated modeling framework.

with moderate energy cooperation. Similarly, natural gas consumption is projected to increase between 5% and 9% in 2030, mainly for the power generation and transport sectors. Gas availability for cooking in households also increases under a cooperative regime, resulting in high human health dividends.

The South Asian region has limited reserves of oil. All countries in the region are major importers of crude oil. Therefore energy cooperation does not change the consumption trajectory of oil significantly. However, there are marginal reductions in oil consumption due to availability of better technologies and greater availability of gas and electricity. Biofuel penetration adds only marginally to reduce oil

Table 12.5. *Benefits of South-Asia regional cooperation (cumulative for 2010–2030).*

Benefit (saving)		% of region's GDP
Energy savings (direct benefits)		
Energy	59 exajoule	0.47
Investment in energy supply technologies		0.19
Investment in energy demand technologies		0.19
Environment (indirect benefits)		
CO_2	5.1 billion ton	0.09
Sulfur dioxide (SO_2)	50 million ton	0.03
Total direct and indirect benefits		0.97
Spill-over benefits		
Water	16 GW additional hydropower capacity	
Irrigation/flood control	From additional dams	
Competitiveness	Reduced per unit energy and electricity costs	

Adapted from Heller and Shukla (2006).

dependence due to cooperation alone, since this is more driven by national-level developmental policies and national biofuel production is projected to be consumed domestically by the producing country, without any regional trading.

Hydroelectricity capacity is projected to increase between 9% and 15% in 2030, up to 15 GW. This increase, however, is below 10% of the combined economic hydropower potential of Nepal, Bhutan, Pakistan, and India; and is therefore likely to be realized even under adverse scenarios of climate change until 2050.

The total energy consumption in the region therefore decreases and becomes cleaner due to sharing of energy resources and efficiency improvements in supply- and demand-side technologies due to regional cooperation. Also with greater availability of gas in the region there are shifts from coal-based technologies to more efficient gas-based technologies on the supply side. As regional cooperation grows, better technologies become available to less developed countries in the region, improving energy efficiencies of GDP and reducing the CO_2 efficiency of energy use in South Asia. However, these improvements in technology efficiencies on both the demand and supply sides do not take place immediately. The benefits of cooperation are realized only beyond the year 2010.

In this way, regional energy cooperation provides multiple dividends in the form of reduced primary energy requirements for the region (59 exajoules over 2010–2030), reduced CO_2 emissions (over 5.1 Bt-CO_2 over 2010–2030), and reduced SO_2 emissions due to lower coal consumption (50 Mt-SO_2 over 2010–2030). There are other spill-over benefits such as increased competitiveness of industry due to

Table 12.6. *Financial benefits to some individual countries (cumulative for 2010–2030).*

Benefit	$ billion					% of country's cumulative GDP				
	Bhutan	Bangladesh	Nepal	Pakistan	India	Bhutan	Bangladesh	Nepal	Pakistan	India
Savings in										
Energy	0.13	29	1.8	13.8	144	0.45	1.21	0.65	0.36	0.45
Investment in energy supply technologies	0.05	13	0.73	6.8	56	0.17	0.54	0.26	0.18	0.17
Investment in energy demand technologies	0.04	11	0.68	6.9	55	0.14	0.46	0.24	0.18	0.17
Gain from trade										
Natural gas export		9.6					0.40			
Natural gas import				14	15.6				0.36	0.05
Electricity exports	2.2		4			7.62	0.02	1.44		
Electricity imports		0.6			6.8					0.02
Total benefits	**2.42**	**63.2**	**7.2**	**41.5**	**276**	**8.39**	**2.63**	**2.59**	**1.07**	**0.87**

lower energy prices; 16 GW additional hydropower capacity resulting in flood control and fish production; lower health impacts on populations due to less coal combustion, and a number of social impacts. All of these provide direct and/or indirect economic benefits. The cumulative economic value of these benefits over a 20 year period from 2010 to 2030 would be around US$390 billion, i.e., nearly 1% of the region's GDP for the entire period. The regional energy and electricity cooperation would add 1% economic growth each year to the region sustained over a 20 year period. The benefits would be larger in smaller countries (Table 12.6). These savings, if invested properly into the respective economies for social sector improvements, would deliver multiple benefits in South Asia where the largest numbers of the world's poor reside.

Numerous barriers to South Asia cooperation exist, most being rooted in history. The development and climate demand the political will to forge cooperative alliances in a region where countries have co-existed cooperatively over several millennia. This by itself would accrue great economic and environmental benefits to the region and climate benefits to the global humanity at little or no cost (Table 12.7).

12.6 CO_2 and SO_2 emission dynamics

A key issue related to regional energy cooperation relates to its impact on local air quality. As we have already noted, South Asia has the dubious distinction of having some of most polluted mega-cities in the world. The story is no better in 40 other cities with a population of over a million each. Over the last decade, some of the mega-cities have gradually switched over to natural gas as a preferred fuel for public transport, and also all road transport. Delhi and Dhaka are prime examples. Some cities are also revamping their urban transport infrastructure, such as the introduction of a metro-rail system in Delhi, while cleaner diesel and gasoline have been introduced in almost 80% of them. These measures have decelerated the growth of SO_2 emissions from the transport sector, which is the second largest emitter of SO_2 emissions after coal-based power plants.

We also analyzed whether it is possible to combine local and global environmental policies so that countries, while pursuing high priority local environmental concerns, for example in relation to local air quality (e.g., SO_2 emissions), can also support CO_2 emission reduction policy objectives. It should here be recognized that CO_2 and SO_2 emission control policies have various interesting links and disjoints. Starting from SO_2 emission control as the major policy priority, it can in many cases be cheaper to install various cleaning techniques that control SO_2 emissions rather than to implement general efficiency improvements or fuel switching that both reduce SO_2 and CO_2 emissions. In contrast, starting with CO_2

Table 12.7. *Overview of economic, social, and environmental impacts of South Asian energy cooperation.*

Parameters	Impacts of South-Asia regional energy cooperation
Total GHG emissions	Cumulative mitigation of 5.1 bt-CO_2 between 2010 and 2030 for the South Asia region
Total investment costs	Very large investment in energy infrastructures would contribute to additional 1% economic growth each year between 2010 and 2030 throughout the South Asia region
Cost per ton of CO_2 mitigation	Negative macroeconomic costs
Employment generation	Added employment from energy trade; very high indirect and spill-over effect on employment due to reduced electricity cost and improved competitiveness of the region
Local air pollution (SO_2, NO_x etc.)	Average SO_2 mitigation of 2.5 Mt/year between 2010 and 2030. Economic value of mitigation between 2010–2030 = US$10 billion
Other environmental impacts	Significant indirect and spill-over benefits from enhanced water supply and flood control from hydroelectric projects
Foreign exchange component	Significant saving of foreign exchange for the South Asia region due to reduced oil and gas imports from other regions
Energy access	Significant increase in energy supply and energy security would enhance energy access and consumption. Cost of primary energy and electricity would reduce on average by 5% throughout the South Asia region for 2010–2030
Host country involvement in project implementation or maintenance	Countries in the South Asia region would have significant participation in cross-country energy and infrastructure projects that would be developed in the region. National governments in the region and the investing public would hold a significant stake in the project equity and benefit from the project operations

emission reduction as the major policy priority will often suggest a number of cost effective options that jointly reduce the two types of emissions. However, such policies seen from the SO_2 reduction perspective alone deliver more expensive local air pollution control than cleaning systems.

The relationship between CO_2 and SO_2 emission development is shown in Figure 12.2 for South Asia under regional cooperation and current dynamics regimes for 2000–2030. These are also contrasted with the projections for China and India under dynamics as usual scenarios to provide a perspective (Halsnaes and Garg 2006).

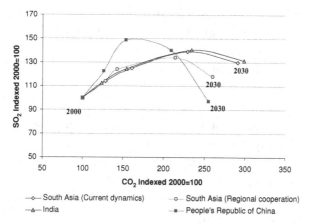

Figure 12.2. Links and disjoints in CO_2 and SO_2 emissions in South Asia under regional cooperation and current dynamics regimes for 2000–2030 (the emissions are indexed separately for each instance to maintain comparability; and dots show the time namely: 2000, 2005, 2010, 2020, and 2030). China and India projections under a reference scenario are shown for comparison (Halsnaes and Garg 2006).

Coal consumption for electricity generation is the major source of CO_2 and SO_2 emissions in South Asia, and coal also is expected to play a major role in the future under current dynamics (Halsnaes and Garg 2006). However, domestic pressure in the countries have led to increasing efforts over time to introduce various policies (Shukla *et al.* 2007) to reform primary energy markets, rationalize electricity generation, and lead to more efficient technology choices along with end-of-pipe solutions, which curb SO_2 and suspended particulate matter (SPM). CO_2 emissions, however, continue to rise but the growth tends to slow down over time under the current dynamics scenario. Road transport emissions are a major source of local air pollution and cleaner road transport technologies, although based on fossil fuels, contribute to reduce SO_2, SPM, NO_x, and CO emissions. CO_2 emissions again continue to rise, since fossil-fuel-based road transport continues to have a major share in all of these countries.

Under regional energy cooperation, coal use declines prompting a reduction in both CO_2 and SO_2 emissions. However, due to larger penetration of natural gas, which emits no SO_2, SO_2 emissions decline faster than CO_2 emissions. Policy dynamics to mitigate local pollution are projected to result in a reduction in SO_2 emissions even though the absolute energy consumption, and therefore CO_2 emissions, would continue to rise.

12.7 Conclusion

The South Asian region is vulnerable to the effects of climate change due to its high dependence on climate-sensitive sectors such as agriculture and forestry and its low financial adaptive capacity. It must, however, be borne in mind that most of the impact studies carried out so far are sectoral in nature and have inherent uncertainties associated with them due to limited understanding of many critical processes in the climate system, the existence of multiple climatic and non-climatic stresses, regional-scale variations, and non-linearity. Uncertainty in projecting socio-economic drivers such as economic growth, demographic patterns, cooperative regimes, and technological growth, adds further uncertainties to these sectoral studies. In any case, there is a need to integrate the diverse scientific assessments and link them with policy making to integrate development and climate change adaptation strategies. Developing regional databases and tools, modeling, and knowledge sharing would go a long way in institutionalizing cooperative climate change research in South Asia.

Among the most important anthropogenic causes for climate change is energy use, which is dominated by fossil fuels worldwide and particularly in the South Asian region. India is the largest country in the region, accounting for almost 90% of regional energy use and related GHG and local pollutant emissions. The benefits of regional energy cooperation (Tables 12.5–12.7) could be gainfully invested in the education, human health, and infrastructure sectors to enhance growth and income generation. Energy cooperation would also enhance the adaptive capacity of millions of people directly or indirectly (Table 12.7) to withstand the adverse impacts of a changing climate.

The above analysis indicates a definite trend of increasing demand for clean and cheap energy for the South Asian region. Alternate supply options are needed to attain the objective of meeting these demands. Cooperation in primary energy and electricity between the South Asian countries could help address the issue of energy security in the region. Supporting and encouraging more robust trade in energy would have far reaching economic, development, and security benefits. Such benefits that arise through cooperation in the energy sector are already being reaped in other regions of the world, such as South Africa, the ASEAN power grid, and the Nordic countries. Their characteristics include: a diverse resource base, consisting of hydropower and thermal-based generation that provides opportunities for significant benefits from regional power trade; significant differences in short-term production costs and availability of supplies that also provide opportunities for trade; and major domestic reform efforts underway within individual countries in the region that contribute to the ability of individual power systems to engage in regional coordination and trading activities. However, to achieve the desired level of cooperation it is imperative to have sustained and high-level political support, and a regional coordinating body.

Under a regional energy cooperation regime for South Asia, sustainable national development automatically gets promoted for each country. The share of renewable and cleaner energy increases, GHG and local pollutant emissions reduce, cleaner energy technologies penetrate faster in all sectors, income generation capacities of the poor increase as energy becomes cheaper and electricity access also improves, while simultaneously enhancing the adaptive capacities of the people to withstand adverse impacts of climate change. Energy cooperation could also provide and sustain confidence building amongst countries in the region, providing the stability necessary for national growth and prosperity.

This chapter highlights the emerging understanding that national energy and climate change futures would benefit from a regional sustainability perspective. These have to coordinate and balance bottom-up-driven national systems such as democratic governance to take people's aspirations and expanding needs into consideration, with top-down-driven processes such as regional cooperation, energy security, and federal structures to ensure regional balance in development and equitable availability of fruits of development such as social and physical infrastructures.

Energy and climate change policies have to be integrated and aligned with national developmental policies and vice versa. A shared regional developmental vision could provide a binding glue. Short-term developmental goals, such as meeting the Millennium Development Goals and national developmental targets up to 2015 have to dove-tail into medium and longer-term developmental visions for South Asia to emerge as a developed and secure region by 2015 and beyond. Energy policies such as energy access and affordability in the short term have to integrate into national visions and the requirements of energy security for each country in the region to achieve their development vision. Environmental integrity at local, regional, and global levels has to be simultaneously ensured so that the fruits of development are enjoyed by all equitably and consistently.

References

ADB 1994. *Climate Change in Asia: India country report.* Manila: Asian Development Bank.

Ahmed, M. 2007. *Pakistan's Energy Sector and Investment Climate: Emerging Investment Opportunities in South Asia Power Sector.* Presentation available at SARI: http://www.sari-energy.org/InvestConfPres3=07/PDF/01%20Mukhtar%20Ahmed,%20Pakistan.pdf

Ahmed, N., M. N. Islam, I. Hossain, *et al.* 2002. *Report of the Committee for Gas Demand Projections and Determination of Recoverable Reserve & Gas Resource Potential in Bangladesh.* Dhaka: Ministry of Energy and Mineral Resources, Government of the People's Republic of Bangladesh.

Ali, M. R. 2005. Energy resources and regional economic cooperation. *Regional Studies,* **23**(2), 14–57.

CEA 2007. *Cooperation with Neighboring Countries.* http://www.cea.nic.in/hydro/Cooperation%20with%20Neighbouring%20Countries.pdf

Chaturvedi, S. 2005. India's quest for energy security: the geopolitics and the geoeconomics of pipelines. In D. Rumley and S. Chaturvedi (eds.), *Energy Security and the Indian Ocean Region*. New Delhi: South Asian Publishers Pvt. Ltd.

CMIE 2007. *Energy: February 2007*, Economic Intelligence Service, Centre for Monitoring Indian Economy Pvt. Ltd., Mumbai.

Cornot-Gandolphe, S., O. Appert, R. Dickel, M.-F. Chabrelie, and A. Rojey 2003. *The Challenges of Further Cost Reductions for New Supply Options (Pipeline, LNG, GTL)*. Paper presented at the 22nd World Gas Conference, Tokyo.

Cruz, R. V., H. Harasawa, M. Lal, *et al.* 2007. Asia. *Climate Change 2007: Impacts, Adaptation and Vulnerability. Contribution of Working Group II to the Fourth Assessment Report of the Intergovernmental Panel on Climate Change*. Cambridge: Cambridge University Press.

EIA 2004. *South Asia Regional Overview*. http://www.eia.doe.gov/emeu/cabs/srilanka.html

EIA 2006. *Online Country Analysis Briefs*. Washington, DC: Energy Information Administration, US Department of Energy. www.eia.doe.gov/international

Ericson, J. P., C. J. Vörösmarty, S. L. Dingman, L. G. Ward, and M. Meybeck 2005. Effective sea-level rise and deltas: causes of change and human dimension implications. *Global and Planetary Change*, 50(1, 2), 63–82.

Faruque, M. A., M. T. R. Khan, M. Nur-e-Alam, and M. Zebunnssa 2002. *Present Structure of Bangladesh Power Sector*. http://www.sari-energy.org/sariresources.asp

Garg A., D. Ghosh, and P. R. Shukla 2001. Integrated modeling system for energy and environment policies. *OPSEARCH* (special issue on energy and environment modelling), 38, 1.

Gautam, U. and A. Karki 2005. Nepal: thermal energy for export. *South Asian Journal*, July–September.

GoI 2006. *Integrated Energy Policy: Report of the Expert Committee*. New Delhi: Planning Commission.

GoP 2007. *Vision 2030*. Islamabad: Planning Commission, Government of Pakistan.

Gupta, S. K. and R. D. Deshpande 2004. Water for India in 2050: first-order assessment of available options. *Current Science in India*, 86, 1216–1224.

Halsnæs, K. and A. Garg 2006. *Sustainable development, energy and climate: exploring synergies and tradeoffs*. A research publication of UNEP Risoe Centre, Denmark.

Hayes, M. H. and D. G. Victor 2006. Politics, markets, and the shift to gas: insights from the seven historical case studies. In D. G. Victor, A. M. Jaffe, and M. H. Hayes (eds.), *Natural Gas and Geopolitics: from 1970 to 2040*. Cambridge: Cambridge University Press.

Heller, T. C. and P. R. Shukla 2006. *Financing the Climate-friendly Development Pathway (with illustrative case studies from India)*. Paper presented at the Financing Integrated Development and Climate Strategies, New Delhi.

IEA 2006. *World Energy Outlook*. Paris: International Energy Agency.

IPCC 2007. *Climate Change 2007: Impacts, Adaptation and Vulnerability. Contribution of Working Group II to the Fourth Assessment Report of the Intergovernmental Panel on Climate Change*. M. L. Parry, O. F. Canziani, J. P. Palutikof, *et al.* (eds.). Cambridge, UK: Cambridge University Press.

Jensen, J. T. 2003. The LNG revolution. *Energy Journal*, 24(2), 1–45.

Jung, N. 2003. TAR: OTH 37018 – *Technical Assistance for the Turkmenistan Afghanistan Pakistan Natural Gas Pipeline Project*. Asian Development Bank.

Kumaraswamy, P. R. and S. Datta 2006. Bangladeshi gas misses India's energy drive? *Energy Policy*, 34(15), 1971–1973.

Nair, R., P. R. Shukla, M. Kapshe, A. Garg, and A. Rana (2003). Analysis of long-term energy and carbon emission scenarios for India. *Mitigation and Adaptation Strategies for Global Change*, **8**, 53–69.

Pandian, S. 2005. The political economy of trans-Pakistan gas pipeline project: assessing the political and economic risks for India. *Energy Policy*, **33**(5), 659–670.

PIP 2007. *Pakistan Energy Outlook 2005–2006*. Downloaded on 13th September 2007 from http://www.pip.org.pk/NewsDetails.aspx?do=32

PTI 2005a. Iran, Pak, Ind pipeline work may start soon. *Economic Times*, August 22, 2005. http://economictimes.indiatimes.com/articleshow/msid-1207040,prtpage-1.cms22/08/2005

PTI 2005b. India, Pakistan to begin work on Iran pipeline by mid '07. *Economic Times*, December 17, 2005. http://economictimes.indiatimes.com/articleshow/msid-1335749,prtpage-1.cms17/12/2005

Ravindranath N. H., N. V. Joshi, R. Sukumar, I. K. Murthy, and H. S. Suresh 2003. Vulnerability and adaptation to climate change in the forest sector. In P. R. Shukla, Subodh K. Sharma, N. H. Ravindranath, A. Garg, and S. Bhattacharya (eds.), *Climate change and India: Vulnerability Assessment and Adaptation*. Hyderabad: Universities Press (India) Pvt Ltd, pp. 227–265.

Rupa Kumar, K., K. Krishnakumar, R. G. Ashrit, S. K. Patwardhan, and G. B. Pant 2002. Climate change in India: observations and model projections. In P. R. Shukla, K. S. Sharma, and P. Venkata Ramana (eds.). *Climate Change and India: Issues, Concerns and Opportunities*. New Delhi: Tata McGraw-Hill, pp. 24–75.

Rupa Kumar, K., K. Krishnakumar, V. Prasanna, *et al.* 2003. Future climate scenarios. In P. R. Shukla, Subodh K. Sharma, N. H. Ravindranath, *et al.* (eds.), *Climate Change and India: Vulnerability Assessment and Adaptation*. Hyderabad: Universities Press (India) Pvt Ltd, pp. 69–127.

Shukla, P. R. 2006. India's GHG emission scenarios: aligning development and stabilization paths. *Current Science*, **90**(3), 354–361.

Shukla, P. R., R. Nair, and A. Rana 2001. Integrated modeling and analysis of long-term energy and emission trajectories for India. *Proceedings of the IGES International Workshop on Climate Policy in Asia*, Tokyo, Japan, December 17–18. Tokyo: Institute for Global Environmental Strategies.

Shukla, P. R., K. Subodh Sharma, N. H. Ravindranath, A. Garg, and S. Bhattacharya (eds.) 2003. *Climate Change and India: Vulnerability Assessment and Adaptation*. Hyderabad: Universities Press (India) Pvt Ltd.

Shukla, P. R., M. M. Kapshe, and A. Garg 2004. *Development and Climate: Impacts and Adaptation for Infrastructure Assets in India*. OECD Global Forum on Sustainable Development: Development and Climate Change, ENV/EPOC/GF/SD/RD(2004)5/FINAL. Paris: OECD.

Shukla, P. R., A. Garg, S. Dhar, and K. Halsnaes 2007. *Balancing Energy, Development and Climate Priorities in India: Current Trends and Future Projections*. UNEP Risoe Centre on Energy, Climate and Sustainable Development, Roskilde, Denmark, September 2007.

Srivastava, L. and N. Misra 2007. Promoting regional energy co-operation in South Asia. *Energy Policy*, **35**(6), 3360–3368.

Swaminathan, M. S. 2002. Climate change, food security and sustainable agriculture: impacts and adaptation strategies. In P. R. Shukla, K. S. Sharma, and P. Venkata Ramana (eds.), *Climate Change and India: Issues, Concerns and Opportunities*. New Delhi: Tata McGraw-Hill, pp. 196–216.

TNN 2006. India, Pak strike peace on Iran gas. *The Economic Times*, August 4, 2006. http://economictimes.indiatimes.com/articleshow/msid-1849889,prtpage-1.cms04/08/2006

Tongia, R. 2005. Revisiting natural gas imports for India. *Economic and Political Weekly*, May 14, 2032–2036.

Tongia, R. and V. S. Arunachalam 1999. Natural gas imports by South Asia: pipelines or pipedreams. *Economic and Political Weekly*, May 1–7, 1054–1064.

UNDP 2006. *Human Development Report 2006, Beyond Scarcity: Power, Poverty, and Global Water Crisis*. New York: United Nations Development Programme.

Verghese, B. G. 2005. It's time for Indus II. *The Tribune*, 25 May, Chandigarh.

Verma, S. K. 2007. Energy geopolitics and Iran–Pakistan–India gas pipeline. *Energy Policy*, **35**(6), 3280–3301.

WWF 2005. An overview of glaciers, glacier retreat, and subsequent impacts in Nepal, India and China. World Wildlife Fund, Nepal Programme.

13

Climate change and regional sustainability in the Yangtze Delta, China

YONGYUAN YIN

13.1 Introduction

Climate change impact assessment has, over the last two decades, evolved from simple expert judgment to simulations of numerous components and processes of the Earth system. Carter *et al.* (1994) provide technical guidelines for climate change impact assessment and option evaluation, and introduce a variety of methods. Various applications of assessment methodologies can be found in the literature of different disciplines. The majority of the applications focus either on physical, biological, or economic aspects of the Earth system. Applications which take a holistic approach are relatively rare. Another important rigidity of these narrow methods is that they often deal with one economic sector in isolation, failing to recognize the importance of intersectoral relations. Given the complex nature of the interactions between sustainability and climate change, integrated analytical methods are desirable.

Although the concept of sustainability lacks a uniform definition, a general consensus holds that sustainability, in principle, represents a long-term goal in planning or development which explicitly incorporates a holistic view of social, economic, and ecosystem values (Robinson and Van Bers 1996; Munasinghe 1998; UN 2004). The desire for sustainability in regional planning and development reflects an increasing public concern over whether the existing resource base can provide goods and services to support a full range of human and ecological values, equitably, in perpetuity (WCED 1987; Pearce and Turner 1990; UNDP 2006). Lack of a universally accepted definition, however, should not be used as a reason to stop working towards a more sustainable future, and/or attempting to ensure sustainable regional development (SRD) while meeting the challenge of climate change. In this case study, SRD is broadly defined as the long-term use of natural resources that is economically viable, socially desirable, and environmentally non-degrading.

Whilst the concept of sustainable development has been discussed worldwide, there seems to be little progress towards linking climate change impact assessment and regional sustainability evaluation. Cohen *et al.* (1998) indicate that while very little attention has been paid by climate change research to sustainable development, the sustainable development research community has not specifically studied the impacts of climate change on societal sustainability. Successful implementation of the concept of sustainable development will require new approaches built upon a foundation of better research into the links between climate change and sustainable development.

There is an increasing concern about the effects of anthropogenic climate change on energy, agriculture, transportation, tourism, and human health. The Fourth Assessment Report (AR4) of the Intergovernmental Panel on Climate Change (IPCC) Working Group II on Impacts, Adaptation, and Vulnerability indicates that climate change will impede progress toward meeting Millennium Development Goals (MDGs). The AR4 report emphasizes the importance of mainstreaming adaptation and mitigation measures into sustainable development strategies to help achieve the MDGs (IPCC 2007).

China, with a large fraction of its population and output dependent on natural resources, is very sensitive to climate change. One challenging issue in evaluating regional sustainability under climate change conditions is to identify the potential impacts on regional sustainability associated with global warming. This can be facilitated by integrated assessment (IA). China needs the development and application of comprehensive and integrated methods to identify impacts of climate change and to ensure sustainable development.

The objective of this chapter is to describe a methodology integrated land assessment framework (ILAF) using the example of the Yangtze Delta of China. The ILAF is designed to identify implications of climate change for sustainable regional development (SRD). The ILAF can provide holistic analysis of climate change and policy responses to improve the scientific understanding of interactions between SRD and climate in the Yangtze River Delta. This IA framework attempts to integrate major physical, biological, and socio-economic components of the region and identify the regional economic–environmental impacts of climate change. Moreover, alternative response options can be evaluated against multiple sustainability goals/indicators, and linkages established between impact assessment and decision making, and between climate change and SRD.

13.2 The integrated land assessment framework (ILAF)

Ideally, a regional IA method would: (1) involve multiple stakeholders, (2) be systematic and holistic, (3) account for multiple objectives and sectors, (4) be able

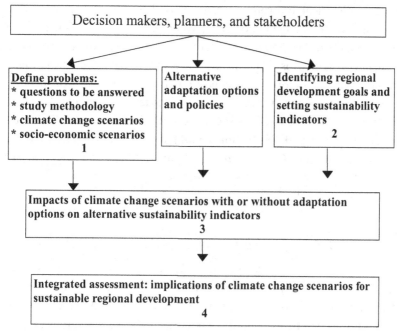

Figure 13.1. The ILAF for the Yangtze Delta study.

to easily identify tradeoffs, and (5) be able to link climate change and SRD. The ILAF adopts a systems analysis approach assisted by simulation modeling and other impact assessment methods, analytic hierarchy process (AHP; Saaty 1980), and goal programming (GP; Ignizio 1976). The ILAF was applied in the Yangtze Delta case study that focused on the identification and specification of regional sustainability goals/indicators and their relationship with climate change impacts and vulnerabilities. The case study identified implications of potential climate change impacts for regional sustainability and possible adaptation policies. The ILAF approach consists of the following main steps (Figure 13.1).

13.2.1 Define problems (step 1)

In conducting climate change impact assessment and SRD evaluation, two essential questions need to be addressed. (1) What are the impacts of climate change scenarios on various sustainability goals/indicators of the study region? (2) What are the effects of the various adaptation options available to reduce the adverse consequences of climate change and to improve sustainability? Finding answers to the two questions can be approached in different ways by applying various methods.

Climate change scenarios are specified in this research and their economic–environmental impacts are examined. Output from general circulation models

(GCMs) and historical information can be used to design scenarios represent-ing different climate change conditions (Xu *et al.* 2003). Other methods can also be used to specify future population increase, economic growth, and technologi-cal advancement scenarios. The purpose of scenario development is to establish a common set of assumptions and conditions to be used by all of the study par-ticipants when they conduct climate change impact assessments. Economic and resource sectors sensitive to climate change in the region should be included for impact assessments. The IPCC Technical Guidelines for Assessing Climate Change Impacts and Adaptation (Carter *et al.* 1994) indicate that the selection of a time horizon for study is influenced by the goals of the assessment. All assumptions should be consistent with this time period selection.

13.2.2 *Specifying sustainability goals/indicators (step 2)*

The research procedure includes the identification of goals/indicators of regional sustainability. In this study, goals and/or indicators are evaluation criteria or stan-dards by which the effects of climate change and/or the efficiency of alternative adaptation options can be measured. Generally, goals are reflections of the prefer-ences and desires of decision makers and indicate some specific target levels to be achieved through resource development. Thus, efforts to measure regional sustain-ability under climate change conditions must deal with the task of identifying and specifying sustainability goals/indicators. The analytic hierarchy process (AHP), a multi-criteria decision making (MCDM) technique (Saaty 1980), is employed in the ILAF to identify the priorities of sustainability goals/indicators. The AHP can provide means by which alternative goals can be compared and evaluated in an orderly and systematic manner (Yin 2001).

13.2.3 *Climate change impact assessment (step 3)*

Various simulation or statistical models can be employed by sectoral impact anal-yses to study the vulnerabilities to or impacts of climate change. A geographical information system (GIS) can be used to provide additional information on the spatial distribution of climate change impacts. For example, the low-lying areas of the Yangtze Delta are susceptible to accelerated sea-level rise (SLR) resulting from climate change, and a GIS can be used to identify ecosystems, coastal infras-tructure, and regional communities that are vulnerable to SLR. Another part of the impact assessment is to determine whether alternative adaptation options or policies can lead to a reduction in damage associated with climate change scenar-ios. To identify the regional environmental and socio-economic impacts of climate change, several practical methods should be used in combination, such as scenario development, methods for identifying impacts of climate change on yields of crops, forests, wildlife, water resources, and other aspects of the region.

13.2.4 Integrated impact assessment and regional sustainability evaluation (step 4)

An integrated analytical system has been developed for linking climate change impact assessment and regional sustainability evaluation, which is the core of the ILAF. The system can be used to relate impact information to regional sustainability using subjective judgment and interpretation, and thus to identify effective and desirable adaptation options among alternatives. In this case study, alternative climate change scenarios were evaluated by relating their various impacts to a number of relevant sustainability goals/indicators. These goals are presented in detail later, and they are used as standards by which the significance of various impacts and the strengths and weaknesses of alternative options or policies can be evaluated. The system uses advanced analytical techniques that are discussed in the case study.

13.3 Applying the ILAF in the Yangtze River Delta

The preceding section has portrayed the conceptual framework of ILAF. The best way to evaluate its capability in integrated climate change assessment is to apply the method in a real-world case. The ILAF was applied to the Yangtze River Delta of China for illustrative purposes. However, it should be noted that in the case study presented below, not all the components of the ILAF are covered with the same level of detail. Rather, the focus is on the two main concerns of the chapter: sustainability indicators/goals identification and integrated climate change impact assessment and adaptation policy evaluation.

The following sections present the importance of indicator/goal setting in regional sustainability research and the approach taken to identify SRD indicators/goals. Sectoral impact assessments of the study are briefly discussed. Then, the IA system assisted by goal programming (GP) is introduced to illustrate how sustainability indicators/goals and climate change impact data can be represented in the analytical system to link climate change impact assessment and regional sustainability evaluation. While the ILAF is a tool that can be used by others to help identify climate change impacts and to evaluate possible response options, the policy analysis is presented in a relatively general way. More detailed consideration of policies should be part of a broad discourse with stakeholders over a much longer term (Yin *et al.* 2006).

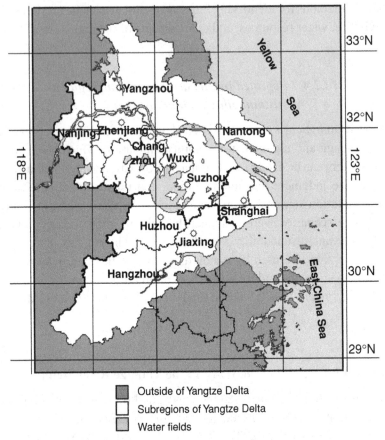

Figure 13.2. Map of the Yangtze River Delta.

13.3.1 The study area

The ILAF was applied to the Yangtze River Delta in China. The Delta is located between latitude 30–32.7°N and longitude 118.2–121.9°E. Figure 13.2 is a map of the study region. The Yangtze Delta is an alluvial plain formed by silt brought down from the upper and middle reaches of the Yangtze River. The Delta includes parts of two provinces (Jiangsu and Zhejiang) and metropolitan Shanghai. With an area of 620 000 km^2 and a population of 63 million people, the Delta accommodates 11 large- and medium-size cities and 61 counties or county-level cities. The study region provides a large amount of rich agricultural land, plentiful water resources, and extensive navigation routes. It also offers diverse recreational opportunities and contains important ecological systems.

The Yangtze River Delta is sensitive to climate. It is the most developed region in China with a high population density, and is critical to the sustainable development of the whole country. While the Delta has fostered the development of many service providers and advanced industrial cities and towns, energy and diversified industries, a broad range of plants, a large number of urban communities, and government agencies, potential climate change may impose considerable economic, social, and environmental impacts. There is already some evidence of changes in the length of the summer extremely hot period, with significant impacts on human health. Potential sea-level rise further complicates the vulnerability of the region to environmental changes, and reinforces the need to evaluate response options (Yin *et al.* 2003; Fan and Li 2006). The Delta offers important opportunities for studies of the human dimensions of global change.

13.3.2 Scenario conditions

One of the distinctive features of climate change impact assessment is the design of scenarios representing different future conditions. Assessing the implications of different response options, policies, or climate change for achieving SRD is as much an art as it is a science. This is because of uncertainties about future conditions, such as the magnitude of warming, the timing of climate change, the vulnerabilities to climate change, and other factors such as future societal demand associated with population growth and income increase, economic development, and institutional and technological changes. In response to these uncertainties, scenarios can be created to represent alternative future conditions. In the case study, three types of scenarios were specified: climate change, future socio-economic conditions, and response options.

In developing climate change scenarios, the study identified a set of baseline assumptions and conditions. Zhao *et al.* (2000) combined results from seven GCMs (GFDL, GISS, LLNL, OSU, MPI, UKMOL, and UKMOH) and suggested that the temperature would rise by about 0.8–3.5 °C in China, and about 1.4–2.5 °C in the Yangtze Delta under doubling of CO_2 conditions. In conducting their sectoral impact assessments, other participants of the Zhao study applied these same assumptions and scenarios. Combining results of GCMs and statistics, and taking uncertainty into account, this study used the following climate change scenarios for the Yangtse Delta:

- temperature rises: 0 °C, 0.5 °C, 1.0 °C, 1.5 °C, 2.0 °C, 2.5 °C
- precipitation varies: +20%, +10%, 0.0%, −10%, −20%.

Changes in socio-economic conditions, such as population, income, technology, and consumption rates were taken into consideration in developing baseline socio-economic scenarios. Based on the environmental and economic development plans of the provincial government, and the 1996–2020 government development

Table 13.1. *Regional development plan of the Yangtze Delta (annual growth rate).*

Item	1995–2000	2000–2010	2010–2020	2020–2030	2030–2050
Population	0.91%	0.50%	0.35%	0.15%	0%
Total grain production	0.90%	0.29%	0.28%	0.27%	0.26%
Yields of grain, cotton, oil seeds	1.92%	0.95%	0.95%	0.95%	0.91%
Aquatics	5.73%	1.60%	1.38%	1.21%	1.03%
Silk production	6.56%	2.25%	1.83%	1.55%	1.26%
Rural town industry	13.25%	12.40%	8.87%	4.34%	2.15%
Urban industry	12.50%	11.99%	8.01%	3.93%	1.95%
Water consumption per industrial production value	−11.00%	−9.50%	−8.44%	−7.84%	−5.28%
Urban domestic water consumption	5.00%	3.00%	2.00%	2.00%	1.00%
Rural domestic water consumption	8.50%	5.00%	3.00%	3.00%	2.00%
Sewage discharge control	37%	50%	59%	72%	90%
Decrease in SO_2	30%	60%	85%	90%	95%
Energy consumption	−6.0%	−5.5%	−4.5%	−3.5%	−3.0%

strategy, the 1996–2050 economic and social development scenario of the Yangtze Delta was specified for the case study. Additional data required for the identification of socio-economic scenarios came mainly from government documents and reports including the population development strategy, the agriculture and rural area modernization strategy, the fishery development strategy, the industry development strategy, the industrial structure and productive forces distribution strategy, and the urbanization plan. The socio-economic scenario for this study is presented in Table 13.1. Readers who are interested in more detailed information on future climate and socio-economic scenarios for the study can refer to Yin *et al.* (2003).

13.3.3 Design indicators/goals to measure regional sustainability

Successful implementation of the sustainability concept will require new approaches, which link climate change impact assessment and sustainability evaluation. In this respect, regional sustainable development goals and indicators must be set and the impacts of climate change on these indicators must be measured in a manner that integrates social, environmental, and economic parameters.

Maclaren (1996) suggests six general frameworks that can be adopted for developing sustainability indicators. The first is a domain-based framework that groups indicators according to the three main pillars of sustainability (economic, environmental, and social). The three-dimensional nature of sustainability and the need to make tradeoffs (e.g., between economic growth and environmental quality) require that these three components are maintained in a dynamic balance. Sustainability indicators should include economic, social, and environmental information in an integrated manner (Tschirley 1996).

Another important framework is a goal-based indicator system. In a study for sustainable agriculture in Australia and New Zealand, goals identified reflected the major concerns of agricultural development. These concerns arise from the national objectives of economic viability, maintenance of the resource base, and minimizing the impacts of agriculture on natural ecosystems. Each goal is composed of a number of attributes or indicators that are measurable in most cases by using existing sources of information (Hamblin 1994).

Other types of indicator frameworks include sector-based, issue-based, cause–effect, and combination ones. Indicators can be developed for each sector or community to measure the condition and trends in each critical sector. Maclaren (1996) describes these indicator frameworks in some detail. No single indicator would be sufficient to determine the sustainability or non-sustainability of a region or a system. A set of goals and/or indicators is required in sustainability evaluation. Notwithstanding the risks in using aggregated indicators, there are also risks in using too many indicators. A large number of indicators may lead decision makers to select only those that support their particular purpose. It may also cause confusion in making tradeoffs among indicators.

The term indicator is not a new concept. Indicators have been used to measure the performance of regional development policies or plans, to identify growth trends, to monitor the social and economic conditions of regions or nations, to inform the general public, to define planning goals or objectives, to guide strategic development options, and to compare different regions. For example, GDP, housing price, unemployment rate, and stock indices are used commonly to measure the social or economic performance of a society or economy. While these indicators are still strongly influencing decision making by government, policy makers, and the general public, they have shortcomings when used for measuring sustainable development. Research has been initiated in developing indicators of societal sustainability. For example, the World Resources Institute (WRI) developed a systematic approach that uses environmental indicators to measure environmental policy performance in the context of sustainable development (Hammond *et al.* 1995).

Indicators may be conflicting in that the achievement of one precludes the achievement of another. Possible tradeoffs between indicators therefore need to be identified. Very often the tradeoff relations are non-linear; creating situations of dramatic changes in the attainment of certain indicator levels once a threshold has

Table 13.2. *Regional sustainability indicators used in ILAF application.*

Goals	Indicator
Economic growth	Economic return, energy consumption, industry productivity
Sustainable resource use	Agricultural production, water use, and forest area protection
Environmental quality	Wastewater control, emission reduction of CO_2, methane (CH_4), and SO_2

been surpassed. Other indicators, however, are complementary; that is, by increasing the attainment of one indicator target, it is possible to increase the attainment of other indicators. It has been suggested, for example, that development and environment are complementary up to some level of resource use. Indicators are also considered compatible when the attainment of one does not sacrifice the attainment of others.

Regional sustainability goals/indicators selected for ILAF

Among the stakeholders in the Yangtze River Delta there is a range of different values or preferences in dealing with climate change and regional development. To select regional sustainability goals or indicators, the first major source of information used for the study was government reports, documents, and other published materials on resource issues (Yin *et al.* 1999). In order to improve the reliability of the information on goals derived from existing sources, an interview with senior decision makers in the region was conducted to discuss major policy issues related to climate change. Based on key policy concerns in the region, indicators for measuring regional sustainability for different climate change scenarios were identified. These indicators are listed in Table 13.2.

The aim of Deng Xiaoping's economic reform was to quadruple GDP in China by the end of the last century. Thus, raising per capita income, the consumption level, and the living standards of people was the primary goal of Chinese development. It is obvious that economic efficiency is one of the most important indicators for measuring development performance. In the climate change study, the economic impacts of climate change scenarios are of particular interest for decision makers. Most adaptation options also attempt to reduce economic costs. Improvements in economic efficiency occur in principle as long as gainers can compensate losers. For this to take place, total benefits must exceed the change in total costs. Assuming that proper account is taken of all resource costs, measures of improvements in economic efficiency are an important element in ensuring that society is reaping the benefits from its resource use.

In China, providing enough food for 1.3 billion people is always a big challenge. There has been an increasing concern about China's food supply and its ability to feed itself (Song 1997). While Brown (1995), on the one hand, warns that the rising dependence on grain imports in China will affect international food prices significantly, others argue that China is capable of producing enough grain to satisfy its needs in the twenty-first century (Rosegrant *et al.* 1995; Song 1997). The provision of adequate food on a continual basis is a major indicator of regional development. The agricultural production indicator reflects the ability of the land base to maintain in perpetuity a given flow of goods and services. Sustainable agricultural production can also be considered as a security goal to achieve higher levels of self-sufficiency and/or it may be used to represent an intergenerational concern to safeguard the resource base for present and future generations.

Many of the industrial and housing developments occur in productive farm-lands, forestry lands, and areas which are perceived to be of natural, historical, cultural, scenic, or scientific importance. Slowing the conversion of farmland to urban and industrial uses is critical for regional sustainability in China. In 1994, the State Council of China issued a Capital Farmland Protection Regulation to control land conversion from agricultural to other uses. To further emphasize the government's concern about the issue of farmland conversion, the Central Commu-nist Party Committee and the State Council published a special document entitled *Further Improving Land Management and Protecting Arable Land* in May 1997. An indicator on protection and conservation of arable land reflects this concern.

It is now generally realized that environmental concerns should be incorporated in decision making in an effort to achieve sustainable development (WCED 1987). Facing rising international pressure on large developing country emitters to take more action on climate change, China is responding positively to reduce CO_2 emission. The Chinese government recently formulated a national action plan in response to climate change according to the UNFCCC (National Development and Reform Commission 2007). In this case study, environmental concern is reflected in the indicators of greenhouse gas (CO_2, CH_4) emission control, wastewater discharge control, and SO_2 reduction.

There is an increasing concern about the implications of climate change for water management (Gleick 1993; Shi 1995). Global warming may change average and extreme high and low river flows, and sediment load in water bodies. Changing water quantity and quality induced by climate warming may increase water use conflicts in the region. Dealing with potential water use conflicts with changing climate is therefore considered to be an important indicator.

Climate change in the region would lead to a series of changes to energy consumption. With a rapid economic growth, energy consumption in China has increased significantly. From 1993 to 2004, China changed from one of the world's largest coal exporters to the second largest oil importer to meet its ever increasing energy demands (Pan 2005). Energy security is another major concern for Chinese decision makers in achieving sustainability. Thus, increasing energy use efficiency is another indicator of regional sustainable development.

Goals/indicators priority setting

Given the fact that not all the goals can be achieved simultaneously, a choice must be made to place different priorities for different goals using multi-criteria decision making. Much of the effort in MCDM has been devoted to constructing the preference relations between goals. An MCDM technique, the analytic hierarchy process (AHP) developed by Saaty (1980), was used to assist goal prioritization in this study. AHP provided a means by which alternative goals identified were compared and evaluated in an orderly manner.

The AHP requires decision makers to provide judgments on the relative importance of each of the goals. The result of the AHP analysis is a prioritized ranking indicating the overall preference for each of the goals. Another feature of the AHP is that it provides a flexible framework for public participation in decision making or problem solving. In an AHP exercise, a decision maker compares goals two at a time, which is termed pairwise comparison. The AHP employs an underlying scale with values from 1 to 9 to score the relative preferences for two items. It should be noted that the scores selected represent linguistic responses and are not strict mathematical ratios. The scores can be determined with qualitative and subjective information. These values were used as input to construct a pairwise comparison matrix (Yin 2001).

Yin and Cohen (1994) presented a systematic approach, assisted by AHP, to identify and prioritize regional sustainability goals/indicators relating to climate change. In the AHP results, a stakeholder's preferences for goals/indicators are expressed by goal ranking orders or the relative importance of goals on an ordinal scale. That is, goals are ranked as "most important" or first priority, "next most important" or second priority, and so on. Thus, the results of priority ranking represent stakeholders' preferences for a set of goals.

The results of the goal priority rankings were later incorporated in the ILAF model to examine the regional impacts of climate change on sustainability. In particular, the objective function of the GP model was set with the identified regional goal priority ranking.

13.3.4 Climate change impacts assessment

Detailed analyses of the social, economic, and environmental impacts (negative and positive) of alternative climate change scenarios for different economic sectors were undertaken in the study for key sectors that were sensitive to climate change. To identify the impacts/vulnerabilities of ecosystems, industry, food and fiber systems, water resources, and energy to climate changes, various simulation and statistical models were employed. These sectoral impact assessments have been presented elsewhere and are only briefly discussed here (Yin *et al.* 2003).

Data required for the integrated impact assessment come from several sources: existing data derived from previous studies on land resource analysis and management, government documents, reports, and scientific literature. The social, economic, and environmental impacts of alternative climate change scenarios for different economic sectors were calculated by a number of individual researchers in the overall project (Yin *et al.* 2003). For example, Zhang *et al.* (1999) and Miao *et al.* (1999) provided data on climate change impacts for the water sector. Impacts of climate change scenarios on crop yields were calculated by Zheng *et al.* (1999). Impact data for the energy and industrial sectors were derived from studies conducted by Wu *et al.* (1999).

Significant computer modeling efforts were employed to fill some of the data gaps for key sectors that would be sensitive to climate change. Computer techniques such as simulation models and statistical methods were used to provide additional information on climate change impacts. For example, CO_2, CH_4, and SO_2 coefficients were calculated using statistical models (Zhou *et al.* 1999).

Ma *et al.* (1999) developed an integrated database for the Yangtze Delta study. Data required for coefficients of various activities include the prices of products, costs of production, average yields, areas of different types of land, pollutant emission rates, and impacts data. The data collected also have spatial and temporal dimensions. The model variables and parameters differ among cities, and vary between the present and the future (changed climate conditions). Thus, the database consists of information for each city under both current and future conditions. A GIS system was used for data manipulation and to display the results.

The information collected was sorted into cities and land-use activities. Economic activities considered for each sensitive sector were consistent with those selected for sectoral analyses. These activities were represented in the model by decision variables. The IA model is flexible enough to incorporate other variables for assessment. With climate change impact data provided by sectoral impact assessments, four sectors were selected in the IA model: agriculture, forestry, fresh water fishery, and manufacturing. Land-use activities considered for the agricultural sector included wheat, rice, cotton, and canola. These crops might be grown only in certain cultivated land areas. In the Yangtze Delta, farmers mainly adopt a double-cropping system: growing rice and cotton in the summer, and wheat and canola in the winter. The activities in the forest sector included trees for timber production, and fruit and tea trees. Fresh water bodies were assumed to be areas used for fish farming activity.

13.3.5 Integrated assessment assisted by goal programming (GP)

It is obvious that decision makers have difficulty in relating to SRD using findings based on a range of impact results from many individual studies. How to determine climate change scenarios that most affect regional sustainability is still unclear. In this respect, the GP model was applied in the case study to identify implications of climate change scenarios for SRD. A brief introduction of the GP model is presented below to show how sustainability goals/indicators and climate change impacts can be represented in the model to study the implications of climate change for regional sustainability.

The basic structure of the GP model

The GP model provides a means for integrating climate change impacts data provided by sectoral studies. For the integrated model, linkages between climate change impact assessment and regional sustainability need to be incorporated in the structure of the model by a clear articulation and reconciliation of objective functions and decision variables. The basic structure of the GP model adopted in the ILAF includes goals and constraints. The specific equations of the model are grouped into the following types: resource and other restrictions, supply–demand balances, goal constraints, and the objective function (Ignizio 1976). The mathematics of the goal programming model designed for this study is provided in the appendix to the chapter.

The objective function or achievement function of the GP model is the minimization of non-attainment of defined target levels of sustainability goals/indicators. The purpose of the model solution is thus to achieve as much of the target levels for indicators as possible given certain conditions. In the case study, regional sustainability indicators identified previously were represented in the objective function and goal constraints of the model.

A comprehensive survey was conducted to collect required data for AHP analysis. Survey design and implementation was tempered by factors such as financial resources and manpower limitations and was conducted among 30 representative samples. While the sample number is limited, considerable knowledge and experience of representatives from several stakeholders was crucial for this survey. Sampled representatives were selected considering three societal groups: government agencies, academic institutes, and public organizations. The specific sample distribution is listed in Table 13.3.

Table 13.3. *Sample representatives of various stakeholders.*

Government agencies	Planning Commission: Long-Term Development Division (1); Societal Division (1) Science Commission (1); Economic Commission (1); Transportation Bureau (1); Family Planning Commission (1); Construction Commission (1); Agriculture Commission (1); Education Commission (1); Health Bureau (1); Environmental Protection Bureau (1); Welfare Bureau (1); Civil Affairs Bureau (1); Labour Bureau (1); Urban Planning Bureau (1)
Academic institutions	Fudan University (3); East China Normal University (3); Tongji University (2); Academy of Social Sciences (2)
Public organizations	Labour Union (1); Women's Organization (1); Rural Community (1); Township Community (1); City Community (1)

Note: numerical numbers in parentheses are numbers of representatives from the agency.

Table 13.4. *The rank ordering of sustainability indicators.*

Goals/indicators	G1	G2	G3	G4	G5	G6	G7	G8	G9	G10	G11
Priority rank	1	2	2	3	4	9	8	6	5	5	7

Note: G1, economic return; G2, grain production; G3, cash crop production; G4, arable land protection; G5, industry; G6, energy consumption; G7, water use; G8, wastewater control; G9, control CO_2 and methane emission; G10, control other air pollution; and G11, forest area protection.

The survey results from each responding person were used to establish a pairwise comparison matrix. An existing software package, Expert Choice (EC), was used to calculate the priority ranks. Results of the AHP application illustrated the rank ordering of sustainability indicators proposed for the study (see Table 13.4), which were used as inputs to represent the priority ranks of indicators in the objective function. Different indicators can have the same priority rank. Government documents and regional strategic plans provided extensive data on target levels as well as other required information (see Table 13.1).

Identify implications of climate change scenarios for sustainability

The primary purpose of the application of the GP model in this study was to identify the implications of various climate change scenarios for regional sustainability. In this connection, the parameters of the GP model were modified to reflect conditions under certain scenarios. The model-solving procedure of the scenario analysis is similar to sensitivity analysis. It is an iterative process and the results of alternative runs representing different scenarios can be compared with baseline scenarios.

Alternative climate change scenarios were specified in the model to explore the possible implications of climate change for regional sustainability. For example, climate change scenarios might significantly affect the quality/quantity of resources available for crop and forest production. Such changes might also be accommodated by altering the yield coefficients of the model from current yields to adjusted yields that reflect conditions under climate change. By proceeding in this manner through a series of scenarios, it is possible to evaluate whether the changes that may occur are in keeping with the stated regional sustainability goals or indicators. Sometimes it is preferable to make only one change at a time, and then obtain a solution before making further changes. This permits identification of the impacts of each individual climate change scenario. Commonly, several changes (climate and other socio-economic changes) are needed to reflect future scenario conditions.

It is recognized that future environmental and socio-economic conditions will change even if the climate were not to change. The implications of socio-economic changes for regional sustainability can be at least as significant as the implications of potential climate change. Thus, the development of socio-economic scenarios is a critical step in the study. To identify the effects of climate change on regional sustainability, it is necessary to separate impacts of climate change from those caused by other independent environmental and socio-economic changes that will occur in the future. In this respect, the IA first ran the GP model with baseline conditions that represent future environmental and socio-economic conditions without taking climate change into consideration. The results of baseline conditions runs were then compared with results of the GP model runs with future environmental and socio-economic conditions coupled with climate change scenarios.

After calibration using 1995 statistical data, the model was run repeatedly with future socio-economic and environmental scenarios without climate changes for 2000, 2010, 2030, and 2050 (Zhou *et al.* 1999). Then, climate change scenarios were incorporated into the model associated with future socio-economic and other environmental conditions. The differences between model run results with and without climate change scenarios indicated the impacts of climate change on the achievement of sustainability. The detailed results of the model runs are presented in Yin *et al.* (1999, 2003). Only the impact results of the model runs for the year 2050 with a climate change scenario of a 2.5 °C temperature increase and 20% increase in precipitation (Table 13.5) are discussed here for illustrative purposes. Negative numbers in Table 13.5 denote reduced quantities under climate change conditions.

Table 13.5. *Climate change impacts in 2050 ($\Delta t = 2.5\,°C$, $\Delta p = 20\%$).*

City	Wheat area (ha)	Wheat production (ton)	Rice area (ha)	Rice production (ton)	Oil seed area (ha)	Oil seed production (ton)	Urban industry production value (10000 Yuan)	Rural town production value (10000 Yuan)
Shanghai	9438	−148 827	0	−387 088	−9438	−28 683	−43 289 707	−20 793 040
Changzhou	273	−136 251	2 001	−244 454	−2 274	−5 760	−3 185 502	−8 051 071
Nanjing	443	−147 728	44 843	268 368	−45 286	−135 671	−10 684 248	−4 679 046
Suzhou	13 453	−198 374	0	−439 265	0	0	−8 373 325	−28 851 640
Wuxi	0	−149 316	0	−235 711	0	0	−5 182 171	−24 953 987
Zhenjiang	239	−122 190	13 229	−23 553	−13 469	−40 744	−1 514 224	−5 042 112
Yangzhou	0	−634 115	0	−708 449	0	0	−8 012 123	−9 348 988
Nantong	−107 377	−683 318	0	−402 899	107 377	445 813	−3 626 378	−5 731 140
Jiaxing	37 196	119 535	0	−404 478	−37 196	−114 081	−2 475 844	−8 571 598
Hangzhou	46 648	126 604	0	−371 726	−46 648	−103 034	−8 343 907	−11 297 275
Huzhou	6 577	1 849	0	−270 658	−6 577	−17 837	−3 984 914	−2 671 482
Total	6 893	−1 972 132	60 074	−3 219 916	−56 514	0	−98 672 349	−129 991 385

Table 13.5. (cont.)

City	Electricity consumption (10000 kWh)	Coal consumption (ton)	Petroleum consumption (ton)	Water resource consumption (10000 ton)	Sewage emission (10000 ton)	CO_2 emission (ton)	CH_4 emission (ton)	SO_2 emission (ton)
Shanghai	−626586	−253279	−74329	2036595	−14273	−842177	0	−402
Changzhou	−77240	−126101	−5047	363502	−2553	−377578	4351	−222
Nanjing	−127584	−235171	−262229	587736	−10928	−1465486	97512	−487
Suzhou	133646	−334213	−14433	1182673	−4107	−1027506	0	−606
Wuxi	175059	−160083	−9665	958845	−2770	−488417	0	−282
Zhenjiang	−56533	−194636	−5149	217921	−1393	−505416	28758	−301
Yangzhou	0	−114552	−18620	891095	−2036	−523074	0	−275
Nantong	−81214	−128725	−2904	305672	−588	−416184	0	−248
Jiaxing	−78104	−103109	−2947	361249	−1152	−3364443	0	−218
Hangzhou	−161041	−155589	−5368	643790	−6036	−383701	0	−227
Huzhou	−55142	−111425	−1344	218714	−690	−254849	0	−155
Total	−954742	−1916887	−402040	7767797	−46531	−6648836	130631	−3429

In the Yangtze Delta, it is estimated that the climate change scenario (a 2.5 °C temperature increase and a 20% increase in precipitation) in the year 2050 might increase the area of wheat production by 7000 ha. But total wheat production would decline by 1.97 million tonnes or 28%. Spatially, production reduction would be mostly experienced in Yangzhou and Nantong City, while wheat production in Hangzhou, Huzhou, and Jaxing City (in Zhejiang Province) would be increased, associated with an increased production area. The results also indicate that while a warmer climate may increase the total area suitable for rice production in the Delta (60 000 ha), wetter conditions under the climate change scenario would reduce rice production by 3.21 million tonnes or 10.7% of the regional total.

A different impact pattern was identified for the growth of vegetable oil seed. Potential production remains unchanged with a reduction of production area under a warmer climate. In addition, the study results suggest that the industrial sector in the region may experience potential risks of production value decline by RMB 986.7 billion or 4.3% under the climate change scenario. Shanghai, Nanjing, and Suzhou would suffer the most.

Referring to the results in Table 13.5, significant impacts are identified under the climate change scenario on net economic return to village and township enterprises (VTEs). The economic loss is RMB 1299.9 billion or 4.4% in the year 2050 compared with the results of the baseline case. It is estimated that climate change would result in a more significant economic return reduction to VTEs in the cities of Suzhou and Wuxi (Jiangsu Province). This shows that in the absence of adaptation the more developed areas would suffer more in terms of declining economic return.

A significant increase in water consumption (doubling) is found for the region under the climate change scenario. While water consumption may increase by 77.6 billion tonnes, wastewater effluent would only decline by 0.4 billion tonnes. This suggests that water quality would get worse under climate change, since wastewater effluent is not reduced at the same rate as the increase in water consumption.

While the CO_2 and SO_2 emissions are cut by 660 000 tonnes or 0.7%, and 3 000 tonnes, respectively, methane emissions increase by 13 000 tonnes or 2.3%. The results of the assessment indicate that air pollutants are not significantly reduced even with a considerable decline of industrial production associated with a warming climate.

Response policy analysis

The IA model can also be used for policy analysis to estimate the likely consequences of potential adaptation policies on regional sustainability goal achievement. This type of information provides a basis for planners or decision makers to determine the adequacy and effectiveness of the policy before it is implemented. In policy analysis, a potential response option can be specified as a policy scenario. The policy scenario can be represented in the model by adjusting the model parameters or structure. The aim of the adaptation policy is to reduce negative impacts of the climate change scenario on regional sustainability.

Since adaptation policies possess different characteristics, implementing them would have various impacts on different locations and on different goal achievements. Each adaptation option may cause both positive and negative impacts. For example, a new irrigation system may reduce negative impacts on crop yield, but may also create negative impacts on the water balance goal.

Adaptation policies will influence resource production, the land availability for each sector, demands for resource products, greenhouse gas emission and water consumption rates, and other factors relating to sustainability. These policy consequences are represented in the GP model during the evaluation process by modifying parameters in the coefficient matrix, the right-hand side (RHS) vector, and the objective function.

In policy evaluation, the effectiveness of alternative policies can be measured by relating their various effects to relevant goals/indicators. The baseline scenario is also used for comparison. Alternative policy scenarios can then be created to reflect conditions coupled with a specific adaptation policy to deal with climate change impacts. The comparison of the results of the policy scenario and the baseline scenario shows the different goal achievements under the two scenarios. If the goal achievements are improved significantly under the policy scenario, then this policy is assumed to be effective.

The model can be run repeatedly with a set of alternative policy scenarios. By proceeding in this manner through a series of scenarios, it is possible to evaluate whether the policies are in keeping with the stated goals or indicators, and the desirable policy options can be identified. Thus the most effective or desirable adaptation options can be identified to improve regional sustainability under climate change conditions. In the Yangtze Delta case study, however, only a limited number of model runs were carried out for policy scenarios (Yin *et al.* 1999). The response policy scenarios were designed mainly to represent different urban growth management options and energy consumption control measures. The results of the policy evaluation showed the effectiveness of alternative policies in terms of improving goal achievement. For a detailed discussion on adaptation policy evaluation, readers can refer to Yin *et al.* (1999, 2003).

13.4 Conclusion

The preceding discussion has illustrated an integrated assessment approach that can be employed to link climate change assessment and sustainability evaluation.

A more detailed discussion on the main techniques employed to form the integrated approach can be found in other articles (Yin and Cohen 1994; Yin *et al.* 1994, 1999, 2003, 2006).

In summary, the chief contribution of this study is not so much to provide information or solutions for improving regional sustainability under climate change conditions. Rather, it is to provide procedures for integrating regional sustainability goals/indicators development and a range of economic sectors, in order to investigate systematically the impacts of climate change scenarios on regional sustainability. In this sense the model developed is more for heuristic purposes. The ILAF model possesses some characteristics of a learning tool and as a means of communication.

Although an extensive effort has been made in model development, there are limitations in the integrated assessment system and thus there is room for improvement. Model testing and validation are critical to ensure that potential users have confidence in the results from the model. In this study, considerable effort has been made in the model construction and application phases to detect possible errors and unreliable aspects in the model. The case study provided a good opportunity for testing the ILAF system. However, the study has not tested the model systematically and comprehensively with respect to model sensitivity. In order to realize the potential of the ILAF system as a means to provide meaningful and reliable guidelines for policy making considering climate change, the model must be tested further.

Acknowledgements

This study was originally funded by the Government of Canada's Green Plan, Environment Canada and the State Environmental Protection Administration of China (SEPA Project Code 95301). The author would like to thank the Research Grants Council of Hong Kong for providing a grant to complete the study. The author received additional funds from Environment Canada to carry out the study. Thanks are also due to Suoquan Zhou, Qilong Miao, Guangsheng Tian, and many other people who participated in the study.

References

Brown, L. R. 1995. *Who Will Feed China? Wake-Up Call for a Small Planet*. New York: Norton.

Carter, T. R., M. L. Parry, H. Harasawa, and S. Nishioka (eds.) 1994. *IPCC Technical Guidelines for Assessing Climate Change Impacts and Adaptations*. London: Department of Geography, University College, London.

Cohen, S. J., D. Demeritt, J. Robinson, and D. Rothman 1998. Climate change and sustainable development: towards dialogue. *Global Environmental Change*, **8**(4), 341–371.

Fan, D. D. and C. X. Li 2006. Complexities of China's coast in response to climate change. *Advances in Climate Change Research*, **2**(suppl. 1), 54–58.

Gleick, P. H. (ed.) 1993. *Water in Crisis: A Guide to the World's Fresh Water Resources*. New York: Oxford University Press.

Hamblin, A. 1994. Indicators of sustainable agriculture in Australia and New Zealand. In *FAO Report, Technology Assessment and Transfer for Sustainable Agriculture and Rural Development in the Asia–Pacific Region*.

Hammond, A., A. Adriaanse, E. Rodenburg, D. Bryant, and R. Woodward 1995. *Environmental Indicators: a Systematic Approach to Measuring and Reporting on Environmental Policy Performance in the Context of Sustainable Development*. Washington, DC: World Resources Institute.

Ignizio, J. P. 1976. *Goal Programming and Extensions*. Lexington, MA: Lexington Books.

IPCC 2007. *Climate Change 2007: Impacts, Adaptation and Vulnerability Working Group II Contribution to the Intergovernmental Panel on Climate Change Fourth Assessment Report – Summary for Policymakers*. Geneva: IPCC Secretariat.

Ma, L., X. Y. Gu, W. Q. Liu, and J. H. Jiang 1999. A management system of climate and socio-economic information in the Yangtze Delta. *Journal of Meteorology, Nanjing Institute of Meteorology*, **22**, 558–564.

Maclaren, V. W. 1996. Urban sustainability reporting APA. *Journal of the American Planning Association*, **62**(2), 184–202.

Miao, Q. L., Y. Q. Zhang, L. Jin, and Y. Y. Xiang 1999. Impacts of climate change on agricultural water consumption in the Yangtze Delta. *Journal of Meteorology, Nanjing Institute of Meteorology*, **22**, 518–522.

Munasinghe, M. 1998. Climate change decision-making: science, policy and economics. *International Journal of Environment and Pollution*, **10**(2), 188–239.

National Development and Reform Commission 2007. *China's National Climate Change Programme*. Beijing: Government of the People's Republic of China.

Pan, J. H. 2005. *China and climate change: the role of the energy sector*, Science and Development Network, June 2005. http://www.scidev.net/dossiers/index. cfm?fuseaction=policybrief&policy=64&dossier=4

Pearce, D. and R. K. Turner 1990. *Economics of Natural Resources and the Environment*. New York: Harvester Wheatsheaf.

Robinson, J. B. and C. Van Bers 1996. *Living within our Means: Foundations of Sustainability*. Vancouver: The David Suzuki Foundation.

Rosegrant, M. W., M. Agcaoili-Sombilla, and N. D. Perez 1995. Global food implications to 2020: implications for investment. In *Food, Agriculture, and the Environment Discussion Paper 5*. Washington, DC: International Food Policy Research Institute.

Saaty, T. L. 1980. *The Analytic Hierarchy Process*. New York: McGraw-Hill.

Shi, Y. F. 1995. *Impacts of Climate Change on Water Resources in North-western and Northern China*. Jinan: Shandong Science and Technology Press (in Chinese).

Song, Jian 1997. No impasse for China's development. *Beijing Review*, May 12–18, 18–22.

Tschirley, J. 1996. *Use of Indicators in Sustainable Agriculture and Rural Development*. A Report by Sustainable Development, Environment and Natural Resources Service (SDRN), Research, Extension and Training Division, FAO, Rome, Italy.

UN 2004. *Implementation of the United Nations Millennium Declaration: Report of the Secretary-General*. United Nations, A/59/282. New York.

UNDP 2006. *Making Progress on Environmental Sustainability: Lessons and Recommendations in over 150 Country Experiences*, United Nations Development Programme, New York.

WCED 1987. *Our Common Future*. The World Commission on Environment and Development. New York: Oxford University Press.

Wu, X., Q. L. Miao, X. Y. Gu, and L. Jin 1999. Climate change impacts on the industry and energy consumption in the Yangtze Delta. *Journal of Meteorology, Nanjing Institute of Meteorology*, **22**, 541–546.

Xu, Y., Y. H. Ding, Z. C. Zhao, and J. Zhang 2003. A scenario of seasonal climate change of the 21st century in Northwest China. *Climatic and Environmental Research*, **8**(1), 19–25 (in Chinese).

Yin, Y. 2001. *Designing an Integrated Approach for Evaluating Adaptation Options to Reduce Climate Change Vulnerability in the Georgia Basin*. Final Report Submitted to Adaptation Liaison Office, Climate Change Action Fund, Ottawa, Canada.

Yin, Y. and S. Cohen 1994. Identifying regional policy concerns associated with global climate change. *Global Environmental Change*, **4**(3), 245–260.

Yin, Y., P. Gong, and S. Cohen 1994. An integrated database for climate change impact assessment. *Canadian Journal of Remote Sensing*, **20**(4), 426–434.

Yin, Y., Q. Miao, and G. Tian (eds.) 1999. Climate change impact assessment and sustainable regional development in the Yangtze Delta. Special Issue of *Journal of Meteorology, Nanjing Institute of Meteorology*, China (in English and Chinese).

Yin, Y., Q. L. Miao, and G. S. Tian 2003. *Climate Change and Regional Sustainable Development: a Case in the Changjiang Delta of China*. New York: Sciences Press.

Yin, Y. Y., G. Cheng, Y. Ding, *et al.* 2006. *Vulnerability and Adaptation to Climate Variability and Change in Western China*. Final Report, AIACC Project No. AS25. Washington, DC: START/AIACC International START Secretariat. www. aiaccproject.org

Zhang, Y. Q., Y. Y. Xiang, Q. L. Miao, and F. C. Zhang 1999. An analysis of the impacts of climate change on water balance of supply and demand in the Yangtze Delta. *Journal of Meteorology, Nanjing Institute of Meteorology*, **22**, 529–535.

Zhao, Zongci, Y. Luo, and X. J. Gao 2000. GCM studies on anthropogenic climate change in China. *Acta Meteorologica Sinica*, **14**, 247–256.

Zheng, Y. F., C. J. Wan, Q. L. Miao, and F. C. Zhang 1999. Influence of climate change on rice photosynthetic yields. *Journal of Meteorology, Nanjing Institute of Meteorology*, **22**, 536–540.

Zhou, S. Q., Q. L. Miao, Y. Yin, and G. Tian 1999. The parameter scheme and prediction of the regional climate change impact assessment model for the Yangtze Delta. *Journal of Meteorology, Nanjing Institute of Meteorology*, **22**, 493–499.

Appendix

The mathematical form of the GP model is expressed as follows:

$$Z_{min} = \left[g_1\left(d^-, d^+\right), g_2\left(d^-, d^+\right), \ldots, g_k\left(d^-, d^+\right)\right] \tag{13.1}$$

$$\sum_p x_{pj} + \sum_j x_{ij}^\otimes - \sum_i x_{ji}^\otimes \leq A_j \tag{13.2}$$

$$\sum_p \sum_j \left(R_{pj} \times x_{pj}\right) + \sum_p \sum_i \sum_j \left(R_{pij}^\otimes \times x_{pij}^\otimes\right) + d_r^- - d_r^+ = b_r \tag{13.3}$$

$$\sum_{p}\sum_{j}\left[Y_{pj}\times\left(x_{pj}+\sum_{i}x_{pij}^{\otimes}\right)\right]+d_{y}^{-}-d_{y}^{+}=b_{y} \tag{13.4}$$

$$\sum_{p}\sum_{j}\left[U_{pj}\times\left(x_{pj}+\sum_{i}x_{pij}^{\otimes}\right)\right]+d_{u}^{-}-d_{u}^{+}=b_{u} \tag{13.5}$$

$$\sum_{p}\sum_{j}\left[T_{pj}\times\left(x_{pj}+\sum_{i}x_{pij}^{\otimes}\right)\right]+d_{t}^{-}-d_{t}^{+}=b_{t} \tag{13.6}$$

$$\sum_{p}\sum_{j}\left[E_{pj}\times\left(x_{pj}+\sum_{i}x_{pij}^{\otimes}\right)\right]+d_{e}^{-}-d_{e}^{+}=b_{e} \tag{13.7}$$

$$\sum_{p}\sum_{j}\left[S_{pj}\times\left(x_{pj}+\sum_{i}x_{pij}^{\otimes}\right)\right]+d_{s}^{-}-d_{s}^{+}=b_{s} \tag{13.8}$$

$$\sum_{p}\sum_{j}\left[W_{pj}\times\left(x_{pj}+\sum_{i}x_{pij}^{\otimes}\right)\right]+d_{w}^{-}-d_{w}^{+}=b_{w} \tag{13.9}$$

$$\sum_{p}\sum_{j}\left[H_{pj}\times\left(x_{pj}+\sum_{i}x_{pij}^{\otimes}\right)\right]+d_{h}^{-}-d_{h}^{+}=b_{h} \tag{13.10}$$

$$\sum_{l}\sum_{j}(EN_{lj}\times x_{lj})+d_{en}^{-}-d_{en}^{+}=b_{en} \tag{13.11}$$

$$\sum_{f}\sum_{j}\left[F_{fj}\times\left(x_{fj}+\sum_{i}x_{fij}^{\otimes}\right)\right]+d_{f}^{-}-d_{f}^{+}=b_{f} \tag{13.12}$$

$$\sum_{j}(V_{cj}\times x_{cj})+d_{v}^{-}-d_{v}^{+}=b_{v} \tag{13.13}$$

$$\sum_{i}x_{ji}^{\otimes}\leq A_{j}^{\otimes} \tag{13.14}$$

$$x,x^{\otimes},d^{-},d^{+}\geq 0 \tag{13.15}$$

where Z is the objective function or achievement function of the integrated model, which is to minimize the non-attainment of defined target levels of goals/indicators. $g_{k}(d^{-},d^{+})$ is a linear function of the deviation variables at priority level $k = 1, 2, \ldots, k$. x is area of land use; x_{pj} is the area of land use p in sector j; $x^{\otimes}$ is land conversion variable; $x_{ij}^{\otimes}$ and $x_{ji}^{\otimes}$ are two decision variables representing, respectively, land areas converted from sector i to j, and land areas converted from sector j to sector i; $x_{pij}^{\otimes}$ is area of land use p on converted land (from sector i) in sector j. A_{j} is the resource availability for sector j; $A_{j}^{\otimes}$ is the total area of land of sector j converted from other sectors. R_{pj} is net return for land use x_{p} in sector *j.* $R^{\otimes}_{pij}$ is the net return from converted land (from other sectors *i*) for land use *p* in sector *j*. Y_{pj} is the yield of land use *p* in sector *j*; U_{pj} is the city industrial output value of land use x_p in sector *j*; T_{pj} is the town-run industrial output value of land use x_p in sector *j*; E_{pj} is the GHG emission rate of land use *p* in sector $j (E = E_{CO_2} + E_{CH_4}).S_{pj}$ is the emission rate of sulfur dioxide for land use x_p in sector *j*; W_{pj} is the discharge rate of wastewater from land use x_p in *j* sector; H_{pj} is the water consumption rate for land use x_p in sector *j*. x_{lj} is land with energy consumption level *l* in sector *j*. EN_{lj} is the energy of *l* consumed by sector *j*. x_{fj} is the land area of tree species *f* (i.e., timber tree, fruit tree, or mulberry tree) planted in sector *j*. $x^{\otimes}_{fij}$ is the land area of tree species *f* (timber tree, fruit tree, or mulberry tree) transferred from sector *i* to *j*; x_{cj} is the water surface area of fresh water fishery in sector *j*. V_{cj} is the fishery output value in sector *j*. b_i is the right-hand side vector representing the target values for goals *y, r, u, t, e, h, s, en, f,* and *v* (resource production, economic return, industrial output, emission control, water and energy consumption control, and fish harvest) respectively; d^+ and d^- are the over-achievement and under-achievement vectors of goal target levels, respectively.

14

From CLIMPACTS to SimCLIM: development of an integrated assessment model system

RICHARD WARRICK

14.1 The context

The challenge of addressing the complex causes and consequences of global environmental changes has prompted researchers to turn increasingly to collaborative, interdisciplinary assessments that seek to integrate at the regional scale. In part, this trend acknowledges that whereas climate change is a global phenomenon, impacts and adaptation are inherently regional and local. For example, in the Third Assessment Report (TAR), the Intergovernmental Panel on Climate Change (IPCC) balanced its more traditional sectoral focus with integrated assessments at the world-regional level in an attempt to attain greater clarity on the extent and distribution of vulnerability to climate change (McCarthy *et al.* 2001). The Fourth Assessment Report followed suit (IPCC 2007a).

The 1990s witnessed the development of a number of integrated assessment models (IAMs) that could potentially be used in support of such regional assessments. In the context of climate change, most IAM developments were concentrated at the global/world-regional scale for purposes of assessing greenhouse gas mitigation options. Other IAMs actually spanned hierarchies of function, complexity, and scale (Weyant *et al.* 1996; Schneider 1997) and some began to capture sufficient detail to be used in integrated regional assessments (Tol and Fankhauser 1998), such as the Asia–Pacific Integrated Model (AIM; Kainuma *et al.*, Chapter 11, this volume).

Nonetheless, even at this sub-global scale such IAMs, due to their resolution and demands for spatial consistency, tended to obscure the richness of environmental and socio-economic characteristics of individual countries and localities. Yet it is precisely this sort of national and local information that is required to support policy and decision making as regards issues of climate change impacts and adaptation. Adaptation is a process that largely takes place at national and local scales. Few IAMs are focused at these scales.

The purpose of this chapter is to describe one integrated modeling system that seeks to do so – *SimCLIM*. The roots of SimCLIM stretch back to the early 1990s. SimCLIM was developed from a more targeted, location-specific model called CLIMPACTS, designed for assessing impacts and adaptation to climate variability and change specifically in New Zealand. CLIMPACTS, in turn, provided a foundation for other specialized applications, both within New Zealand and internationally. This chapter begins by describing the historical development of CLIMPACTS and its derivative models, and then looks at the more recent development of SimCLIM and its applications.

14.2 CLIMPACTS: developing the core structure and functionality

The CLIMPACTS program began in 1993, with funding from the New Zealand Government, and continued for 14 years.[1] The program was led by the International Global Change Institute (IGCI) at the University of Waikato and involved collaboration with other research organizations based in New Zealand.[2] From the outset, the focus of the program was on the development of the CLIMPACTS model system, designed to examine, in a spatial and temporal context, the sensitivity of the managed environment to climate change and variability (Kenny *et al.* 1995, 2000, 2001; Warrick *et al.* 1996, 2001).

14.2.1 Model components

In the context of the different types of IAMs described by Weyant *et al.* (1996), CLIMPACTS was conceived as a simple, vertically integrated "end-to-end" model. The basic model structure, which is retained in the more recent SimCLIM system, is shown in Figure 14.1. The connections between the model components are one-way, with no feedbacks between them, in what has been termed a "linked-model system" by Carter *et al.* (1994). There are three major components to the model system.

At the "top end," the first component contains the outputs of MAGICC (Model for the Assessment of Greenhouse gas Induced Climate Change: Wigley 2003) the energy-balance, box-diffusion–upwelling global climate model that has been used for producing global climate projections for IPCC. The inputs to MAGICC are emissions of greenhouse gases and the principal outputs are projections of

---

[1] From the Foundation for Research, Science and Technology.

[2] The CLIMPACTS program was led by IGCI in collaboration with the National Institute of Water and Atmospheric Research (NIWA), with contributions from Crop and Food Research, AgResearch, HortResearch, Landcare Research and the University of Auckland, all within New Zealand.

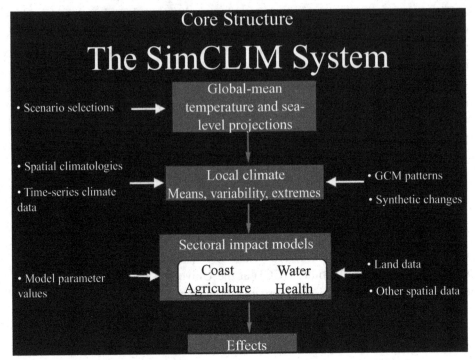

Figure 14.1. The core structure of both the earlier CLIMPACTS model for New Zealand and the more recent, generic SimCLIM system. The systems link models and data from global to local scales for the purpose of assessing impacts and adaptation.

global mean temperature and sea-level changes over the period 1990–2100. MAGICC itself contains models for converting greenhouse gas and other emissions to atmospheric concentrations (e.g., a carbon cycle model), and concentrations to radiative forcing, which drives a simple one-dimensional global climate model. MAGICC was used to convert emission scenarios into climate projections in the first three Assessment Reports produced by IPCC, in 1990, 1995 and 2001. Successive versions of CLIMPACTS, and later SimCLIM, incorporated the updated IPCC projections as they became available.[3]

The second component consists of climate and land data and the capacity for generating scenarios of climate change. The model system contains historical climate data, both time-series site data and spatially interpolated climate normals. In addition, general circulation model (GCM) results for monthly temperature

---

[3] In the early versions of CLIMPACTS, the option was available to run MAGICC "live" and to develop custom scenarios. However, the users interested in climate impacts and adaptation more often favored only the use of the tabulated outputs for the range of IPCC emission scenarios. Thus, later versions of CLIMPACTS and its derivative models did not have MAGICC itself linked to the system.

and precipitation are downscaled to the country or region of concern. For the New Zealand CLIMPACTS model, statistical downscaling techniques and limited area model (LAM) results were used for this purpose. These patterns are "normalized" to account for the "climate sensitivity" of the GCM in order to give a spatial pattern of climate change per degree of global warming.[4] These change patterns are scaled by the projections of global-mean temperature change in order to give time-dependent, spatial patterns of climate change for New Zealand. This "pattern-scaling" method of climate change scenario generation, which is central to CLIMPACTS and subsequently to SimCLIM, was first described by Santer *et al.* (1990; and later by Carter and La Rovere 2001) and applied, for example, by IPCC in 1990 (Mitchell *et al.* 1990) and by the MAGICC/SCENGEN system (Hulme *et al.* 2000). The scaled patterns of change can be used to perturb the present climatology, thus providing scenarios for future climates (spatially and for individual sites).[5]

Operationally, in generating scenarios of climate change the user takes account of three main areas of uncertainty: the regional variations of climate change as produced by GCMs; the sensitivity of the climate response to changes in greenhouse gas emissions due to climate feedback effects; and greenhouse gas emissions as they affect the rate of global warming. As illustrated in Figure 14.2, the user selects the future year of interest, a GCM pattern, a GHG emission scenario, and a value for the climate sensitivity. Thus, there is considerable scope for the user to exercise creativity in the development and application of scenarios.

The third component consists of a suite of impact models. The emphasis in the initial development of CLIMPACTS was on the agricultural sector – soils (soil carbon and moisture), grasses (invasive sub-tropical species and pasture production), fruit (kiwifruit and apples), and arable crops (grain maize, sweet corn, wheat, and barley). Variations of these models were incorporated that allowed both spatial and site-specific analyses (see below).

In developing CLIMPACTS, the "game plan" was to concentrate initially on the development of biophysical impact models. The intent was to first develop and test models to gain confidence in the ability to assess biophysical effects of climate change and variability on agriculture, and then to turn attention to other sectors and the human dimensions of change.

---

[4] The *climate sensitivity* conventionally refers to the equilibrium change in global-mean surface temperature for a doubling of carbon dioxide.

[5] In addition, in the Pacific region, the influence of ENSO and the Interdecadal Pacific Oscillation (IPO) are important components of the natural variability of the climate. Patterns of El Niño Southern Oscillation (ENSO) and the Interdecadal Pacific Oscillation (IPO) were therefore incorporated into the CLIMPACTS system (Mullan *et al.*, 2001). This provided the capacity to examine the present or future (in conjunction with GHG-induced effects) influence of such sources of natural variability on climate and sectoral impacts.

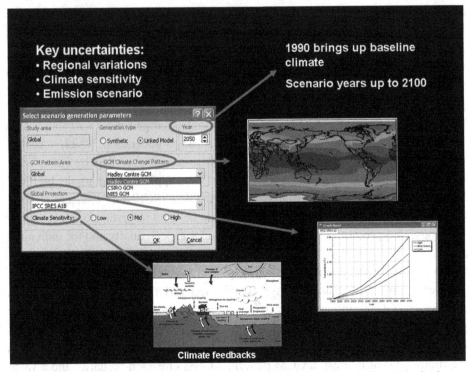

Figure 14.2. Generating scenarios using the method of pattern-scaling. Standard-ized (per degree of global warming) GCM change patterns are scaled by global-mean temperature change projections. In choosing amongst a range of GCM patterns, global emissions/warming projections, and different values for the cli-mate sensitivity (due to feedbacks in the climate system) within SimCLIM, the user takes account of a wide range of scientific and socio-economic uncertainties.

14.2.2 *Spatial and temporal modes of analysis*

One of the main aims of the CLIMPACTS program during the 1990s was to develop a "nested" capacity for multi-scale spatial and temporal analyses – nationally and sub-nationally, and at individual sites (Kenny *et al.* 2001a). As envisioned, this capacity would extend to the functions for generating scenarios of climate change, as well as to impact and adaptation analyses based on those scenarios. These functions provided the foundation for later models. They are depicted in Figure 14.3 which shows (for the current SimCLIM system – see Section 14.3) the main menu for selection of the mode of scenario generation and some illustrative output. With the spatial mode, one is able to generate "time-slice" comparisons of spatial changes in climate means (yearly, seasonally, or monthly). With the site-specific mode, one is able to generate projections of changes in climate variables or perturb existing time-series data (monthly, daily, or hourly).

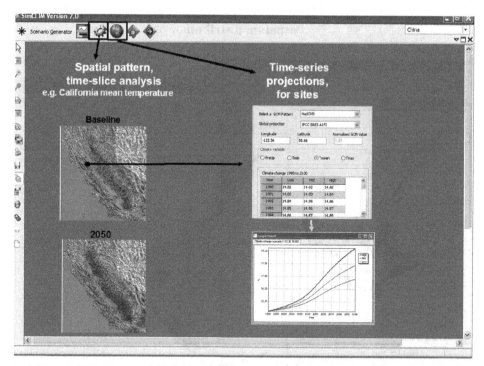

Figure 14.3. Two forms of climate change outputs (shown for SimCLIM system): (a) spatial, "time-slice" comparisons (left). In this example, the spatial pattern of the baseline mean-annual temperature (top left) is perturbed by the spatial pattern of temperature change for 2050 (bottom left), for California, USA. (b) site-specific, time-series projection (right). In this example, the mean-annual temperature (for low, mid, and high climate sensitivity values) for a user-selected location in California is projected forward to the year 2100 under the scenario of climate change.

Spatial "time-slice" analyses

During the 1990s, the initial effort of the CLIMPACTS program focused on the development of a *national-scale* capacity for spatial analyses that would allow "broad-brush" spatial analyses to be conducted over the North and South Islands of New Zealand. For this purpose, a basic Geographic Information System was developed, along with the scenario generator, described above, and monthly climatologies based on the period 1951–1980, spatially interpolated to a 0.05° latitude–longitude grid (approximately a 5 km grid). In addition to climate, the CLIMPACTS database also included an eight-category land classification, which allowed the user to filter land quality in relation to particular agricultural activities, and, later, digital elevation models. A range of agricultural impact models for spatial analyses related to pastoral activities, arable cropping, horticulture, and soils were developed for CLIMPACTS (descriptions of these models are available in Warrick *et al.*

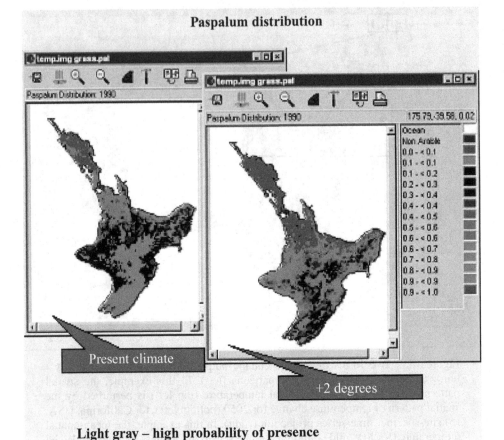

Figure 14.4. The distribution of Paspalum (*Paspalum dilitatum*), an invasive sub-tropical grass species found in North Island, New Zealand, as modeled under current climate and a two-degree climate warming scenario. The mapped values, as generated by the original CLIMPACTS model, indicate the probability of Paspalum being present. See Campbell *et al.* (1999) for methods.

2001).[6] These impact models, which were either developed purposefully for CLIMPACTS or modified from existing models in order to be run in a computationally efficient manner in this spatial context, were linked to the spatial climate scenario generator for national-scale analyses.

For example, Figure 14.4 shows the results of a model for Paspalum (*Paspalum dilitatum*), an invasive sub-tropical grass species that has gained a foothold in

---

[6] Additional "generic" tools were also developed for CLIMPACTS, including a *degree–day model* and an *atmospheric water balance* model, the results of which are calculated and displayed spatially. Tools are also available for mapping *threshold values* of climate variables or model outputs (e.g., a critical value of accumulated degree–days needed to mature a crop) and their changes over time. Functions are now available for manipulating or combining the spatial data with GIS-like tools.

New Zealand, particularly in the warmer regions of North Island. Under the current climate, there is a high probability of finding Paspalum in pastures in the far north of the Island. The selection of a synthetic scenario of climate change, a 2 °C warming applied uniformly over the year, results in a substantial southward migration of the areas subject to invasion by Paspalum. Such a prospect would be of serious concern to the major areas of dairying in the central (Waikato) and southern (Taranaki) portions of North Island.

At this national scale, it was envisioned that CLIMPACTS could provide end-users (e.g., central government) with broad-scale sensitivity analyses and impact assessments for national and international reporting purposes. Indeed, CLIMPACTS did serve as a methodological cornerstone of New Zealand's statements on impacts and adaptation in, for example, its national communications to the Parties to the UN Framework Convention on Climate Change (in Ministry for the Environment 1994, 1997, 2001b, 2002, 2006), its national report on climate impacts (Ministry for the Environment 2001a) and the Annual Reports of its National Science Strategy Committee for Climate Change (1997, 2001).

A complementary need for integrated assessment modeling was found to exist at the regional and local scale. In New Zealand, this scale is where much of the practical decision making regarding the potential impact of climate variability and change is focused.[7] Issues of climate variability and extremes – e.g., floods and droughts – as they pertain, for example, to water quantity and quality, biodiversity and land use, are often high on the agenda of regional councils and local government authorities. Consequently, the opportunities for climate change adaptation are often most perceived as incremental risk-reducing measures to cope with climatic hazards. This notion, from the user perspective, accords well with the viewpoint that human adaptation to climate change is largely a matter of adjusting to changes in variability, especially in the frequency and/or magnitude of extreme events (e.g., Smit and Pilifosova 2001; Warrick 2006).

Thus, a complementary *regional-scale* capacity for spatial analysis was subsequently developed within CLIMPACTS for New Zealand applications. Detailed land data and vector files (e.g., roads and property boundaries) could also be imported into the system for spatial overlays which are important for impact assessment and planning. The spatial resolution, currently at a 100 m grid, is much finer than that of the national-scale model, but the range of modeling capabilities is similar.

---

[7] New Zealand is divided into regions roughly following natural watersheds, each under the jurisdiction of a regional council, together with district councils and local governments, for the purpose of resource planning and policy.

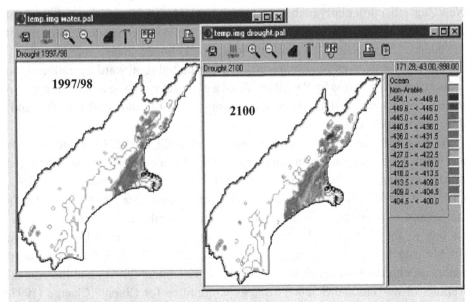

Black – high moisture deficit

Figure 14.5. Possible change in drought intensity in the Canterbury region of South Island, New Zealand. Drought conditions are indicated by soil moisture deficits using a simple atmospheric water-balance model within the original CLIMPACTS system. With a scenario of warmer future conditions in 2100, the evaporative demand intensifies drought conditions, as compared to the 1997/98 El Niño induced drought as the base case (from Kenny *et al.* 2001b).

To illustrate, one analysis examined the possible change in the intensity of drought occurrence in Canterbury using the CLIMPACTS system (Kenny *et al.* 2001b).[8] The growing season of 1997/98, a strong El Niño year, witnessed very severe drought and consequent soil moisture deficits in the Canterbury region. How might future climate change affect such droughts? As shown in Figure 14.5, this question was addressed using an atmospheric water-balance model within CLIMPACTS. The left-hand panel in Figure 14.5 shows the soil moisture deficits as estimated for the year 1997/98. The right-hand panel shows the effect of super-imposing a scenario of climate change using a Hadley Centre GCM climate change pattern (HadCM2; Mitchell and Johns 1997) and a projection of global warming based on the SRES A1 emission scenario (Nakicenovic *et al.* 2000) with a mid-range value for the climate sensitivity (2.5 °C). Under this scenario, both the intensity and the spatial extent of such soil moisture deficits would be greatly exacerbated if the same drought were to recur in the year 2100.

[8] The Canterbury region is located on the drier, eastern side of the South Island.

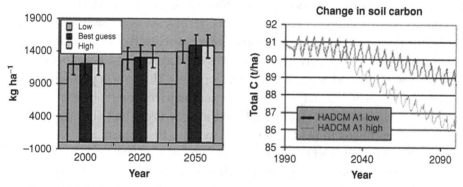

Figure 14.6. Analyses of pasture yields and soil carbon using the site-specific capacity within the original CLIMPACTS model. The left panel indicates an increase in annual dry matter yields obtained using the Hadley Centre GCM (HadCM2) and three different emission scenarios (mean of five New Zealand sites; Clark *et al.* 2001). In contrast, the right panel indicates the possibility of a decline in soil carbon level and thus soil quality (Parshotam and Tate 2001).

Temporal site-specific analyses

A *site-specific* modeling capacity was also developed within the early CLIMPACTS system. While non-spatial, this capacity allows for the inclusion of time-series data and more in-depth analyses, especially as regards variability and extremes, both in terms of climate itself and its present and future impacts. The individual site files contain monthly, daily, and/or hourly time-series climate data. For the early CLIMPACTS applications, these data were used to drive process-based crop simulation models that were modified and "wired" into the CLIMPACTS system. Such simulation models typically give outputs related to the responses, developmental stages, and production indicators of commercial crops, such as dates of flowering and maturation, yield, quality indicators and so on. The user interface also allowed for the examination of some adaptation measures, such as changing cultivars or crop type, altering planting dates, or irrigating.

For example, as part of a national assessment of climate change and agriculture (Warrick *et al.* 2001), CLIMPACTS was used to assess the effects of climate change on pasture grass yields and soil carbon (see Figure 14.6).[9] Using a mechanistic physiological model of pasture growth that takes into account the influence of atmospheric CO_2 concentration, temperature, radiation, and moisture supply on vegetative growth and development, Clark *et al.* (2001) conducted an assessment using site-specific scenarios superimposed on historical daily climate data. The

---

[9] Pastoral agriculture is the major activity on more than half of New Zealand's land, and together dairy, meat, and wool exports make up a large percentage of the value of the country's exported goods. Thus, a critical issue for the country is how climate change could affect pasture productivity.

daily data were perturbed according to the downscaled HadCM2 GCM results and three different emission projections. The results showed that annual pasture yields (the mean of five sites) tended to *increase* as a result of the climate change (Figure 14.6, left-hand panel). This increase is due largely to the benefits of higher CO_2 concentrations and warmer winter temperatures. However, a parallel analysis of soil organic matter, a major conditioner and nutrient reservoir in most soils, showed that changes in soil quality could be a major constraint. Using a soil carbon turnover model linked to the CLIMPACTS system, Parshotham and Tate (2001) showed that total soil carbon at the site, assuming constant land management, tends to *decrease* under the same climate change scenario (Figure 14.6, right-hand panel). This suggests that in order for pasture production to realize the full potential benefits of future climate change, the various land uses and their management must be conducive to the promotion of sustainable soil resources upon which pasture production depends.

A set of generic tools were also developed to enhance the capacity to analyze time-series site data (see Ye *et al.* 1999 for details). A *data browser* was developed that allows the user to examine in detail the time-series climate data at a site, providing the tools to group data (e.g., by days, months, seasons, or years), to sort data (e.g., according to climate variables), and to calculate means and standard deviations. Four different *weather generators* were incorporated, the parameters of which were linked to the climate scenario generator, thus allowing a high degree of flexibility for running and testing impact models under a large variety of weather conditions that otherwise would not be possible with only the observational datasets.[10]

Importantly, an *extreme event analyzer* was also developed that allows the user to fit a generalized extreme value (GEV) distribution to observed (and perturbed) extreme values from daily or hourly time-series data. As shown in Figure 14.7 (the recent user interface from SimCLIM) this function allows the user to estimate the return periods of extreme events and their exceedance probabilities. The user can perturb the observed data with scenarios of future climate change in order to estimate the changes in return periods for specified extremes. The extreme event tool is especially valuable for supporting risk-based assessments of impacts and adaptation.

14.2.3 *"Clones" of CLIMPACTS*

Over the period 1995–2001, the CLIMPACTS model structure was "cloned" for other problems and to other geographical settings and countries. The relatively

---

[10] Three of the weather generators were based on methods developed by Richardson (1981), and one was developed by Semenov and Barrow (1997).

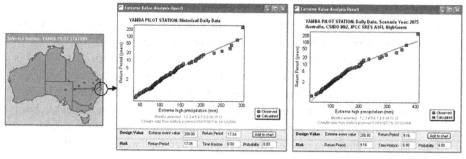

Figure 14.7. Example of the application of the Extreme Event Analyzer for extreme daily rainfall events from a station in New South Wales, Australia. The left graph shows the GEV curve fit to observed extremes for 130 years of record. For these data, the return period for a 200 mm event is estimated to be 1-in-17 years. The right graph shows a re-analysis after applying a scenario of climate change for the year 2075, perturbing the entire dataset and re-fitting the GEV curve. The return period for the 200 mm event is now estimated to be about 1-in-9 years – nearly a doubling of its frequency of occurrence.

simple, end-to-end structure was especially appealing for the purpose of developing, in the first instance, a platform for assessing the first-order, biophysical impacts of climate variability and change upon which more sophisticated tools could be added. Some examples of these extensions include: the model *OzCLIM,* developed for Australia in collaboration with the CSIRO Division of Atmospheric Research; *BDCLIM*, developed for examining climate change and flooding in Bangladesh, with funding from the Ford Foundation; and *VANDACLIM*, a training tool developed initially for the United Nations Institute of Training and Research. These and some other "clones" that were built on the core framework of the CLIMPACTS system are briefly described in Box 14.1.

Box 14.1

"Cloning" the CLIMPACTS System: 1995–2001

During the period 1995–2001, the CLIMPACTS model, created for New Zealand, was replicated for a number of other areas and contexts. These include the following.

- *OzCLIM*. One of the first extensions of CLIMPACTS was to Australia. In collaboration, the IGCI and the CSIRO Division of Atmospheric Research (DAR) created the model OzCLIM. The initial, collaborative effort focused on building the basic system platform, including the scenario generator, in order to allow OzCLIM to be developed according to the specific requirements for impact analyses in Australia. Impact model developments carried out by CSIRO include the incorporation of watershed-scale hydrological models and a coral reef bleaching model (CSIRO 2007).

- *BDCLIM* was developed collaboratively with the Bangladesh Unnayan Parishad with support from the Ford Foundation in the mid-1990s. In Bangladesh, one of the recurrent environmental problems is widespread over-bank flooding from the Ganges, Brahmaputra, and Meghna river system. The unique feature of BDCLIM is its capacity for investigating the possible effects of climate change on the depth and areal extent of flooding within the country, taking into account changes in climate and runoff over the vast river basins that feed into Bangladesh. The model has been used, for example, by the World Bank in examining how climate change could impinge on development projects in the country over the long term (Smith *et al.* 1998).

- *VANDACLIM* was created for a fictitious country – the "Republic of Vanda" – as a software training tool for climate change vulnerability and adaptation assessment. Two versions were developed, for continental countries and small islands, as a cooperative endeavor between IGCI, the United Nations Institute for Training and Research (UNITAR), and the South Pacific Regional Environment Programme (SPREP), with UNDP-GEF funding. VANDACLIM includes a suite of models for coastal, water resource, human health, and agricultural impact assessments (Warrick *et al.* 1999).

- *PACCLIM* and *FijiCLIM*. PACCLIM was created as a regional climate change scenario generator for the Pacific Island region, as part of the Pacific Island Climate Change Assistance Programme (PICCAP), a UNDP-GEF funded project. The idea was for PACCLIM to serve as a means for generating consistent, comparable scenarios for the many island countries scattered across the Pacific. Individual island models, like FijiCLIM (with additional funding from the World Bank) were developed to examine in detail the implications for individual islands. FijiCLIM underpinned an initial World Bank study of adaptation in the Pacific (World Bank 2000).

- *HOTSPOTS*. Funded by the NZ Health Research Council and developed jointly by IGCI and the Wellington School of Medicine, HOTSPOTS is designed to examine and characterize the risks of dengue, a vector-borne disease, occurring in New Zealand (it is not currently present). In order to address this issue spatially, the model overlays components dealing with the mosquito vector (*Aedes albopictus* and *Aedes aegypti*) population, human population, and virus introduction. Applications of the model indicate that climate change could substantially increase the risks of survival and persistence of an effective vector (especially *Aedes albopictus*) in the North Island (De Wet *et al.* 2001).

All of these national and international applications are built on the core framework of the CLIMPACTS system. In effect, these models served as evolving platforms for integrating scientific knowledge and data for purposes of supporting policy and planning in the context of climate change and variability.

14.2.4 Achievements and drawbacks of the initial CLIMPACTS model

In summary, the CLIMPACTS system filled a gap that existed at the national-to-local level with respect to tools for climate impact assessment. CLIMPACTS provided the user with the possibility to use a range of linked models, from simple to complex, and to examine impacts in multi-dimensional temporal and spatial contexts. In short, it provided a capacity to:

- examine spatial variations in baseline values for mean monthly, seasonal, or annual temperature (minimum, maximum, and mean values) and precipitation;
- construct scenarios of climate change based on greenhouse gas emissions and create "new" climates for the future (up to the year 2100);
- examine historical time-series data (monthly, daily, or hourly) for climate variables;
- apply climate change scenarios to create perturbed time-series data;
- simulate time-series of daily weather;
- conduct risk analyses of extreme events, with and without climate change;
- run impact models forced by either spatial climate variables or time-series data; and
- analyze distributions of model outputs (e.g., for means, variability, or extremes).

Despite these notable achievements, the early CLIMPACTS system had three major limitations. First, while serving its originators well in the context of research and producing outputs for others, the main motivating factor behind the development of the CLIMPACTS system was to eventually provide a flexible tool that could actually be put in the hands of planners, resource managers, educators, and policy makers, as well as researchers. This was not happening. The system was not quite "accessible" enough for these end-user groups.

Second, the intent of CLIMPACTS was to provide an evolving platform on which such new data and models could be easily incorporated and updated. However, the "hard-wired" structure of CLIMPACTS meant that only the model developers themselves could make such modifications, and, as the system grew, the system became increasingly difficult to maintain.

Third, the system emphasized biophysical impact assessment. There was little explicit provision for socio-economic data and risk-based assessments of adaptation measures, especially at the local scale.

If CLIMPACTS was to transition effectively from the hands of its developers to the end-users at large, it had to be more open, easily maintainable, user-friendly, and relevant. This led to the development of the open-framework SimCLIM system, as described below.

14.3 SimCLIM: the "next-generation" model

The original CLIMPACTS model was envisaged as a tool for bridging science, policy, and planning. Evolution of the model system should thus be driven by end-user needs as well as by science, and should be created in a manner accessible to planners and policy makers. With these principles in mind, steps were taken, beginning around the year 2002, to create the "next-generation" model. The aim was to build on the existing functionality of CLIMPACTS and to nudge it into being an active support tool for evaluating options for adapting to climate variability and change.[11] The main improvements are described below.

14.3.1 An "open-framework" structure

As mentioned above, CLIMPACTS and its associated family of models were "hard-wired." That is, they were developed and programmed by its developers[12] and therefore could only be updated, expanded, or modified by them. In order to increase the flexibility and the accessibility of the system to a wider range of users, a generic "open-framework" structure was developed. The new system was re-named *SimCLIM*.[13] As shown schematically in Figure 14.8, the open structure of SimCLIM and its interfaces allow users to define geographical boundaries and spatial resolution, enter spatial data and downscaled GCM patterns, import time-series climate data for sites, and attach "SimCLIM-compatible" impact models. In effect, with the open-framework structure SimCLIM has made significant strides in providing a modifiable system that can be put in the hands of analysts, planners, educators, and decision makers, and customized to meet user needs.

14.3.2 Human dimensions elements

The original version of the model contains no components or data related to people, activities, or property. Thus, there was no provision for estimating the number of people, land uses, or infrastructure potentially at risk from climate variability and change. Improvements were thus made to allow the incorporation of geo-referenced demographic, land-use, and infrastructural data, and to explicitly link them to biophysical impacts arising from climate and sea-level variability and change. For example, Figure 14.9 shows a simple spatial relationship between low-lying areas and property boundaries in an area of the Queensland, Australia coast subject to inundation and storm surges.

[11] The "next-generation" improvements were initiated through a new phase of the CLIMPACTS Programme for New Zealand, funded by the New Zealand Foundation for Research, Science and Technology, as well as through the AIACC program (Assessment of Impacts and Adaptation to Climate Change), funded by the Global Environment Facility and implemented by the United Nations Environment Programme, and executed jointly by START and the Third World Academy of Sciences (TWAS). The particular AIACC project (SIS09) under which model improvements were made was carried out jointly by the University of the South Pacific (Fiji) and IGCI, University of Waikato (New Zealand) and focused on small island developing states.

[12] IGCI, University of Waikato.

[13] The version of SimCLIM used in New Zealand is still called "CLIMPACTS" for reasons of consistency and name recognition.

SimCLIM open-framework structure

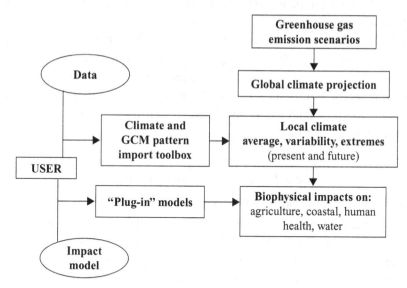

Figure 14.8. The open-framework structure of SimCLIM. With this structure there is greater flexibility for the user to customize the system by adding or deleting data, creating new areas and sub-areas (from global to local) and attaching impact models.

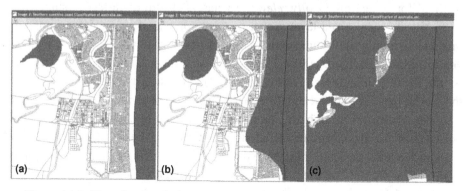

Figure 14.9. Elevation in relation to property titles using SimCLIM for a southern, low-lying portion of the Sunshine Coast, Queensland, Australia. Gray indicates areas that are: (a) less than 1m above mean sea level; (b) less than 1.5 m; and (c) less than 2.0 m.

14.3.3 Land-use scenario generator

While the earlier model allowed the user to construct scenarios of future climates, there was no comparable scenario generator for non-climatic factors as they affect vulnerability. For this reason, a stochastic land-use scenario generator was developed for SimCLIM, particularly for structures at risk from coastal or riverine flooding (Warrick *et al.* 2005). There are two sets of parameters which drive the changes and which can be manipulated by the user. First, the user can specify the model settings or rules that determine the characteristics of buildings (for example, the lifetime of a house or commercial building type). Secondly, the user can assign weightings to the various building types in order to influence the change in the future mix of buildings (for example, by increasing the fraction of, say, commercial buildings at each time-step). With both the user-specified rules and weightings, the spatial pattern of land use is generated stochastically in the simulation on a cell-by-cell basis over time.[14]

14.3.4 Sea-level scenario generator

In many parts of the world, like the Asia–Pacific region with many small island states and heavily populated coastlines, issues of sea-level variations and change are a major concern. For scenarios of change, the vast majority of coastal impact and adaptation assessments conducted within such regions have simply used the global-mean values of future sea-level rise. Regional variations in the rate of rise (due to regional differences in the rates of oceanic thermal expansion and dynamic atmospheric and oceanic effects on sea level, as projected by coupled atmosphere– ocean GCMs (AOGCMs)), as well as local factors such as vertical land movements as they affect *relative* sea level, were excluded. Generally, there has been a lack of methods for generating *locally relevant* scenarios of future sea-level change.

To fill this gap, a sea-level scenario generator was developed for SimCLIM that takes account of all three components of sea-level change – global, regional, and local (Warrick *et al.* 2005). The *global-mean* projections are those produced for the IPCC Third Assessment Report (TAR).[15] For any user-selected location in the world, these are scaled by normalized *regional patterns* of sea-level changes (i.e., per centimeter of global-mean change) produced by AOGCMs for the TAR (seven normalized patterns are included in SimCLIM).[16] Lastly, the *local observed trend* in sea level (as from tide-gauge data) can be entered and added to the future projection as well.[17] The user interface for generating a sea-level scenario is shown in Figure 14.10.

[14] However, this function is still "hard-wired" for particular places and is not yet generic. It was developed under the AIACC project (see footnote 11) and, in the first instance, implemented for a community in the Cook Islands as part of an Asian Development Bank study (ADB, 2005).

[15] These global-mean sea-level projections were generated by MAGICC for IPCC and were kindly provided by Tom Wigley. Additional projections for IPCC AR4 are now included.

[16] The AOGCM results were kindly provided by the Hadley Centre, UK. AOGCM results from IPCC AR4 are now included.

[17] If using the total trend as obtained from tide-gauge data, SimCLIM estimates the portion of the past observed trend presumably due to climate change (based partly on the users' selection of model parameters). This portion is then subtracted from the estimated trend, leaving the non-climate-change-related part of the trend (due to local factors like vertical land movements) which is then added onto the future projection. This procedure avoids "double-counting" of the effects of global warming on local sea-level changes.

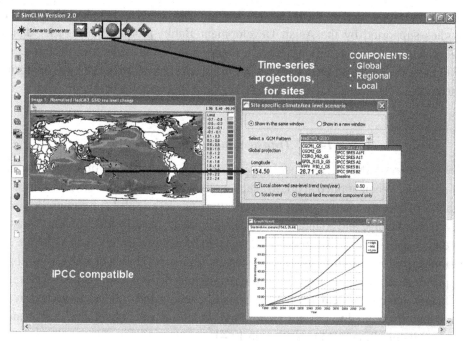

Figure 14.10. The user interface for generating scenarios of future sea-level changes in SimCLIM. Global, regional, and local components of sea-level change are taken into account. For a user-selected location, GCM pattern, SRES emission scenario, and local trend, the program generates high, mid, and low projections of future sea-level change.

14.3.5 Adaptation options

The first-generation CLIMPACTS model was designed principally for examining climate change impacts, not adaptation. Some adaptation options could be examined, but usually in an ad hoc fashion (e.g., changing the crop variety within a specific crop model). For SimCLIM, a more explicit, systematic treatment of adaptation was developed in order to give the user the opportunity to examine and evaluate a range of options. An important part of the work was the development of adaptation response functions that link specific adaptation measures to the impact models and socio-economic baselines for the purpose of creating the capacity for assessing their effects under scenarios of future change. (Examples of adaptation analyses using SimCLIM are given in Section 14.4.)

14.3.6 Economic tools

The first-generation CLIMPACTS model did not possess the capacity for economic evaluation of impacts or adaptation. Subsequent modeling developments for

SimCLIM (supported by the AIACC[18] program and tested and applied under pilot studies supported by the Asian Development Bank) produced first-order damage calculators and functions for estimating adaptation costs and benefit–cost comparisons.[19] This allowed damages from extreme events such as floods or storm surges to be estimated and aggregated over time (see Section 14.4).

14.3.7 Transient mode of analysis

The previous CLIMPACTS model concentrated on developing a "time-slice" mode of analysis. When applied spatially, such analyses lend themselves well to visualization of biophysical impacts. However, they are less useful for comprehensive economic damage and adaptation assessments under scenarios of evolving climate and land-use changes, especially in the context of extreme events such as floods or storm surges. For this purpose, a time-step, or "transient," mode of analysis is more appropriate in which the array of extreme events can be simulated at each time-step, damages and costs calculated, and streams of benefits and costs accumulated over time and discounted to present value. Such functions were developed, tested and applied within SimCLIM (see Section 14.4).

14.3.8 Impact model linkages

Improvements were made in SimCLIM to expand the capacity to attach or link various models that are forced by climate variables for the purposes of impact and adaptation assessment. For example, collaboration with the Danish Hydraulic Institute (DHI) has led to a DHI-compatible version of SimCLIM that interfaces directly with DHI data files. This development allows the suite of DHI hydrologic and hydraulic models to be run under scenarios of climate change in a seamless manner (Warrick and Cox 2007). Similarly, there is a new DSSAT-compatible version of SimCLIM that allows DSSAT crop models to be run under SimCLIM-generated scenarios. Optional "plug-in" models can be installed and registered by users depending on their requirements; for example, there are plug-in models for coastal shoreline change, domestic water tank systems, degree–day calculations, and atmospheric water balance. The concept is to provide SimCLIM as a core system that can be expanded and customized by end-users.

---

[18] Assessments of Impacts and Adaptations to Climate Changes (http://www.aiaccproject.org/aiacc.html).
[19] These tools were developed in the context of coastal and riverine flood risk for particular locations and, at this stage, are not yet "generic" within SimCLIM. Nonetheless, they can still be modified for other SimCLIM applications.

Figure 14.11. The 3-D visualization tool in SimCLIM. This example shows current annual-mean precipitation in northern California (San Francisco Bay on the left).

14.3.9 Other improvements

Other improvements in SimCLIM that are designed to make it more accessible to end-user groups include: system-building tools including "wizards" for importing and exporting site and spatial data and images; map calculation tools that allow manipulation of spatial data in multiple maps; and improved visualization, including three-dimensional "fly-over" tools (see Figure 14.11).

14.4 Examples of SimCLIM applications

As mentioned in the introductory section, it has been increasingly realized that, in terms of benefiting planning, policy, and action, assessments of climate change impacts and adaptation need to focus more at the community and local scale. Moreover, at this scale the importance of climate change is how climatic variability and extremes – and thus risks of floods, droughts, heat waves, and the like – will change in the future. Many of the improvements in SimCLIM have been directed to providing tools to assist in risk-based assessments. Two examples serve to illustrate this kind of application of SimCLIM: adaptation to storm surge risks in the community of Avatiu in the Cook Islands; and adaptation to the risks of drought and requirements for domestic water tank systems in the Brisbane region in Queensland, Australia.

14.4.1 Storm surge risks in the community of Avatiu, Cook Islands

Many of the new SimCLIM simulation tools were used in pilot studies for a risk-based approach to "climate-proofing" development, a project of the Asian Development Bank (ADB 2005). One case study involved storm surges from tropical cyclones in relation to the community of Avatiu on the island of Rarotonga, Cook Islands (ADB 2005; also see Warrick *et al.* 2005; Warrick and Cox 2007). How might climate change exacerbate the risks of tropical cyclones in the vulnerable community of Avatiu. How can the costs and benefits of adaptation options be compared?

The island of Rarotonga has a long history of dealing with tropical cyclones. For example, the storm surge (including wave run-up as well as barometric effects and wind and wave set-up) from Cyclone Sally in 1987 caused extensive damage to the port of Avatiu and overtopped the beach ridge along that segment of coast, causing considerable damage to residential and commercial structures and infrastructure equivalent to 66% of Cook Islands GDP (Kirk and Dorrell 1992). Although an unusual spate of five tropical cyclones in 2005 heightened awareness of the risks, ongoing development along the exposed coastal strip continues to increase exposure to such extreme events. Climate change is threatening to exacerbate the risks. Sea-level rise will add to the storm surge height, and the intensity of severe tropical cyclones is likely to increase with global warming (IPCC 2007b).

To assess the risks, the Avatiu area was modeled within SimCLIM at a spatial resolution of 5 m. A digital elevation model (DEM) was developed for this resolution and data on residential and commercial structures, roads, streams, and the like were collected and entered into the system. For simulating coastal flooding events, a simple reduced-form coastal flood model was developed based on outputs from engineering studies for Avatiu that used complex wave models and storm surge models (JICA 1994). A chain of relationships – from wind speed, wave height to total water run-up elevation, and their associated return periods – was developed and related to the potential wave-overtopping of the beach ridge.[20] After overtopping, water is distributed over the study area using a simple distance–decay function calibrated on the areal extent and the depth of flooding during Cyclone Sally (Figure 14.12).

---

[20] The overtopping height is the difference between total run-up elevation and the height of the beach ridge or protection structure at a site.

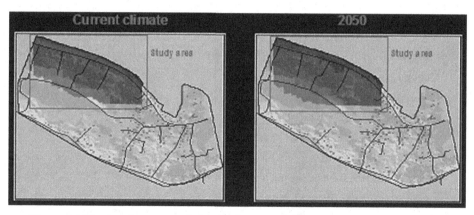

Figure 14.12. Areal extent and depth of coastal flooding from a severe tropical cyclone with an estimated return period of 1-in-50 years, as simulated in SimCLIM for a custom-built application for a study area at Avatiu, Rarotonga, Cook Islands (from Warrick *et al.* 2005).

In addition to the scenarios of sea-level rise (which adds to the total water run-up elevation), the model is driven by scenarios of changes in cyclone intensity. For this purpose, SimCLIM uses a simple scaling technique related to global-mean temperature change: a range of 2.5–10% increase in cyclone intensity per degree of warming (based on Henderson-Sellers *et al.* 1998; Giorgi and Hewitson 2001). Damages from storm surge flooding are simulated for the range of flood frequencies at each step as sea level and cyclone intensities change. For each building type, functions were developed for SimCLIM that relate flood heights to dollar damages (i.e., "stage-damage" curves). At each time step, the expected damages are summed over the range of return period floods. The yearly simulated damages are aggregated and discounted to present dollar value.

In terms of adaptation, functions were developed for examining three options for the Avatiu community: (1) raising the minimum floor heights of new structures built in the hazard zone over time; (2) constructing a protective sea-wall; and (3) modifying land use over time to avoid the most hazardous areas. The SimCLIM user specifies the design and enters the additional unit cost that is entailed in meeting the requirement, as shown in Figure 14.13 (top panel). This provides a basis for comparing the benefits of adaptation (i.e., damages prevented) to their costs. These adaptation options can be run individually or in combination. Importantly, multiple simulations with and without climate change, and with and without adaptation, provide a basis for identifying the *incremental* benefits and costs associated specifically with climate change (as compared to natural climate variability).

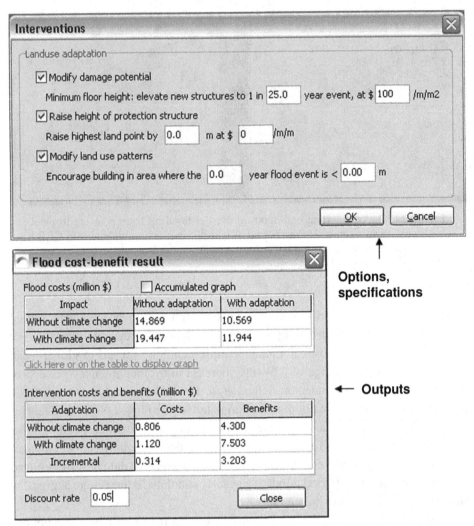

Figure 14.13. Analyzing adaptation options with a custom-built flood damage model within SimCLIM. Single or multiple options can be selected along with their specifications in the user interface (top panel). The output (bottom panel) gives flood damages (with and without adaptation, with and without climate change) as well as the costs and benefits (damages prevented) of adaptation. Values are annual direct costs (in dollars) to the year 2050, discounted (at user-specified rate) and summed to present value. (Modified from Warrick *et al.* 2005.)

An example of output from SimCLIM is shown in Figure 14.13 (bottom panel). This particular simulation used a scenario that resulted in a sea-level rise of 20 cm and a change in cyclone intensity of +8% by the year 2050 (a mid-range scenario). The adaptation option chosen was the raising of floor heights on new structures to a height sufficient to avoid the 1-in-25 year flood, at a nominal cost of $100 per m/m^2.[21] The top half of Figure 14.13 (bottom panel) shows simulated flood damages, in dollars. It can be seen that without adaptation, damages increase considerably, by about 30% (from $14.9 M to $19.5 M) as a result of the climate and sea-level changes. With adaptation, raising floor heights is effective in reducing damages with and without climate change, by 29% and 39%, respectively.

How do the benefits of adaptation compare to the costs? The bottom half of Figure 14.13 (bottom panel) compares the benefits (damages prevented) with the costs of adaptation (discounted at 5% and summed to present value). In this example, the benefit of raising floor heights in the absence of climate change is $4.3 M (i.e., $22.4–21.5 M), which compares favorably to a cost of $0.8 M. With climate change, the benefits are even greater, $7.5 M, as compared to a cost of $1.2 M. In other words, the benefits of raising floor heights exceed their costs, with and without climate change – a "no regrets" option.

Finally, the output in the lower half of Figure 14.13 (bottom panel) also shows the *incremental costs and benefits* of adaptation. These are the adaptation costs and benefits associated solely with climate change after subtracting the effects from natural variability.[22] In this example, the incremental benefits of raising floor heights are $3.2 M as compared to the incremental costs of $0.3 M.

14.4.2 Drought risk and domestic water tank systems, Queensland, Australia

At the time of writing, Southeast Queensland is experiencing one of its worst droughts on record. By mid-2007, high-level restrictions on water use were in place in a number of areas, and issues of water supply and demand topped the political agenda. In a region with a rapidly growing population and economy, there is now serious consideration being given to requirements for new developments to be self-sufficient and sustainable in terms of water supply. Domestic water tanks, once banned within the region, are now back in favor. The adequacy of domestic tank systems to supply water is largely dependent on reliable amounts and timing of rainfall in relation to tank size, water consumption, and catchment area (usually the roof surface). How reliable is the current climate and what are the risks of water shortage? How might the risks change under future climate change? What adaptation measures are viable options for reducing the risks?

---

[21] This value is arbitrary (as is the choice of design risk and scenario selection and consequently the output tables) and is used here solely to demonstrate the simulation methods and should not be interpreted as Cook Island government preference.

[22] This identification of the *additional* impacts and adaptation costs associated with climate change, as opposed to natural variability, is an important feature of the simulation, for practical considerations. For example, the Global Environment Facility's main funding mechanism only provides funds to developing countries for reducing vulnerability to climate change, not natural climate variability.

SimCLIM is used here to address these questions for a typical situation that might exist in the Brisbane area (Warrick 2007). One of the "plug-in" models available for SimCLIM is a "rainwater tank model," which is driven by daily time-series rainfall data. The input parameter values for the model include: *daily water consumption* per household (assumed here to be 600 liters); *roof area catchment* (assume 250 m^2); *initial water storage* (at the start of the simulation assumed to be 50% full); *tolerable length of a dry period* (assumed here to be 2 days before the tanker is called out);[23] and *tank size* (in liters). The key outputs of the model are: *longest dry period* (number of days in which the tank is empty) and the *number of dry periods* (which exceed the "tolerable length").

One way of performing the simulation is, in the first instance, to first determine an "acceptable level of risk" in relation to the outputs. Then, after setting other model input parameters, alter the tank storage size to achieve the acceptable level of risk. For our simulation, it is assumed that over 30 years, running out of water about six times (once every 5 years on average), with a single dry period of no more than 3 weeks, is a risk that can be tolerated. In order to determine required tank storage, the model is run with 30 years (1961–1990) of daily rainfall as recorded at the Brisbane Aero station. With a single 45 000 liter storage, the tank runs dry 86 times – clearly unacceptable. Doubling the storage to 90 000 liters reduces the failure rate to five times, with 22 days being the longest dry period, which is within the acceptable limits.

How might changes in climate projected for 2050 alter the risks? In order to address this question, a "worst-case" scenario of climate change was generated within SimCLIM that shows regional drying.[24] The changes in monthly rainfall are used to perturb the entire 30-year record daily rainfall values. The rainwater tank model was re-run using the perturbed dataset. The results showed that the rate of failure tripled, from five to 16 failures (or about once in every 2 years on average) – a presumed "unacceptable" level of risk.

What are the viable options for adaptation to reduce the added risks from climate change? Given that retro-fitting is often costly, there is some merit in considering measures to reduce the risks at the outset. Four options were considered:

- an additional 10 000 liter tank;
- reduce daily water consumption by 5%;
- include the garage roof area for water catchment, an extra 40 m^2; and
- raise the tolerable threshold to 4 days without tank water, through better emergency preparedness.

---

[23] This is the number of days that can be tolerated after the water runs out and before the emergency tanker truck arrives. The length of this period is a function of such factors as emergency water supplies and individual preferences and behavior.

[24] The scenario used the Hadley Centre (UK) GCM (HadCM3) results, assuming a high sensitivity of climate to changes in greenhouse gas concentrations and a high rate of future greenhouse gas emissions (SRES A1F1).

Table 14.1. *Assessment of adaptation options for reducing the incremental risks of rainwater tank system failure arising from climate change, using a domestic water tank model attached to SimCLIM. With no climate change, there were five tank failures, with the longest dry period of 22 days. With a scenario of climate change, the failures increase to 16 if there is no adaptation (i.e., "do nothing"). Each of the adaptation options can be assessed in terms of its reduction of risk (Warrick 2007).*

Adaptation option	No. tank failures	Longest dry period
Do nothing	16	22
Add extra 10 000 liter tank	11	22
Reduce daily consumption by 5%	4	22
Add garage roof to catchment area (40 m²)	4	10
Raise critical threshold level (extra 2 days)	13	22

Each of the options was simulated separately and the results are shown in Table 14.1. The most obvious option, adding an extra tank, was only moderately effective, reducing the number of failures from 16 to 11, and would probably be the most expensive. By comparison, lowering water consumption by a mere 5% dropped the number of failures to only four, and is probably a very low-cost option, although it has little effect on the longest dry period (unchanged at 22 days). Overall, the most technically effective option appears to be the inclusion of the garage roof area into the water catchment (four failures, 10 days for the longest dry period), although the costs are likely to be moderately high. The least effective (but low cost) option in reducing the risks was raising the tolerable threshold.

In general, it is important to note that the rainwater tank system itself is one way of adjusting to current climatic variability. The modifications of the system to take account of the additional – or *incremental* – risks from climate change is *adaptation*. As demonstrated here, SimCLIM can help in assessing adjustments to natural variability as well as in identifying the incremental risks and evaluating adaptation to future climate change. Final decisions are a matter of weighing and balancing a number of factors such as risk acceptance, costs, future discounting, uncertainties, and perceived efficacy of the actions.

14.5 Summary and conclusions

The original CLIMPACTS model was developed in, and for, New Zealand during the 1990s as a tool to contribute to integrated regional assessments of the effects of climate change and variability. The core structure of the system was designed as an "end-to-end" vertically linked set of models and data. A key feature of the system was the capacity to quickly generate scenarios of climate change from which to examine climate variations over time and space, and their effects. In the first instance, it focused solely on biophysical impacts. The unique aspects of the early CLIMPACTS model included its:

- focus at national-to-local scales of analysis, which filled a gap left by existing integrated assessment models focused at world-regional scales;
- inclusion of climate variability and extremes and their relation to climate change; and
- user-friendliness as a way of engaging planners and policy makers.

The model system was soon replicated for other countries and contexts. Once implemented, CLIMPACTS and its off-shoots provided new tools for regional assessments that were: quick running and flexible; capable of both spatial and temporal analyses; multi-scale – national, regional, sites; and useful for awareness raising and training, as well as for assessments.

However, there were several major drawbacks. First, CLIMPACTS was essentially a "researcher's" tool that was used mainly by those who developed it. Second, the system was "hard-wired" – that is, the program could only be expanded and maintained by the model developers themselves. Third, the primary focus was on biophysical impacts, with minimal explicit attention to the human dimensions of climate change, the risks of variability and extreme events, and the adaptation options for reducing the risks. In short if CLIMPACTS was to transition effectively from the hands of its developers to the end-users at large, it had to be more open, easily maintainable, user-friendly, and relevant.

This led to the development of the SimCLIM system. SimCLIM has a much more open structure; like a GIS, it can be customized and maintained by the user. Other software models can now be linked to the system (as with the linkage to Danish Hydraulic Institute models or DSSAT crop models). Images and data layers can be exported to, or imported from, GIS systems. The scenario generating capacity has been expanded to include a unique sea-level scenario generator. There is an enhanced capacity for local-scale modeling of impacts and adaptation, including the incorporation of environmental, social, and economic data. Functions to assess extreme events and risks have been improved. Initial developments of economic impact and evaluation tools have been introduced. Overall, with such modifications SimCLIM has become accessible to end-user groups, such as regional councils, environmental consultants, and universities, and has recently been made available to them.

The aim in developing CLIMPACTS and SimCLIM has been to provide technical capacity for facilitating integrated assessments. Ultimately, the goal is to promote adaptation and reduce present and future risks from climate variability and change. Integrated assessments of climate change, even those conducted by scientific committees or working groups, can be cumbersome. They typically require multiple sets of data, models, and tools, all of which require updating as scientific understanding and information improve. A customizable, evolving SimCLIM system is a way of integrating such data, models, and tools for assessment purposes, not by climate impact analysts, but by planners, policy makers, and environmental and resource managers who ultimately make decisions about adapting to climate change. In this way, the model system becomes a "tool-box" for integrated assessments and, in effect, becomes part of the adaptation process itself.

Acknowledgements

The original development of the CLIMPACTS model was driven by the International Global Change Institute (IGCI, University of Waikato) in collaboration with the National Institute of Water and Atmospheric Research (NIWA) and contributions from AgResearch, Crop and Food Research, HortResearch, Landcare Research, and the University of Auckland. For this early development, the author is particularly grateful for the hard work provided by Wei Ye, Graham Sims, Gavin Kenny, and Jacqui Harman (at IGCI) and Brett Mullan and Jim Salinger (at NIWA). This work would not have been possible without the consistent funding provided by the New Zealand Foundation for Research, Science and Technology (FRST) since 1993. Additional financial support for various extensions of the CLIMPACTS system to places and contexts was provided variously by the Ford Foundation, the UNDP-Global Environment Facility, World Bank, UNITAR, and the New Zealand Health Research Council, amongst others.

The transition to the open SimCLIM system was also funded by FRST. In addition, important support was provided by the AIACC project (executed by START-TWAS, implemented by UNEP/WMO/IPCC and funded largely by GEF). The modeling contributions of Wei Ye, Peter Kouwenhoven, Yinpeng Li, Xianfu Lu, and Matthew Dooley (at IGCI) and Gary Brightwell are gratefully acknowledged. The pilot case studies in the Cook Islands provided for the modifications, testing, and application of a number of new SimCLIM features as reported herein; for this, the author acknowledges the support provided by the Asian Development Bank and the Government of the Cook Islands, the project leadership provided by John Hay, and the economic modeling by Chris Cheatham. The coordination of recent SimCLIM modifications and the distribution of SimCLIM to end users has been accomplished through the hard work of Peter Urich (IGCI and CLIMsystems Ltd). Professor Tom Wigley and the UK Hadley Centre kindly provided data for use in SimCLIM.

References

Asian Development Bank (ADB) 2005: *Climate-proofing: a Risk-Based Approach to Adaptation*. Manila: Asian Development Bank.

Campbell, B. D., N. D. Mitchell, and T. R. O. Field 1999. Climate profiles of temperate C3 and subtropical C4 species in New Zealand pastures. *New Zealand Journal of Agricultural Research*, **42**, 223–233.

Carter, T. R. and E. L. La Rovere 2001. Developing and applying scenarios. In J. J. McCarthy, O. F. Canziani, N. A. Leary, *et al.* (eds.), *Climate Change 2001: Impacts, Adaptation, and Vulnerability. Contribution of Working Group II to the Third Assessment Report of the Intergovernmental Panel on Climate Change*. Cambridge: Cambridge University Press, pp. 146–190.

Carter, T. R., M. L. Parry, H. Harasawa, and S. Nishioka 1994: *IPCC Technical Guidelines for Assessing Climate Change Impacts and Adaptations*. University College, London, United Kingdom, and Centre for Global Environmental Research, Tsukuba, Japan.

Clark, H., N. D. Mitchell, P. C. D. Newton, and B. D. Campbell 2001. The sensitivity of New Zealand's managed pastures to climate change. In R. A. Warrick, G. J. Kenny, and J. J. Harman (eds.), *The Effects of Climate Change and Variation in New Zealand: an Assessment Using the CLIMPACTS System*. Hamilton: International Global Change Institute (IGCI), University of Waikato, pp. 65–77.

CSIRO 2007. *OzClim*. Commonwealth Scientific and Industrial Research Organisation. Clayton South, Australia. http://www.csiro.au/ozclim/home.do

De Wet, N., W. Ye, S. Hales, *et al.* 2001. Use of a computer model to identify potential hotspots for dengue fever in New Zealand. *New Zealand Medical Journal*, **114**(1140), 420–422.

Giorgi, F. and B. Hewitson 2001. Regional climate information – evaluation and projections. In J. T. Houghton, Y. Ding, D. J. Griggs, *et al.* (eds.), *Climate Change 2001: The Scientific Basis. Contribution of Working Group I to the Third Assessment Report of the Intergovernmental Panel on Climate Change*. Cambridge: Cambridge University Press, pp. 581–638.

Henderson-Sellers, A., H. Zhang, G. Berz, *et al.* 1998. Tropical cyclones and global climate change: a post-IPCC assessment. *Bulletin of the American Meteorological Society*, **79**, 19–38.

Hulme, M., T. M. L. Wigley, E. M. Barrow, *et al.* 2000. *Using a Climate Change Scenario Generator for Vulnerability and Adaptation Assessments: MAGICC and SCENGEN Version 2.4 Workbook*. Norwich: Climatic Research Unit.

IPCC 2007a. *Climate Change 2007: Impacts, Adaptation and Vulnerability. Contribution of Working Group II to the Fourth Assessment Report of the Intergovernmental Panel on Climate Change*. Cambridge: Cambridge University Press.

IPCC 2007b. Summary for policymakers. In S. Solomon, D. Qin, M. Manning, *et al.* (eds.), *Climate Change 2007: the Physical Science Basis. Contribution of Working Group 1 to the Fourth Assessment Report of the Intergovernmental Panel on Climate Change*. Cambridge: Cambridge University Press, pp. 1–18.

Japan International Cooperation Agency (JICA) 1994. *The Additional Study on Coastal Protection and Port Improvement in the Cook Islands. Final Report for Ministry of Planning and Economic Development – The Cook Islands*. Pacific Consultants International (PCI) and The Overseas Coastal Area Development Institute of Japan (OCDI).

Kenny, G. J., R. A. Warrick, N. D. Mitchell, A. B. Mullan, and M. J. Salinger 1995. CLIMPACTS: an integrated model for assessment of the effects of climate change on the New Zealand environment. *Journal of Biogeography*, **22**(4/5), 883–895.

Kenny, G. J., R. A. Warrick, B. D. Campbell, *et al.* 2000. Investigating climate change impacts and thresholds: an application of the CLIMPACTS integrated assessment model for New Zealand agriculture. *Climatic Change*, **46**, 91–113.

Kenny, G. J., J. J. Harman, and R. A. Warrick 2001a. Introduction: the CLIMPACTS Programme and method. In R. A. Warrick, G. J. Kenny, and J. J. Harman (eds.), *The Effects of Climate Change and Variation in New Zealand: an Assessment Using the CLIMPACTS System.* Hamilton: International Global Change Institute (IGCI), University of Waikato, pp. 1–10.

Kenny, G. J., J. J. Harman, T. L. Flux, R. A. Warrick, and W. Ye 2001b. The impact of climate change on regional resources: a case study for Canterbury and Waikato Regions. In R. A. Warrick, G. J. Kenny, and J. J. Harman (eds.), *The Effects of Climate Change and Variation in New Zealand: an Assessment Using the CLIMPACTS System.* Hamilton: International Global Change Institute (IGCI), University of Waikato, pp. 85–98.

Kirk, R. and D. E. Dorrell 1992. *Analysis and Numerical Modelling of Cyclone Sea-states: Avarua and Nikao Areas.* Report commissioned by Government of the Cook Islands, September 1992.

McCarthy, J. J., O. F. Canziani, N. A. Leary, D. J. Dokken, and K. S. White (eds.) 2001. *Climate Change 2001: Impacts, Adaptation, and Vulnerability. Contribution of Working Group II to the Third Assessment Report of the Intergovernmental Panel on Climate Change.* Cambridge: Cambridge University Press.

Ministry for the Environment 1994. *Climate Change: the New Zealand Response.* September, Ref. ME147. Wellington: Ministry for the Environment.

Ministry for the Environment 1997. *Climate Change: the New Zealand Response II.* June, Ref. ME241. Wellington: Ministry for the Environment.

Ministry for the Environment 2001a. *Climate Change Impacts on New Zealand.* Wellington: Ministry for the Environment.

Ministry for the Environment 2001b. *National Communication 2001: New Zealand's Third National Communication Under the Framework Convention on Climate Change.* Wellington: Ministry for the Environment.

Ministry for the Environment 2002. *National Communication 2001: New Zealand's Third National Communication under the Framework Convention on Climate Change.* January, Ref. ME419. Wellington: Ministry for the Environment.

Ministry for the Environment 2006. *New Zealand's Fourth National Communication under the United Nations Framework Convention on Climate Change.* March, Ref. ME745. Wellington: Ministry for the Environment.

Mitchell, J. F. B. and T. C. Johns 1997. On the modification of global warming by sulphate aerosols. *Journal of Climate*, **10**, 245–267.

Mitchell, J. F. B., S. Manabe, T. Tokioka, and V. Meleshko 1990. Equilibrium climate change. In J. T. Houghton, G. J. Jenkins, and J. J. Ephraums (eds.), *Climate Change: the IPCC Scientific Assessment.* Cambridge: Cambridge University Press, pp. 131–172.

Mullan, A. B., M. J. Salinger, C. S. Thompson, and A. S. Porteous 2001. The New Zealand climate: present and future. In R. A. Warrick, G. J. Kenny, and J. J. Harman (eds.), *The Effects of Climate Change and Variation in New Zealand: an Assessment Using the CLIMPACTS System.* Hamilton: International Global Change Institute (IGCI), University of Waikato, pp. 11–31.

Nakicenovic, N., J. Alcamo, G. Davis, *et al.* 2000. *Emissions Scenarios. A Special Report of Working Group III of the Intergovernmental Panel on Climate Change.* Cambridge: Cambridge University Press.

National Science Strategy Committee for Climate Change 1997. *Annual Report 1997.*
    Wellington: Ministry for the Environment.
National Science Strategy Committee for Climate Change 2001. *Annual Report 2000.*
    Wellington: Ministry for the Environment.
Parshotham, A. and K. R. Tate 2001. The impacts of climate change on soils and land
    systems in New Zealand. In R. A. Warrick, G. J. Kenny, and J. J. Harman (eds.), *The
    Effects of Climate Change and Variation in New Zealand: an Assessment Using the
    CLIMPACTS System.* Hamilton: International Global Change Institute (IGCI),
    University of Waikato, pp. 79–84.
Richardson, C. W. 1981. Stochastic simulation of daily precipitation, temperature and
    solar radiation. *Water Resources Research,* **17,** 182–190.
Santer, B. D., T. M. L. Wigley, M. E. Schlesinger, and J. F. B Mitchell. 1990. *Developing
    Climate Scenarios from Equilibrium GCM Results.* Max Planck Institute für
    Meteorologie, Report No. 47, Hamburg, Germany.
Schneider, S. H. 1997. Integrated assessment modeling of global climate change:
    transparent rational tool for policymaking or opaque screen hiding value-laden
    assumptions? *Environmental Modeling and Assessment,* **2**(4), 229–248.
Semenov, M. A. and E. M. Barrow 1997. Use of a stochastic weather generator in the
    development of climate change scenarios. *Climate Change,* **32,** 293–331.
Smit, B. and O. Pilifosova 2001. Adaptation to climate change in the context of sustainable
    development and equity. In J. J. McCarthy, O. F. Canziani, N. A. Leary, *et al.* (eds.),
    *Climate Change 2001: Impacts, Adaptation, and Vulnerability. Contribution of
    Working Group II to the Third Assessment Report of the Intergovernmental Panel on
    Climate Change.* Cambridge: Cambridge University Press, pp. 877–912.
Smith, J. B., A. Rahman, and M. Q. Mirza 1998. *Considering Adaptation to Climate
    Change in the Sustainable Development of Bangladesh.* Report to the World Bank by
    Stratus Consulting Inc., Boulder, Colorado, USA.
Tol, R. S. J. and S. Fankhauser 1998. On the representation of impact in integrated
    assessment models of climate change. *Environmental Modelling and Assessment,* **3,**
    63–74.
Warrick, R. A. 2006. Climate change impacts and adaptation in the pacific: recent
    breakthroughs in concept and practice. In R. Chapman, J. Boston, and M. Schwass
    (eds.), *Confronting Climate Change: Critical Issues for New Zealand.* Wellington:
    Victoria University Press, pp. 189–196.
Warrick, R. A. 2007. SimCLIM: recent developments of an integrated model for
    multi-scale, risk-based assessments of climate change impacts and adaptation.
    *Proceedings of the 2007 ANZSEE Conference on Re-inventing Sustainability: a
    Climate for Change,* 3–6 July 2007, Noosaville, Queensland, Australia.
Warrick, R. A. and G. Cox 2007. New developments of SimCLIM software tools for
    risk-based assessments of climate change impacts and adaptation in the water
    resource sector. In M. Heinonen (ed.), *Proceedings of the Third International
    Conference on Climate and Water,* Helsinki, Finland, 3–6 September 2007. Helsinki:
    SYKE, pp. 518–524.
Warrick, R. A., G. J. Kenny, G. C. Sims, *et al.* 1996. Integrated model systems for
    national assessments of the effects of climate change: applications in New Zealand
    and Bangladesh. *Journal of Water, Air and Soil Pollution,* **92,** 215–227.
Warrick, R. A., G. J. Kenny, G. C. Sims, W. Ye, and G. Sem 1999. The VANDACLIM
    simulation model: a training tool for climate change vulnerability and adaptation
    assessment. *Environment, Development and Sustainability,* **1**(2), 157–170.

Warrick, R. A., G. J. Kenny, and J. J. Harman (eds.) 2001. *The Effects of Climate Change and Variation in New Zealand: an Assessment Using the CLIMPACTS System.* Hamilton: International Global Change Institute (IGCI), University of Waikato.

Warrick, R. A., W. Ye, P. Kouwenhoven, J. E. Hay, and C. Cheatham 2005. New developments of the SimCLIM model for simulating adaptation to risks arising from climate variability and change. In A. Zerger and R. M. Argent (eds.), *MODSIM 2005, Proceedings of the International Congress on Modelling and Simulation.* Modelling and Simulation Society of Australia and New Zealand, December 2005, pp. 170–176.

Weyant, J., O. Davidson, H. Dowlatabadi, *et al.* 1996. Integrated assessment of climate change: an overview and comparison of approaches and results. In J. P. Bruce, H. Lee, and E. F. Haites (eds.), *Climate Change 1995: Economic and Social Dimensions of Climate Change, Contribution of Working Group III to the Second Assessment Report of the Intergovernmental Panel on Climate Change.* Cambridge: Cambridge University Press, pp. 367–396.

Wigley, T. M. L. 2003. *MAGICC/SCENGEN 4.1 Technical Manual.* Boulder, CO: National Center for Atmospheric Research.

World Bank 2000. *Cities, Seas, and Storms: Managing Change in Pacific Island Economies. Volume IV: Adapting to Climate Change.* Washington, DC: The International Bank for Reconstruction and Development, The World Bank.

Ye, W., G. J. Kenny, G. C. Sims, and R. A. Warrick 1999. Assessing the site-scale effects of climate variability in New Zealand: developments with the CLIMPACTS integrated model system. *Proceedings of MODSIM 99 Conference*, University of Waikato, Hamilton, New Zealand, 6–9 December 1999.

15

Why regional and spatial specificity is needed in environmental assessments

RIK LEEMANS

15.1 Introduction

During the last two decades model-based global environmental assessments have become established as tools to evaluate the potential consequences of different environmental and societal trends. Such assessments integrate different disciplinary perspectives in understanding and seeking a solution to a problem. Generally, these assessments aim to inform policy makers and assist in policy development (e.g., the Intergovernmental Panel on Climate Change, IPCC: IPCC 2001; Metz *et al.* 2007; Parry *et al.* 2007; Solomon *et al.* 2007; and the Millennium Ecosystem Assessment: Reid *et al.* 2005). As such, these assessments bridge between three different domains: research, monitoring, and policy development (Figure 15.1). Any convincing assessment will combine scientific insights with data from monitoring networks to determine if, how, and why trends evolve. Scientists have an obvious role: they initiate and develop the monitoring networks, interpret the data, and communicate their findings in the assessments. They thus develop adequate tools, methods, and indicators to enhance understanding of complex problems. The maintenance and quality control of long-term monitoring networks is generally done by dedicated agencies. Both research and policy development use results of these networks but in different ways. Researchers use the raw data to analyze processes to enhance understanding, while policy makers use highly aggregated indicators to detect and analyze trends and evaluate the effectiveness of policy measures. Unfortunately, too often assessment is solely seen as a scientific enterprise, and the importance of the bridging role between research, monitoring, and policy is neglected.

This chapter briefly discusses the history of global environmental assessments. It focuses on global climate change and the methodologies with which I have been involved over the last 15 years. This focus is not to endorse these specific approaches and criticize others, but to use them as examples to present general requirements

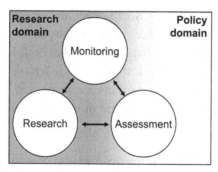

Figure 15.1. The relation between research, monitoring, and assessment in assisting policy development.

for global environmental assessments. My in-depth knowledge of these approaches allows a comprehensive highlighting of their strengths, weaknesses, and limitations. Throughout, the term "environmental assessment" is used to indicate the description of an environmental problem from several different perspectives integrated into a (semi-)quantitative framework (a model) to analyze and project ongoing trends in order to inform policy makers. Often the adequacy and efficiency of proposed policy measures are evaluated with the same framework. These assessments are often labeled *integrated assessments* and the mathematical frameworks, *integrated assessment models.* It can be argued, however, that these terms falsely provide a sense of completeness and comprehensiveness, while most so-called integrated assessments only apply simplistic cause–effect chains, which do not describe the complex behavior of a system with all relevant interactions, such as feedback and synergistic processes, and time lags in responses.

After this review, the chapter discusses the linkage of different dimensions and presents one of the state-of-the-art models, describes its application, and discusses the largest uncertainties and their origin. Some of them are just a consequence of lack of (disciplinary) understanding; others, the lack of comprehensive integration. Many stem from an inappropriate selection of dimensions and resolutions. Every assessment (or assessment model) has a specific objective and generally focuses only on a limited aspect of the Earth system. The rationale to develop such a model is, however, to describe the essence of a problem or system and to reduce or remove redundant aspects. What is essential or redundant and thus the desired level of simplification depends mainly on the questions that have to be addressed. Thus the result is always a "model" that represents the studied system but with different levels of realism. Therefore, it is only fair when criticizing an assessment to critically evaluate its achievements against its specific objective(s) and not against one's personal expectations or biases. Uncertainties must also be seen in relation to such objectives. Assessments have traditionally focused largely on the natural science aspects of environmental problems. Socio-economic aspects are often ignored. This is identified as one of the main sources of uncertainty in environmental assessments (e.g., van der Sluijs 1997). This chapter shows that high-resolution models allow for effective aggregations, communicating results in a more robust manner. Finally, some general but concluding observations are made that could set the scene for productive and successful future global environmental assessments.

15.2 A short history of environmental assessment models

Environmental assessment logically evolved from environmental impact assessments when problems grew more complex and covered larger regions or more sectors. Central tools in these assessments were simulation models. Most of these models followed the cause–effect paradigm. A broad range of causal factors, also called drivers,[1] change the state of the studied system. Often these drivers were based on empirical correlations at high levels of aggregation. The IPAT identity (Impact = Population × Affluence × Technology; Commoner *et al.* 1971; Ehrlich and Holdren 1971) and derived identities such as KAYA (Kaya 1989), for example, have often been used to estimate pollution or emission levels globally and regionally. The KAYA identity estimates CO_2 emissions by the product of population (P), income per capita (GDP/P), energy intensity (E/GDP), and carbon intensity (CO_2/E). The resulting identity ($CO_2 = P \times$ GDP/$P \times E$/GDP $\times CO_2/E$) works very well for determining regional and global emissions from energy use, but neglects, for example, emissions from deforestation and cement production. However, most identities perform relatively well for specific aggregated problems but provide no insight for solutions or adequate policy responses. These identities are too simple to capture the broad range of possible policies.

In the 1980s these identities were also applied to deforestation. Here, they did not work at all. The causes of deforestation were much more diverse. Geist and Lambin (2002) have shown that only 3% of deforestation cases can be explained unambiguously by the IPAT factors. The real causes of deforestation are regionally specific (e.g., expansion of agricultural land in Brazil; wood extraction in southeast Asia), multiple, and interacting. Often the underlying drivers of these impacts are not locally determined but relate to distant markets. For example, deforestation in Indonesia and eastern Siberia are driven by the Japanese demand for forest products. In these more complex situations, generally several drivers interact simultaneously and simple cause–effect chains cannot be distinguished.

---

[1] A challenge is to find terms that mean the same thing to many different users. The term "driver" is, for example, widely used in the ecological and other natural sciences but seldom used by economists.

Recently, however, Kauppi *et al.* (2006) have developed a successful identity to determine a simple development threshold below which deforestation occurs.

The OECD's InterFutures Study Team (1979) put forward the Drivers-Pressure-State-Impact-Responses (DPSIR) framework to improve the capture of such complexity in their environmental assessments. In a DPSIR scheme there is more room for interactions and feedbacks between regions and sectors. This emerging complexity assisted in developing more advanced system analysis approaches in environmental assessments, recognizing different drivers.

Two frequently used sets of drivers are discussed here[2] – "direct" versus "indirect" and "underlying" versus "proximate." Direct drivers unequivocally influence processes and can therefore be easily identified and measured (cf. Leemans *et al.* 2003). Their systemic effect can often be quantified by experimentation (especially when the system studied is relatively small and closed). Indirect drivers operate more diffusely (i.e., from a distance), often by altering one or more direct drivers. An indirect driver can seldom be observed; its influence is established by understanding its effect on a direct driver. Often the indirect drivers operate at coarser-scale levels. For example, a farmer can relate to local market prices and can probably influence his/her income by properly timing entry into this market. The national and international institutions that govern local markets also influence farmer behavior but these institutions are certainly exogenous to a farmer's influence. Such institutions do not even intend to govern a specific local market: their impact is truly indirect.

The underlying and proximate driver terminology is widely used in the land-use change literature (e.g., Lambin and Geist 2006). Proximate drivers have an immediate influence on processes in the system studied, while underlying drivers strongly alter one or more proximate drivers. Proximate and underlying drivers are conceptually similar to direct and indirect drivers but tend to be used when analyzing the spatial and temporal processes in which the human intent, such as the need for food, is linked with actually occurring activities, such as burning, sowing, and harvesting. Proximate drivers thus define the hands-on land-management activities that, for example, lead to a land-use conversion (e.g., forestation) or modification (e.g., irrigation and fertilization) of ecosystems. These drivers are linked to a specific locality (e.g., an upland field) and timing (e.g., in the spring). Underlying drivers influence the nature and strength of the proximate drivers and include many diverse and often diffuse factors (e.g., food demand, population structure, market constitution, and property rights) that often operate at more distant different scale

---

[2] There are many other typologies (e.g., anthropogenic versus biophysical; dependent versus independent; endogenous versus exogenous; and primary versus secondary) that can be used but most of them can be mapped into the two discussed here.

levels than the ensuing proximate drivers. Proximate drivers are likely to be under the control of local ecosystem managers or decision makers, while underlying drivers tend to be exogenous.

Recent comprehensive analyses of environmental problems (e.g., Petchel-Held *et al.* 1999; Geist and Lambin 2002; Ostrom *et al.* 2002) have shown that analyzing causes of environmental change requires a multi-scale and multi-dimensional assessment of major components of the system, their dynamics and interactions. An appreciation of the feedbacks, synergies, and tradeoffs among these components in the past improves our understanding of current conditions and enhances our ability to project future outcomes and intervention options. This more comprehensive approach to relating cause and effect nowadays forms the basis of many environmental assessment models that dynamically integrate sectoral processes (e.g., emissions from energy generation, transport, industrial activities, and land use), physical processes (e.g., atmospheric chemistry and transport), and ecological processes (e.g., critical loads).

The approach was first applied to the acid rain problem. RAINS, an acidification model developed at the International Institute for Applied Systems Analysis (Alcamo *et al.* 1987), combined insights from different disciplines and was the first that was used for policy development in a region with many countries (Europe). Its impact components were already spatially explicit. Since then, environmental assessment has excelled especially in the domain of climate change (Weyant *et al.* 1996), biodiversity (Sala *et al.* 2000), desertification (Stafford Smith and Reynolds 2002, Reynolds *et al.* 2007) and also sustainable development (Rotmans and de Vries 1997). Over the last decade many models and approaches have been developed (see the review by Costanza *et al.* 2007).

Most of the earliest models were highly aggregated. For example, some of the early climate change models simulated emissions and impacts globally (Weyant *et al.* 1996). This was rapidly seen to be too great a limitation, because differences between regions or countries were neglected and positive and negative impacts were averaged out. Therefore, the next generation specified different regions. Initially, only developed and developing regions were recognized but currently state-of-the-art models use up to 25 different regions globally. Some regional models, like RAINS, simulate countries. In each region different sectors for impact models and ecosystems were recognized. These sectors were characterized by their share of the gross domestic product (GDP), specific emission factors and expected development.

Ecosystems were first characterized by their extent (i.e., surface area but without their specific distribution or location), their sensitivity, and their productivity (i.e., CO_2-uptake). The sectors and ecosystems were thus generally parameterized by average characteristics, which were often calibrated against coarse global data sets. For example, one of the earliest global carbon cycle models (Goudriaan and Ketner

1984) was based on six typical ecosystems (including a simple land-use class), each characterized by a specific productivity level and allocation towards different compartments (e.g., leaves, stems, roots, humus, and charcoal) with specific residence times. The parameterization was highly calibrated towards changes in atmospheric CO_2 concentration, using the CO_2-fertilization factor as the main tuning knob. For many years, this led to overstating of the CO_2-fertilization process in many other models and assessments.

The disadvantage that the actual processes cannot be properly parameterized by aggregated models led to the development of higher-resolution models (e.g., more ecosystems) and spatially explicit models in the early 1990s. This was made possible by the development of a series of gridded global databases of environmental properties (e.g., land cover by Olson *et al.* 1985; climate by Leemans and Cramer 1991; terrain by the National Geophysical Data Center (NOAA) 1998; and soil by FAO and CSRC 1987). These databases allowed the implementation of different climate classifications (e.g., Leemans *et al.* 1996) and carbon cycle models (e.g., Melillo *et al.* 1993) in a spatially explicit way. This approach allowed the geographical patterns of ecosystems and their productivity to be captured in a much more realistic way. The actually occurring environmental drivers could now be simulated and assessed. This led to a boost of continental and global ecology (Kuhn *et al.* 2008), which was only matched by the earlier eras of von Humboldt and Köppen. The spatially explicit approach also allowed for a much-improved validation of the models with local measurements and satellite-derived patterns.

In the mid-1990s many ecosystem models became spatially explicit, while land use and land-use change was still modeled as a single economic sector for different countries (e.g., Fischer *et al.* 1988). This made it difficult to assess, for example, the causes and consequences of deforestation. Deforestation was then commonly modeled by extrapolating the observed rates, which did not discriminate between different causes and intensities. Skole and Tucker (1993) showed that by using a satellite-derived spatial analysis of deforestation in the Amazon, the actual process was more regionally heterogeneous. The deforestation process should be described by forest conversion to agricultural land, abandonment, and regrowth of forests. The simple aggregated deforestation rates did not capture these spatial processes, which change in intensity over regions and time. Their paper and their simple model was then the start of the development of spatially explicit land-use models.

Leemans and Solomon (1993) applied the land-evaluation approach, agro-ecological zoning developed by FAO, to estimate global patterns and potential productivity of temperate and tropical crops. Later, this approach was linked to a simple agricultural-demands model (Zuidema *et al.* 1994). Their model reconciled regional demand for different crops with current and future crop patterns and productivity. In some regions more agricultural land was needed to satisfy demand, in others less land was needed. This dynamic drove deforestation and forestation patterns and was used to determine the consequences for the terrestrial carbon cycle and land-use related emissions. This approach became the kernel of the land-use modeling in IMAGE 2 (Alcamo 1994), which was the first environmental assessment model that dynamically integrated sectoral, regional, and spatial dimensions. Since then, several other models have adopted the IMAGE-2 approach (e.g., AIM: Matsuoka *et al.* 1994; Kainuma *et al.* 2002; Masui *et al.* 2006; Kainuma *et al.*, Chapter 11, this volume).

15.3 Integrating the different dimensions

The discussion above already highlighted that relevant processes should be parameterized at the scale level where they operate. Often, this means that proper dimensions and resolutions should be selected. This is not straightforward! The different assessments are often typified according to scale level: local, regional, and global assessment. This is misleading because few environmental issues operate only at single regional or the global level. Globally, many models therefore rely on empirical relationships between assumed global proximate indicators for a given pressure, state, or impact indicator. This was done earlier in the IPAT approaches but it is still practiced in many global top-down models (e.g., Tol 1999). Although those simple models provide some insights for well-articulated problems, they surely do not provide a panacea for broader global-change problems.

When a problem is explored, one of the first steps should be the definition of the relevant dimensions. This not only means the spatial and temporal dimensions, which are often clearly specified in textbooks, but also the institutional, legal, organizational, ecological, atmospheric, and other environmental dimensions and their inherent scale levels. Figure 15.2, for example, depicts two important dimensions in land-use change research (cf. Turner *et al.* 1995): the organizational and ecological dimension.

The basic component of the ecological dimension is the plant. It photosynthesizes and produces biomass. This process forms the basis of all productivity and is influenced by local soil, water, and climatic constraints and human management. The latter is not generally done at the plant level, but at the field level. Several fields or stands comprise an ecosystem, several ecosystems a landscape, and several landscapes a region. All ecosystems in a recognizable climate zone define biomes. All biomes comprise the global ecosystem. The regional diversity of ecosystems can best be captured in a spatially explicit way on a high-resolution grid. In this way it can easily be linked to climatic, soil, and terrain characteristics. Characterizing ecosystems by extent alone does not disclose these important patterns. Many ecosystem properties can be scaled linearly. Global productivity, for

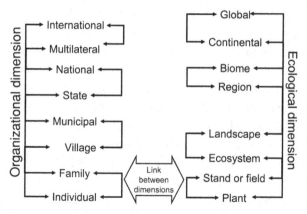

Figure 15.2. The relations between the organizational and ecological dimensions and the different scale levels.

example, is the aggregated sum of all plants and ecosystems. There are, however, also several emergent properties of plant and ecosystem interactions at the higher-scale levels (e.g., ecological succession and disturbance) that do not scale linearly. Scaling ecosystem processes appropriately from the individual plant through the ecosystem towards the global level is important for assessing credible ecosystem responses. The thin arrows in Figure 15.2 visualize the aggregating character of many ecological processes.

The smallest functional unit in the organizational dimension is the individual, who manages plants, fields, and ecosystems. The context of this individual is set by his/her family circumstances, the village, and municipality, where they have their social contacts and have their children educated. Local individuals generally do not influence the state and national levels. Although government is often based on local representation, it strongly defines the institutional and legal context in which individuals operate. In integrated assessments, individuals are rarely recognized; national governments are the prime component of the assessments. These assessments therefore aggregate up to the global level. National representatives populate even the international conventions, such as the United Nations Framework Convention on Climate Change and the United Nations Convention on Biological Diversity. The interactions up and down scales are often more complex in the organizational dimensions than in the ecological dimension. In the ecological dimension clear hierarchies can be distinguished (Allen and Starr 1982). There are top-down and bottom-up influences. This recognition invalidates the single scale-level approach and explains why the recently developed multi-actor oriented approaches have become so popular. The organizational dimensions do not necessarily need to be modeled in a spatially explicit way. More important are appropriate sectoral and institutional classifications.

In environmental assessment many of these dimensions and their interactions meet and it is important that the proper levels at which their interactions occur are recognized. This is actually one of the most important steps in modeling them. For example, land-use interactions occur at the lowest scale levels between the land-use manager (an individual farmer, forest ranger) and his/her plants in fields or stands. The wheat farmer in the Great Plains produces for international markets and uses (inter)nationally developed technologies. The global market and national subsidies define the context in which he/she makes decisions on managing the local soils, as does the landscape-level water availability in a regional climate. A small farmer in Tanzania, in contrast, produces primarily to feed the family. Some surplus will be traded in local markets. Maybe the farmer profits from a foreign-aid project. But similar to the Great Plains wheat farmer, the small farmer in Tanzania depends on local and regional environmental conditions. In principle, there is no global or continental land-use process. The land-use activities of each individual lead to specific impacts on biodiversity and greenhouse gas emissions. These environmental impacts become globally significant through their cumulative effect. This is especially true for greenhouse gas emissions that lead to climate change, which will affect all land users. An individual land user, however, is not impacted by his/her own emissions (in contrast, for example, to erosion) but by all other land use and other emissions worldwide.

Modern environmental assessment models are nowadays composed of different interlinked dimensions. Ecological processes and land-use activities on land are simulated geographically explicitly, economic processes by different sectors in (groups of) countries, the oceans as boxes representing different ocean bodies, and the atmosphere as an array of layers and so on.

15.4 The use of a current assessment model

IMAGE 2 (Integrated Model to Assess the Global Environment; IMAGE Team 2001; Bouwman *et al.* 2006) is an integrated assessment model that calculates the environmental consequences of human activities worldwide. The model carefully linked the different dimensions at appropriate scale levels. IMAGE 2 represents interactions between society, the biosphere, and the climate system to assess environmental issues such as climate change, biodiversity decline, and human development. The objective of the IMAGE 2 model is to explore the long-term dynamics of global environmental change. The model is designed to compare business-as-usual scenarios with specific mitigation and adaptation scenarios in order to be able to compare the effectiveness of such measures. Scenarios are "what if" representations of how the unknown future might unfold. They form an accepted and valuable tool in analyzing how different comprehensive sets of driving forces may

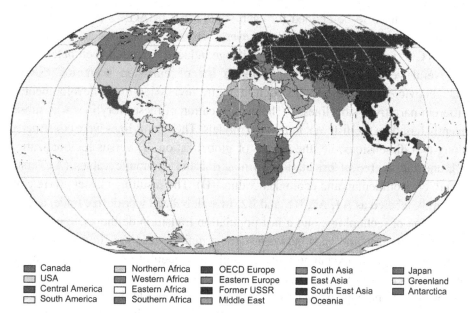

■	Canada	☐	Northern Africa	■	OECD Europe	■	South Asia	■	Japan
☐	USA	■	Western Africa	■	Eastern Europe	■	East Asia	☐	Greenland
■	Central America	☐	Eastern Africa	■	Former USSR	■	South East Asia	■	Antarctica
☐	South America	■	Southern Africa	■	Middle East	■	Oceania		

Figure 15.3. The different regions for which socio-economic processes are modeled in IMAGE 2.2.

influence future emissions, concentrations, climate change, and impacts. No specific likelihood can be defined for any scenario due to major uncertainties in the society–biosphere–climate system.

The socio-economic and energy-use calculations are performed for 17 regions (plus Antarctica and Greenland) (Figure 15.3). Countries that form a region are assumed to have similar socio-economic characteristics and geographically border each other. Interactions between regions and data availability limit the number of modeled regions. A relatively high number of 17 regions were distinguished in IMAGE 2, because of applications for UNEP's Global Environmental Outlook (2002), in which IMAGE 2 defined the global context. Most other global models distinguish fewer regions.

The land-use and terrestrial-carbon calculations are performed on a high-resolution grid of $0.5° \times 0.5°$ longitude and latitude. This relatively high resolution was needed to simulate distinct changes in land-cover patterns and the terrestrial carbon cycle. The atmosphere and climate are simulated by a one-dimensional climate model and oceans are modeled by a box model, each box representing a major ocean body. The innovative aspects of IMAGE 2 consisted in bringing together in a single modeling framework these highly different dimensions and scale levels. The IMAGE-2 simulations are calibrated with observed data for the period 1970–1995. Future trends in socio-economic scenario variables (e.g., demography, wealth, and technology) have to be defined for each simulation. IMAGE 2 represents major global dynamic processes, including several natural interactions and feedbacks, such as CO_2 fertilization and land-use change induced by changed climate.

Several years ago, IPCC published a set of emission scenarios (SRES; Nakícenovíc *et al.* 2000). A thorough literature review led to the development of different narrative "storylines."[3] The quantification of these storylines was subsequently based on six different integrated models. These storylines were constructed along two dimensions, i.e. the degree of globalization (1) versus regionalization (2), and the degree of orientation on material and economic values (A) versus social, environmental, and ecological values (B). The resulting clusters were consequently labeled as A1, A2, B1, and B2. In a globalized world, free trade, a rapid conversion of technology, and a fast population transition define socio-economic developments. Income gaps between regions rapidly decrease. In a regionalized world there is little exchange between and consequently more diversity among regions. There is little conversion of technology and the demographic transition is slower. Income gaps between regions remain. In a world that focuses on material and economic values, energy and resource use rapidly increases, while in a world with a social, environmental, and ecological focus resources are used much more efficiently and a service economy dominates. The original SRES implementation focused mainly on the emissions from energy use. Atmospheric, climatic, and environmental processes that could feed back on emissions were neglected. They were also only implemented for four regions (OECD, countries in transition, Asia, and rest of the world). This coarse aggregation concealed important regional differences in developmental, socio-economic, and environmental conditions.

In IMAGE 2 the SRES storylines were implemented for more regions and the whole emissions-concentrations–climate change-impacts chain was comprehensively simulated, taking several interactions and feedbacks into account. Basic demographic and economic information is derived from the original SRES data and disaggregated for 17 world regions. The data was checked for consistency by using a population model and the world-economy model, WorldScan, into the linked sub-systems of IMAGE 2 (Figure 15.4).

In IMAGE 2 relatively simple but broadly accepted models are used to simulate the dynamics of the sub-systems. Therefore, a more complete coverage of many feedbacks is included. For example, climate change induces shifts in global vegetation patterns; this changes the yield of crops and hence changes land-use patterns

---

[3] The innovative aspect of using narratives in the SRES storylines lies in the consistency of future trends in population, wealth, technology, equity, and energy use within each scenario. The earlier IPCC IS92 emission scenarios were based on expert judgment or literature surveys for every scenario assumption independently, returning a high, middle, and low scenario. In the SRES approach each narrative defines characteristic, consistent trends in the input assumptions, recognizing explicitly the correlations between the different assumptions.

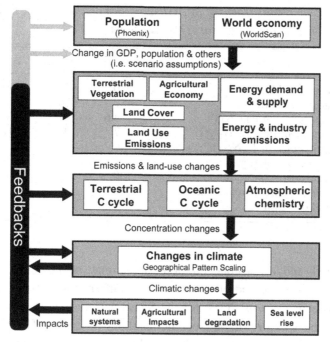

Figure 15.4. The structure of IMAGE 2.2.

and emissions, which influences carbon uptake by the biosphere and oceans. In other words, the strength of the IMAGE-2 model is found in the comprehensive coupling of the anthropogenic influences, the biosphere, ocean, and atmosphere (see Figure 15.4) and not in the complexity of the different sub-models. IMAGE 2 is calibrated on historical data from 1970 to 1995. This calibration is performed on a number of data sources, concerning energy use, land use, and national emission inventories. The global carbon cycle model required a longer calibration period, which is based on data for the atmospheric increase and energy and industry emissions of CO_2 for the period from 1765 to 1995.

The SRES results clearly show that different patterns can evolve from similar assumptions that are part of a SRES narrative (summarized in Table 15.1; more detailed regional, sectoral, and geographical information is on the IMAGE 2 CD-ROM by the IMAGE team (2001) and on http://www.mnp.nl). For example, the population assumptions in B1 and A1 were identical. The land-use dynamics, driven by shifts in demography, technological development, diets, and trade, were shown to be very important in shaping future worlds.

In implementing the IMAGE-2 SRES storylines, much attention has been paid to identifying major sources of uncertainty. The different storylines result in largely different sets of energy use and energy carriers, land-use and land-cover patterns,

Table 15.1. *Some results from the IMAGE-2 implementation of the SRES storylines after 100 simulated years (IMAGE team 2001).*

	A1	A2	B1	B2
Global forest area (in Mha)	980	420	750	800
Global crop area (in Mha)	480	600	260	660
Global CO_2 emissions (in Pg C/yr)	3.6	9.0	0.7	0.1
Global CH_4 emissions (in Tg CH_4/yr)	67	102	91	96
Global N_2O emissions (in Tg N/yr)	0.7	7.4	7.5	0.8
Global CO_2 concentration (ppmv)	55	70	15	5
Global CO_2-equivalent concentration (ppmv)	30	315	40	20
Global mean temperature increase since pre-industrial times (in °C)	0.4	0.7	0.3	0.9

emission trends and levels, concentrations, and climate change (Table 15.1). Consequently, sea-level rise and other impacts also differ. A sensitivity analysis was performed (Leemans *et al.* 2003) for the SRES narratives with different climate sensitivities and with or without specific feedback processes. The results of this sensitivity analysis indicate a possible global mean temperature increase, ranging from 1.6 to 5.5 °C at the end of the twenty-first century.

Another uncertainty stems from the pattern scaling of global-mean temperature changes to geographically explicit temperature and precipitation change by a given climate change pattern obtained from state-of-the art, three-dimensional climate models (i.e., GCMs). To assess such uncertainty, the climate change pattern scaling in IMAGE 2 has been carried out with six different GCM climate change patterns. Between each GCM pattern there were large differences in impacts. At a global mean temperature increase of 2 °C, 33–42% of all ecosystems will change (i.e., will display a shift in biome). All simulations showed that large shifts occurred in high-latitude areas due to relatively large temperature increases. Most disagreement occurred in the tropics where the change was generally determined by changes in moisture availability (i.e., resulting from changes in precipitation and evaporation). Similar patterns were seen in crop productivity. All simulations showed positive and negative impacts. At the regional scale, contradictory results were obtained. For example, using one GCM pattern, crop productivity increases strongly in India, while with another it rapidly decreases. Some consistent positive impacts were seen in all scenarios: crop productivity increased in many of the northern areas, which are currently marginal. Also crops could be grown in the future at higher altitudes. All other regions were sometimes winners and sometimes losers, depending on the specific crop and the GCM used. Although there were large regional differences, impacts did not differ largely when they were globally aggregated. The magnitude

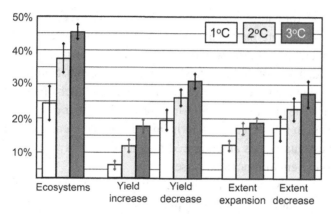

Figure 15.5. The aggregated impacts on ecosystems and crop distributions for different increases in global mean annual temperature. Impacts are expressed as the percentage area affected. The reference is the current extent of ecosystems or the current area for crops. The errors bars are determined by using the spatial patterns of temperature and precipitation change from four different GCMs.

of impacts between the different GCM climate change patterns is very similar for all and increases with increasing temperatures (Figure 15.5). In general, the negative crop impacts (decreases in potential yield and extent) are larger. In principle, the potential for negative impacts almost everywhere should provide enough motivation to limit climate change.

All of these sensitivity experiments provide some insights into the uncertainty of quantification of these storylines. In summary, the SRES narratives largely determine emissions, the final concentrations and thus climate change and impacts are strongly dependent on several feedback processes. This is consistent with the recent conclusions of the IPCC (Solomon *et al.* 2007). The outcome of these processes can only be adequately simulated when the appropriate dimensions and scales are considered. The sensitivity analysis highlights the importance of a proper parameterization of the specific systemic interactions between the different components of the system. The consequence of this finding is that environmental assessment models that solely focus on energy production and the different sectors that use energy should only be used to calculate emissions. The often simple highly aggregated linked carbon-cycle and climate change models do not provide a sound basis for impact assessment. These components need a highly integrated approach to cover the spatially heterogeneous processes.

These IMAGE 2 results from the SRES storylines, however, provide sufficient detail for researchers interested in the causes and consequences of many aspects of global change for higher-resolution studies. RIVM and Stichting DLO (2002), for example, have applied the SRES storylines to develop four reliable maps for different future land uses in the Netherlands. The resulting land-use patterns show large differences for nature conservation. The A-scenarios emphasize multiple use of landscapes. There are many parks but few strict nature reserves. All landscapes are used for recreation. In the wealthy A1 scenario, mansions emerge again, as if the Dutch Golden Age of the seventeenth century reappears. Nature conservation is best guaranteed in the B-scenarios. In the B2 scenarios much of the conservation is implemented at the provincial level, which results in a poor connective network of nature reserves. In B1 such a network is implemented on a national and European scale. This best guarantees an enhanced resilience of ecosystems against climate change and other environmental stresses.

The SRES narratives were also implemented for Europe on a 10×10 km grid and EU-nuts-2 regions (cf. states or provinces) by the ATEAM team (Schröter *et al.* 2005). They interpreted socio-economic consequences of each narrative further and created consistent scenarios for land use (Rounsevell *et al.* 2006), atmospheric composition, climate change, and N-deposition. Although IMAGE determined the global context, other approaches were developed for the European level (Ewert *et al.* 2005). These scenarios were used to determine the vulnerability of different ecosystem services (Metzger 2005) and biodiversity (Thuiller *et al.* 2005). The ATEAM concluded that Mediterranean and mountain regions were the most vulnerable.

These two examples illustrate the power of developing comprehensive narratives to systematically develop local and regional assessments from the results of global models. Although originally implemented for only four regions in the original SRES and for 17 regions in IMAGE 2, both studies show that it is possible to enhance the resolution of both the environmental and socio-economic dimensions. This can only be done when one realizes that consistency and coherency at the different levels continues to be determined by the narrative. The local or regional results, however, are qualitatively different from those of the global models. Additional detail has been added, which would never have been possible with just increasing, for example, the socio-economic and environmental resolutions of IMAGE 2. For example, in the Dutch study the consequences for many more sectors, land uses and land-cover classes were quantified than those available in IMAGE 2. IMAGE 2 just characterized the necessary context.

The above examples strongly focus on qualitative results stemming from narratives developed by a multi-disciplinary team of experts. Quantitative nesting approaches with global and regional models have been developed (e.g., Mearns *et al.* 2001). The disadvantage of these approaches, however, is that the global model provides the boundary conditions for the regional models. This one-way approach neglects important interactions and feedbacks between the local and regional levels (see Figure 15.2). This drawback is very difficult to solve. Only one GCM used an elegant approach with a variable grid: a high resolution on the region of interest and a coarse resolution for the rest of the world (Déqué and Piedelievre 1995). UNEP's Global Environmental Outlook (2002) used an iterative approach based on global IMAGE-2 scenarios, which were used and further interpreted by regional expert groups. Important regional trends, characteristics, and restrictions on trade, productivity, etc. were collected for a new IMAGE 2 simulation. This process was repeated until a consensus was reached. This process stressed the importance of socio-economic consistency between regions (on issues like trade) but less on the environmental interactions. The approach led to a very rich qualitative regional description of the scenarios, illustrated by global environmental trends.

15.5 Concluding remarks

There are still many uncertainties in developing environmental assessment models and plausible scenarios of global change. The narrative approach has increased consistency and will assist to improve broader and more precise applications of the scenarios, such as mitigation, adaptation, and vulnerability assessments. Each narrative will logically have its own set of effective policy responses to mitigate and adapt. The policy goal to stabilize atmospheric greenhouse gas concentrations, for example, is easier to reach in a B1 world than in an A1 world. Enhancing energy efficiency, however, is not likely in the B1 scenario because energy use is already highly efficient. The distinct differences in the narratives clearly illustrate that different sets of responses to climate change are needed depending on the societal and environmental context. Also the exposure, the systemic sensitivity, and adaptive capacity are different for each narrative, which results in largely different vulnerabilities. An appropriate context is defined by specifying narrative bridges between global, regional, and local assessments.

The differences between narratives will in the near future surely facilitate an advancement of climate change risk analyses and policy response assessments. The coarse top-down broadbrush approaches will likely be linked with regionally specific, bottom-up approaches that realistically capture the diversity of possible responses at all discrete scale levels of the organizational dimension (see Figure 15.2). To achieve this, local, regional, and global assessors have to collaborate closely. This will enhance the understanding of the different approaches and will enrich narrative scenario development and make scenarios much more relevant for various audiences.

Although the four SRES narratives illustrate plausible future developments, they remain simple caricatures of possible realistic developments. The future will probably evolve as a combination of all SRES narratives (together with many other plausible futures). One of the most important lessons of the SRES scenarios is, however, that similar sets of coarse IPAT driving forces can lead to largely different emissions pathways. Highly aggregated sets of drivers in environmental assessments are no longer valid. New approaches to link different dimensions of global change have to be developed. Although there probably remains a significant role for traditional system-analysis approach models, such as IMAGE 2, other approaches have to be explored. Despite the fact that some models already include some stochastic effects, such as interannual variability in climate forcing, these traditional models generally simulate only smooth transitions. Recent insights show that many societal and environmental dynamics show rapid transitions, phase shifts, and multiple stable stages caused by such gradual changes (Scheffer *et al.* 2001; Costanza *et al.* 2007). Higher-resolution models could capture such behavior, but then the available data (of reliable quality) rapidly become limited. New innovative approaches are therefore called for. Candidates include actor-oriented modeling (e.g., Stephen and Downing 2001; Anderies *et al.* 2002), the syndrome approach (Petchel-Held *et al.* 1999), and advanced cellular automata that link human behavior with the development of spatial patterns (e.g., Costanza *et al.* 2002).

Traditional environmental assessors could benefit substantially from these emerging approaches, when they are combined with the more traditional approaches. It will surely more strongly emphasize the appropriate regional and spatial specificity that describes the appropriate dimensions of environmental problems. From the above discussion, it also becomes clear that the methodology gap between regional and global environmental assessments will be reduced when these new approaches are adopted.

References

Alcamo, J. (ed.) 1994. *IMAGE 2.0: Integrated Modeling of Global Climate Change.* Dordrecht: Kluwer Academic Publishers.

Alcamo, J., M. Amann, J.-P. Hettelingh, *et al.* 1987. Acidification in Europe: a simulation model for evaluating control strategies. *Ambio*, **16**, 232–245.

Allen, T. F. H. and T. B. Starr 1982. *Hierarchy: Perspectives for Ecological Complexity.* Chicago, IL: University of Chicago Press.

Anderies, J. M., M. A. Janssen, and B. H. Walker 2002. Grazing management, resilience, and the dynamics of a fire-driven rangeland system. *Ecosystems*, **5**, 23–44.

Bouwman, A. M., T. Kram, and K. Klein Goldewijk (eds.) 2006. *Integrated modelling of global environmental change.* Bilthoven: Netherlands Environmental Assessment Agency.

Commoner, B., M. Corr, and P. J. Stamler 1971. The causes of pollution. *Environment*, **13**, 2–19.

Costanza, R., A. Voinov, R. Boumans, *et al.* 2002. Integrated ecological economic modeling of the Patuxent river watershed, Maryland. *Ecological Monographs*, **72**, 203–231.

Costanza, R., L. J. Graumlich, and W. Steffen (eds.) 2007. *Sustainability or Collapse? An Integrated History and Future of People on Earth.* Cambridge, MA: MIT Press.

Déqué, M. and J. P. Piedelievre 1995. High resolution climate simulation over Europe. *Climate Dynamics*, **11**, 321–339.

Ehrlich, P. and J. Holdren 1971. Impact of population growth. *Science*, **171**, 1212–1217.

Ewert, F., M. D. A. Rounsevell, I. Reginster, M. J. Metzger, and R. Leemans 2005. Future scenarios of European agricultural land use. I. Estimating changes in crop productivity. *Agriculture, Ecosystems and Environment*, **107**, 101–106.

FAO and CSRC 1987. *Soil map of the world*, 1.5M, UNESCO, Paris. 1/2 degree digitization. Complex Systems Research Center, University of Hew Hampshire, Durham.

Fischer, G., K. Froberg, M. A. Keyzer, and K. S. Pahrik 1988. *Linked National Models: a Tool for International Food Policy Analysis*. Dordrecht: Kluwer.

Geist, H. J. and E. F. Lambin 2002. Proximate causes and underlying driving forces of tropical deforestation. *Bioscience*, **52**, 143–150.

Goudriaan, J. and P. Ketner 1984. A simulation study for the global carbon cycle, including man's impact on the biosphere. *Climatic Change*, **6**, 167–192.

IMAGE team 2001. The IMAGE 2.2 implementation of the SRES scenarios: a comprehensive analysis of emissions, climate change and impacts in the 21st century. *RIVM CD-ROM Publication 481508018*. Bilthoven: National Institute of Public Health and the Environment.

IPCC 2001. *Climate Change 2001: Synthesis Report*. Cambridge: Cambridge University Press.

Kainuma, M., Y. Matsuoka, and T. Morita. 2002. *Climate Policy Assessment: Asia–Pacific Integrated Modeling*. Tokyo: Springer-Verlag.

Kauppi, P. E., J. H. Ausubel, J. Fang, *et al.* 2006. Returning forests analyzed with the forest identity. *PNAS*, **103**, 17574–17579.

Kaya, Y. 1989. Impacts of carbon dioxide emission on GWP growth: interpretation of proposed scenarios. In *Report of the IPCC Response Strategies Working Group Intergovernmental Panel on Climate Change*. Geneva: IPCC.

Kuhn, I., K. Bohning-Gaese, W. Cramer, and S. Klotz 2008. Macroecology meets global change research. *Global Ecology and Biogeography*, **17**, 3–4.

Lambin, E. F. and H. J. Geist (eds.) 2006. *Land-use and land-cover change. Local Processes and Global Impacts*. Berlin: Springer-Verlag.

Leemans, R. and W. P. Cramer. 1991. The IIASA database for mean monthly values of temperature, precipitation and cloudiness on a global terrestrial grid. *Research Report RR-91–18*. Laxenburg: International Institute of Applied Systems Analyses.

Leemans, R. and A. M. Solomon 1993. The potential response and redistribution of crops under a doubled CO_2 climate. *Climate Research*, **3**, 79–96.

Leemans, R., W. P. Cramer, and J. G. van Minnen 1996. Prediction of global biome distribution using bioclimatic equilibrium models. In J. M. Melillo and A. Breymeyer (eds.), *Effects of Global Change on Coniferous Forests and Grasslands*. New York: J. Wiley and Sons, pp. 413–450.

Leemans, R., E. Lambin, A. McCalla, *et al.* 2003. Drivers of change in ecosystems and their services. In *Millennium Ecosystem Assessment, Ecosystems and Human Well-being: a Framework for Assessment*. Washington, DC: Island Press, pp. 85–106.

Masui, T., T. Hanaoka, S. Hikita, and M. Kainuma 2006. Assessment of CO_2 reductions and economic impacts considering energy-saving investments. *Energy Journal*, **27**, 175–190.

Matsuoka, Y., M. Kainuma, and T. Morita 1994. Scenario analysis of global warming using the Asian–Pacific Integrated Model (AIM). *Energy Policy*, **23**, 1–15.

Mearns, L. O., W. Easterling, C. Hays, and D. Marx 2001. Comparison of agricultural impacts of climate change calculated from high and low resolution climate change scenarios: Part I. The uncertainty due to spatial scale. *Climatic Change*, **51**, 131–172.

Melillo, J. M., A. D. McGuire, D. W. Kicklighter, *et al.* 1993. Global climate change and terrestrial net primary production. *Nature*, **363**, 234–239.

Metz, B., O. R. Davidson, P. R. Bosch, R. Dave, and L. A. Meyer (eds.) 2007. *Climate Change 2007: Mitigation. Contribution of Working Group III to the Fourth Assessment Report of the Intergovernmental Panel on Climate Change*. Cambridge: Cambridge University Press.

Metzger, M. J. 2005. *European vulnerability to global change: a spatially explicit and quantitative assessment*. PhD thesis. Wageningen University, Wageningen.

Nakícenovíc, N., J. Alcamo, G. Davis, *et al.* 2000. *Special Report on Emissions Scenarios*. Cambridge: Cambridge University Press.

National Geophysical Data Center (NOAA) 1988. *10-minute topography data base*. Computer tape US Department of Commerce, Washington DC, US.

OECD's InterFutures Study Team 1979. *Mastering the Probable and Managing the Unpredictable*. Paris: Organisation for Economic Co-operation and Development and International Energy Agency.

Olson, J. S., J. A. Watts, and L. J. Allison 1985. *Major World Ecosystem Complexes Ranked by Carbon in Live Vegetation: A Database. Report NDP-017*. Oak Ridge, TN: Carbon Dioxide Information Center, Oak Ridge National Laboratory.

Ostrom, E., T. Dietz, N. Dolsak, P. C. Stern, S. Stonich, and E. U. Weber (eds.) 2002. *The Drama of the Commons*. Washington, DC: National Academy Press.

Parry, M. L., O. F. Canziani, J. P. Palutikof, C. E. Hanson, and P. J. Van Der Linden (eds.) 2007. *Climate Change 2007. Impacts, Adaptation and Vulnerability. Contribution of Working Group II to the Fourth Assessment Report of the Intergovernmental Panel on Climate Change*. Cambridge: Cambridge University Press.

Petchel-Held, G., A. Block, M. Cassel-Gintz, *et al.* 1999. Syndromes of global change: a qualitative modelling approach to assist global environmental management. *Environmental Modelling and Assessment*, **4**, 295–314.

Reid, W. V., H. A. Mooney, A. Cropper, *et al.* 2005. *Millennium Ecosystem Assessment Synthesis Report*. Washington, DC: Island Press.

Reynolds, J. F., D. M. S. Smith, E. F. Lambin, *et al.* 2007. Global desertification: building a science for dryland development. *Science*, **316**, 847–851.

RIVM and Stichting DLO 2002. *Natuurverkenning*, **2**, 2000–2030.

Rotmans, J. and B. de Vries (eds.) 1997. *Perspectives on global change: the TARGETS approach*. Cambridge: Cambridge University Press.

Rounsevell, M. D. A., I. Reginster, M. B. Araujo, *et al.* 2006. A coherent set of future land use change scenarios for Europe. *Agriculture, Ecosystems and Environment*, **114**, 57–68.

Sala, O. E., F. S. Chapin, III, J. J. Armesto, *et al.* 2000. Global biodiversity scenarios for the year 2100. *Science*, **287**, 1770–1774.

Scheffer, M., S. R. Carpenter, J. A. Foley, C. Folke, and B. Walker 2001. Stochastic events can trigger large state shifts in ecosystems with reduced resilience. *Nature*, **413**, 591–596.

Schröter, D., W. Cramer, R. Leemans, *et al.* 2005. Ecosystem service supply and vulnerability to global change in Europe. *Science*, **310**, 1333–1337.

Skole, D. L. and C. J. Tucker 1993. Tropical deforestation and habitat fragmentation in the Amazon: satellite data from 1978 to 1988. *Science*, **260**, 1905–1910.

Solomon, S., D. Qin, M. Manning, *et al.* (eds.) 2007. *Climate Change 2007: the Physical Science Basis. Contribution of Working Group I to the Fourth Assessment Report of the Intergovernmental Panel on Climate Change.* Cambridge: Cambridge University Press.

Stafford Smith, D. M. and J. F. Reynolds (eds.) 2002. *Integrated assessment and desertification.* Berlin: Dahlem University Press.

Stephen, L. and T. E. Downing 2001. Getting the scale right: a comparison of analytical methods for vulnerability assessment and household-level targeting. *Disasters*, **25**, 113–135.

Thuiller, W., S. Lavorel, M. B. Araujo, M. T. Sykes, and I. C. Prentice 2005. Climate change threats to plant diversity in Europe. *Proceedings National Academy of Sciences*, **102**, 8245–8250.

Tol, R. S. J. 1999. Time discounting and optimal emission reduction: an application of FUND. *Climatic Change*, **41**, 351–362.

Turner, B. L., II, D. L. Skole, S. Sanderson, *et al.* 1995. Land use and land-cover change: science/research plan. *IGBP Report* No. 35 and *HDP Report* No. 7. Stockholm: International Geosphere–Biosphere Programme and the Human Dimensions of Global Environmental Change Programme.

UNEP 2002. *Global Environment Outlook 3.* London: Earthscan publications.

Van der Sluijs, J. P. 1997. *Anchoring amid uncertainty. On the management of uncertainties in risk assessment of anthropogenic climate change.* PhD Thesis. University of Utrecht, Utrecht.

Weyant, J., O. Davidson, H. Dowlatabadi, *et al.* 1996. Integrated assessment of climate change: an overview and comparison of approaches and results. In J. P. Bruce, H. Lee, and E. F. Haites (eds.), *Climate Change 1995. Economic and Social Dimensions of Climate Change.* Cambridge: Cambridge University Press, pp. 367–396.

Zuidema, G., G.-J. Van Den Born, J. Alcamo, and G. J. J. Kreileman 1994. Simulating changes in global land cover as affected by economic and climatic factors. *Water, Air and Soil Pollution*, **76**, 163–198.

16

The SARCS integrated study of Southeast Asia: research and assessment with networks

16.1 Introduction

The main purpose of this chapter is to evaluate some of the limitations and benefits of networks for conducting integrated regional studies. The analysis is based on the experiences of the Southeast Asian Regional Committee for START[1] (SARCS) in developing and promoting research and assessment of global environmental change in the Southeast Asia region (Fuchs *et al.* 1998). A key strategy of the committee was the formulation of an "integrated study" around which a network of organizations carrying out several projects and planning others could contribute towards much broader scientific goals of understanding larger-scale biophysical and social processes in the region and their links with the Earth system. The Science Plan for the SARCS Integrated Study (Lebel and Steffen 1998) placed a strong emphasis on synthesis and integration activities aimed at development policy, providing a knowledge base and framework for integrated regional assessments.

The chapter is organized in three parts. The first provides some essential background to environmental changes and economic developments in Southeast Asia, and thus lays out the scope and major issues for research and assessment. Some of this understanding is in itself an outcome of the assessment process. The middle part is a brief history of the origins, goals, and achievements of the first iteration of assessments under the SARCS Integrated Study framework. Key features of the supporting organizational structure are also described. The final section draws conclusions about the strengths and limitations of research networks as a foundation for regional research and assessment.

16.2 Southeast Asia

The transformation of Southeast Asia over the past several decades has been breathtakingly quick, often dirty, and incomplete (Lebel 2005b). For several decades the

---

[1] START denotes the Global Environmental Change Programme "The Global SysTem for Analysis, Research and Training." See www.start.org

newly industrializing economies in the East Asia region have registered some of the fastest changes in health, education, and other indicators of well-being. Economies have grown at dizzying rates, agricultural productivity has soared, and the supply of sanitation, electricity, education, health services, and other basic infrastructure has greatly improved (Lebel 2002a).

Industrialization has led economic growth. Between 1970 and 1993 the contribution by industry to the ASEAN region gross domestic product (GDP) has increased from 25% to 40% and industrial output has increased by 25 times during the same period. Manufacturing contributed more than two-thirds of the GDP in 1994, having expanded at an annual rate of 19% since 1980. East Asia became the *"manufacturing belt of the world"* (Lebel 2004a; Rock and Angel 2005).

The economies of the Southeast Asian newly industrialized economies became more industrial, diversified, and integrated into the global economy than their counterparts elsewhere (Knight 1998). Between 1965 and 1995, for example, openness ratios (exports plus imports divided by GDP) increased from 12% to 49% for Indonesia, 37% to 91% for Thailand, and 79% to 195% for Malaysia. Rapid expansion of croplands has been associated with large-scale deforestation (Lepers *et al.* 2005).

The World Bank, the International Monetary Fund (IMF), and the Asian Development Bank (ADB) have played crucial roles in developing infrastructure and transforming economies through various "restructuring" programs and conditional loans, most recently following the Asian financial crisis in 1997–98. Today, the region is a mixture of small, transitional, fast-growing economies, and a few established larger economies around the edges, in particular Japan, the Island of Taiwan, Korea, and the omni-present USA, which they serve, and for which Singapore acts as a service hub. Cambodia, Lao PDR, and Myanmar, for example, have just begun engaging the regional and global market system and still have small agrarian economies, whereas Singapore's has gone beyond manufacturing and is dominated by the information service sector.

Rapid growth has had its consequences. Air and water quality pollution became serious problems before efforts were made to tackle them; and what were once thought of as just local-scale problems are increasingly recognized as having transboundary causes and consequences (ASEAN 1997; Angel & Rock 2000; Rock 2002). Growing disparities in incomes and the violence of modernization projects, initially carefully kept hidden from view, are emerging into the light (Bello *et al.* 1998). Each country has struggled to maintain diverse mixtures of indigenous and transformed cultures and social structures against a backdrop of international interventions, wars, corruption, military dictatorships, and footloose foreign capital. The financial crisis of 1997–98 was also a poignant reminder of the strong economic connections between countries and led to a substantial amount of re-examination of development models (Root 1996; Jomo 1998; Dixon 1999; Siamwalla 2003).

Moreover, for the least developed and newly industrializing economies in the region, most of the investments in industrialization are still to come (Angel and Rock 2000; Rock 2002). How these investments are guided will have major consequences for the regional environment and the Earth's system.

Incompleteness is most visible in the cities (Marcotullio 2003; Lebel 2005a). Scattered across capital cities is a complex jumble of glass-walled skyscrapers, slums, housing estates, factories, and vegetable gardens. Environmental health hazards from polluted waterways and expressways are juxtaposed with air-conditioned shopping malls and five-star hotel lobbies. The connections among capitals and between cities and their rural hinterlands are strong and cross national boundaries. Decisions by Malaysian firms to invest in oil palm plantations in Indonesia contributed to simultaneous clearing of large areas of forest with fires that resulted in significant haze in Kuala Lumpur and Singapore. The reality of a common destiny but great diversity in capacity is reflected in the modest achievements of the Association of Southeast Asian Nations (ASEAN; Tay *et al.* 2000), the Mekong River Commission (Dore 2003; Hirsch 2006), and regional institutions to tackle trans-boundary environmental pollution (Lebel 2002b; Murdiyarso *et al.* 2004).

There are several economic, institutional, and biophysical linkages that argue for considering the region as whole, even though this means encompassing a great variety of ecosystems, cultures, and development pathways. Setting the boundaries of a region is always arbitrary but the Association of Southeast Asian Nations made sense as a starting point at least from the perspective of international relations in the 1990s (Tay *et al.* 2000).

16.3 The SARCS integrated study

From a global Earth systems perspective, the Southeast Asian region is of particular interest because of its rapid rates of social, economic, and environmental transformation, its archipelagic and mountainous geography with high levels of endemic biodiversity, and the direct dependence of many livelihoods on the Asian monsoon and the El Niño–Southern Oscillation (Lebel and Steffen 1998). From a regional perspective there is an interest in better understanding the implications of global changes, both social and environmental, for future development, so that appropriate adaptation and coping strategies can be pursued (Lebel 2005b, 2007).

16.3.1 *Origins in capacity building*

One of the challenges for the *developing* countries in the region is that the capacity to conduct primary research and make assessments of existing knowledge is often limited. A common solution is to *"fly in the experts."* There are clear dangers with

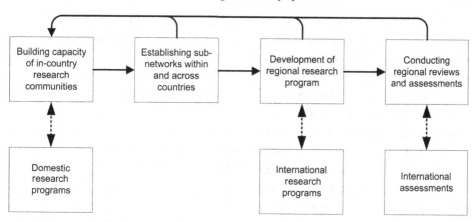

Figure 16.1. Capacity and network building underlined the SARCS approach to shaping a knowledge system capable of carrying out integrated regional studies.

this strategy in terms of the vested interests and unintentional biases that come as carry-on baggage with each person's culture. In the longer term it may also be counterproductive as it could stifle development of in-house analytical capacities (Lebel and Tang 2006). At the same time, international cooperation and exchange can often be very beneficial if it is seen as a two-way dialogue where developing and industrialized countries have things they can learn from each other about transitions to sustainability. A balance of perspectives is highly desirable when making assessments (Lebel 2006b). This is especially true when dealing with incomplete understanding of the complex interactions between human activities, policy, and environment that form the core of integrated regional assessments.

The international global change science programs sponsored the START initiatives to create networks in developing regions capable of conducting research on regional aspects of global change and providing support to policy makers and governments based on the regional findings (Fuchs *et al.* 1998). In the early stages, many of the activities were directed to help meet basic capacity-building needs (Figure 16.1). These included building awareness about how global environmental change, especially climate change, could confound more local ecological and adaptive processes, and providing training on data gathering, organization, and analysis. As the network has grown in strength it has become a significant contributor to basic scientific research in the international programs.

16.3.2 Distributed nodes and supporting networks

Since 1993, the START program in Southeast Asia has been coordinated by SARCS. The committee itself does most of its work through a network of affiliated organizations and a small secretariat. The chair has rotated every few years and the committee has changed in composition. I have been personally involved with the international global environmental change research programs in Southeast Asia since 1995, and for a substantial part of that time have had a role in the coordination of scientific research in the region. This has included helping design research programs, running training workshops, finding funds for collaboration, and synthesizing research findings across projects. In this chapter I endeavor to step back and reflect on those experiences relevant to integrated regional assessment activities.

In its initial phases the programs of the network were largely funded by a single GEF-UNDP grant, and in-kind support from participating organizations, mostly universities. Under this grant the coordination node shifted early on from the National University in Singapore to Chulalongkorn University in Bangkok. Over time, the number of sponsors and the number of actively affiliated scientists and organizations has grown tremendously. Although initially the idea of a central expert or data center was prominent, this was progressively replaced with the view of "multiple centers," or contributing nodes, in a network (Figure 16.2). This philosophical shift was important in getting wider ownership for the science programs in the region, because it provided opportunities for many different groups to play a role in coordination of regional science and assessment activities. It was also crucial for exploring sensitive policy issues, for example, related to transboundary drivers and consequences of land or water resources development.

Several nodes have made substantial contributions to coordination, research, and assessment activities since the mid-1990s. The first fully operational node the SEA START Regional Centre based at Chulalongkorn University in Bangkok continues to be very active in coordinating research on trace gas emissions, aerosols, water resources, coastal, and ocean processes. The University of Kebangsaan Malaysia in Kuala Lumpur has coordinated most of the land-use and land-cover change activities associated with SARCS, under the label of the Southeast Asia Regional Research and Information Network (SEARRIN). The Impacts Centre for Southeast Asia (ICSEA), in Bogor, Indonesia, contributed primarily to developing capacity to analyze the consequences of global environmental change on terrestrial ecosystems (Figure 16.2).

After 2000 additional nodes became more active (The Unit for Social and Environmental Research at Chiang Mai University, USER) and some of the original nodes and networks became less prominent (ICSEA) or relatively independent (SEARRIN). USER has focused primarily on the integration of social and natural sciences. Several regional networks have been established, for example, the U-TURN network on urbanization and carbon management, the IFA network on institutional dimensions of flood disaster management. The Central Taiwan University in Chung-li has played a leading role in the development of the regional

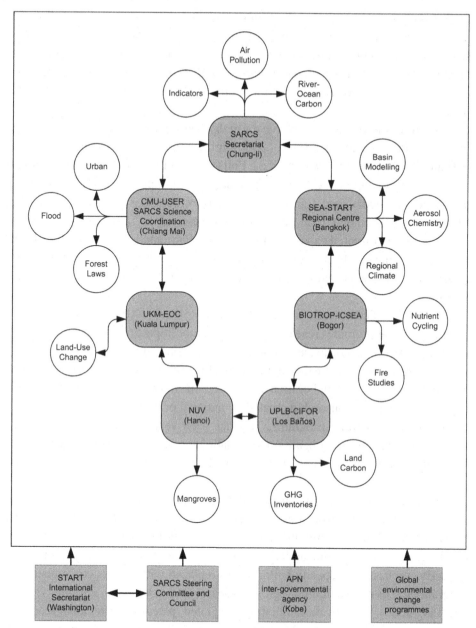

Figure 16.2. Summary of the network of networks structure of the mature SARCS Integrated Study science program in about 2004–5. Nodes are shaded rectangles and more specific project sub-networks or working groups are open circles.

project on sustainable development indicators, and with other centers in Taiwan on land–ocean interactions in the carbon cycle. It currently hosts the SARCS secretariat. The network continues to evolve with different nodes taking on more or less responsibility for regional research collaboration in different areas, fading away as funding of projects comes to an end or key individuals move on to new organizations.

Apart from these nodes which have usually had representatives in the SARCS committee there are also significant contributing nodes at the National University in Hanoi, Vietnam, the University of the Philippines at Los Baños, and the National University in Singapore. Although often targeted for capacity-building activities, groups in Cambodia, and Laos have, apart from activities associated with the Mekong River, been only modestly involved in research and assessment activities, in part because of limited capacity and resources. Participation of researchers from Burma/Myanmar has been even more limited. From time to time researchers and findings from Yunnan Province of China are included in "Southeast Asian" regional activities.

The Asia Pacific Network for Global Change Research (APN),[2] an intergovernmental body that promotes research, has had a major impact on the development of scientific networks and capacity in Southeast Asia since it began funding projects in 1997 (Figure 16.2). Since the beginning the focus of its funding support has been on activities involving collaboration among organizations in different countries filling a critical gap in sponsorship created by national or international agencies with country-specific funding programs. This has been very conducive to the formation, expansion, and strengthening of international scientific networks, the primary mode of functioning for global environmental change scientific programs around the world.

Over time the organization of the SARCS network has transformed from a single central research unit into a loose cluster of nodes that coordinate their actions through formal annual committee meetings, collaborative grant applications, and regular electronic communications. At the same time, much like the parent body, the early emphasis on capacity building was replaced with a more comprehensive program of activities including regional research and assessment activities (Lebel and Tang 2006). A focus on facilitating integrated, place-based regional research in developing parts of the world, especially Asia and Africa, has been one of the primary roles of the START networks.

As the capacity and confidence of the networks within the region grew, and, at the same time, priority needs of decision makers within the region became

---

[2] See www.apn-gcr.org

Table 16.1. *National breakdown of formally acknowledged contributions to the SARCS Integrated Study Science Plan and contributing authors to the Global Change Assessment Report by place of normal residence and work.*

Nation or region	SARCS Integrated Study Science Plan	Contributing authors to the Global Change Assessment Report
ASEAN	48	22
Philippines	12	4
Thailand	15	6
Malaysia	6	5
Indonesia	8	1
Singapore	2	4
Vietnam, Myanmar, Laos, Brunei, and Cambodia	5	2
Other parts of Asia, Australia, and the Pacific	25	3
Europe and North America	22	2

better understood within the science community, the need for a framework to guide research and assessment activities became widely accepted.

16.3.3 The Science Plan framework

The SARCS committee began seriously considering ideas for the Science Plan for an integrated study in late 1995 following an initial concept paper from Will Steffen, then Executive Officer of GCTE. The idea was that a comprehensive plan would provide a framework for an integrated regional study that would support policy. Development of the Science Plan involved contributions from a large number of people. In 1996 SARCS ran an initial planning workshop with 40 participants from 13 countries. In 1997 and 1998 I led a broad consultation process (see Table 16.1) with the scientific community and carried out reviews of regional development documents and literature to complete the Science Plan (Lebel and Steffen 1998).

The over-arching questions for the Integrated Study (Lebel and Steffen 1998) were listed as follows.

(1) What are the major demographic, socio-economic, and institutional driving forces for environmental change in the region and how do they interact?
(2) How might regional climate variability change?
(3) What are the environmental changes, in terms of patterns of land use and land cover, atmospheric composition, and the condition of marine resources, that result from these human and biophysical drivers of regional change?

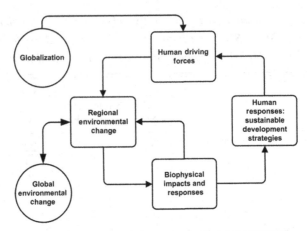

Figure 16.3. The simple conceptual framework of the SARCS Integrated Study emphasizing the feedbacks between development and environmental change at regional and global scales. Modified after Lebel and Steffen (1998) to include "globalization" influence explicitly (cf. Lebel 2002a).

(4) What are the implications of these environmental changes for biophysical processes on land, atmosphere, and the coastal zone, and how do they interact?
(5) What are the consequences of these environmental changes and biophysical responses for human welfare and sustainable regional development?
(6) What are the consequences of regional environmental change in Southeast Asia to the Earth system?

The Science Plan emphasized the importance of integrating, across multiple countries and interest groups, policies that support transitions towards sustainability. It also stressed the importance of viewing the Earth system as a coupled social–ecological system, with processes operating at different scales and incorporating several important biophysical and socio-economic feedbacks (Figure 16.3).

The Science Plan was an unusual departure from most of the earlier global change science planning documents for it placed the concerns of societies in the region, with regards to development, at the central focus of the study rather than at the periphery. At the same time it was part of a trend within the international research programs of striving towards better integration between biophysical and social sciences (cf., Koch *et al.* 1995; LBA 1996; Desanker *et al.* 1997; Scholes and Parsons 1997). This is reflected in the thematic structure of the Science Plan that builds upon more fundamental research themes towards the ultimate goals of synthesis and integration, and the exploration of sustainable development strategies.

The Science Plan provided a framework for an integrated study that could be implemented through a large number of smaller contributing projects over several years to a decade. This was seen at the time as the only feasible way for SARCS to tackle the ambitious integrating research questions about regional–global linkages in the Earth system. Since 1998 the Science Plan has been used by SARCS and its networks to help define priorities for research and assessment.

16.3.4 The START-APN assessment

The first iteration of an integrated regional assessment was organized around the idea of a written report that would address the policy implications of global change for development in Southeast Asia. This started soon after publication and distribution of the Science Plan with the recognition that this was a "time-zero synthesis" likely to be the first of several iterations. The scope of the initial assessment was comprehensive, covering a wide range of interacting global change issues, from changes in biogeochemical cycles, land-use and land-cover changes, degradation of large-scale resources, through to loss of biodiversity and climate change.

The starting point of analysis was development. Thus there were chapters devoted to urbanization, industrialization, expansion, and intensification of agriculture, use and protection of forested lands, exploitation and degradation of coastal and marine resource systems, globalization of trade and investment, and institutional and political transformation.

Each chapter began with a review of the development process including an assessment of major trends and underlying causes. It then proceeded to consider both how global environmental changes will influence that development process as well as how that process will contribute to further global environmental change. Throughout we were concerned with both describing current dynamics as well as assessing the capabilities of societies (individuals, social groups, or nations) to respond to future stresses. We were trying to understand the major processes and their interactions rather than to make quantitative predictions. The assessment process then built on this understanding to explore policy implications of global environmental change in terms of development strategies.

The assessment was completed through working group meetings of authors (Table 16.1) and electronic communications. A significant number of the authors had direct experience as senior bureaucrats and policy makers, or were active in policy think-tanks serving or challenging governments in the region. Without this policy-relevant experience an assessment of this type would not have been feasible. Apart from the core group of authors, the critical feedback of peer reviewers of the individual chapters and additional critical reviewing and editing by the science coordinator helped refine the findings of the assessment.

The assessment activity was financially supported by APN and START. The original plan was for the project to complete publication of a single final report within 2 years, but in the end a comprehensive report containing all the drafted chapters was never published because of the inconsistent quality of the original assessment chapters despite substantial efforts at editing and corresponding with authors to fill gaps. Instead a number of shorter, often more narrowly focused, products, synthesizing major findings were published (Tyson *et al.* 2001; Lebel 2002; Hsiao *et al.* 2002). Some authors also revised their work and published their findings in separate articles, for example, on forest and land management (Lasco 2002; Contreras 2003; Lebel *et al.* 2004; Murdiyarso and Lebel 2006).

Four general findings of the assessment process are highlighted to illustrate its scope. The first was delineating the importance of globalization of trade and liberalization of investment for land- and sea-use changes in the region (Lasco 2002; Lebel 2002a, 2004a; Lebel *et al.* 2002; Rock 2002; Giri *et al.* 2003). The second was the recognition that, while climate change should be of profound concern to the policy community in Southeast Asia, most of the important risks involve interactions with other components of environmental change and are strongly mediated by social structures and processes (Lebel 2004b, 2006). The third was that urbanization has been and will continue to be a profoundly important process in the region, and needs to be much more carefully studied in analyses of the Earth system–regional change linkages (Global Carbon Project 2003; Dhakal 2004; Lebel 2005a, 2006). The fourth was that humans are now an integral part of virtually all ecosystems and their behavior and social institutions need to be better understood if ecosystems are going to be managed in ways that continue to deliver needed goods and services (Lebel 2005c; Millennium Ecosystem Assessment 2005).

There were several other major problems with this first iteration that underline the challenges of regional integration. Many authors struggled to stay focused on how global environmental change confounds or otherwise interacts with more local changes, either remaining "global" or becoming too "local." This reflects the considerable conceptual difficulty of handling multi-scale drivers and responses. Another recurrent difficulty was to integrate quantitative findings from models and statistical analysis of historical trends with more context-rich case studies from which, for example, many institutional insights are derived. Insufficient attention was given to uncertainties or disagreements among findings in background reviews, and as a result the consequences of uncertainties in understanding for policy could not be considered. The failure to deliver a coherent integrated assessment document to the region's decision makers meant that the information gathered and synthesized probably had less impact than it otherwise might have had although this is an arguable point given the ad hoc but frequent interactions between scientists and policy makers in the Southeast Asia region.

16.3.5 The MAIRS SEA assessment

In the latest iteration the concept of an integrated study is being recast at one geographical level higher, that of the Monsoon Asia Region (Fu and de Vries 2006; see also Chapter 10, this volume). This is a logical progression, as it would allow much better engagement with large-scale climate, aerosol, and land-surface interactions than was possible within Southeast Asia. The Monsoon Asia Integrated Study (MAIRS) is a contribution to the Earth System Science Partnership of the international global change science programs. With a much narrower focus than the original SARCS Integrated Study but a much wider geographical scope this iteration is perhaps most significant for the opportunities it could create for growing scientific and political collaboration between China and India along with the other small countries in South and Southeast Asia.

In the lead up to drafting the Science Plan for MAIRS, working groups around the region were formed in South Asia, Temperate East Asia, and Southeast Asia to undertake a so-called "rapid assessment" of the state of scientific knowledge relevant to global environmental change. The Southeast Asian review drew on some of the authors from the previous round as well as adding new scientists. One of the main differences relative to earlier assessment was the much stronger set of reviews about land-surface, hydrology, and coastal zone interactions, in part as the result of additions of new groups of authors from Singapore and Taiwan. The assessment book was published in 2009 (Lebel *et al.* 2009).

The MAIRS collaboration also provides a framework for the continuing interest in interactions between land-use management practices and their impacts on hydrology and aerosols and interactions with climate (e.g., Sidle *et al.* 2006; Thanapakpawin *et al.* 2006; Murdiyarso and Lebel 2007). Interest in water resources management continues to expand. It also encourages further integration with human dimensions research, as for example, work on floods and flood-related disaster management has undertaken (Lebel *et al.* 2006a; Manuta *et al.* 2006; Nikitina 2006).

16.4 Networks

The START approach in Southeast Asia to integrated regional assessment has been based on the building and maintenance of networks of researchers, of coordinating nodes and of contributing projects, punctuated with occasional forays into policy, practitioner, and business communities. Was the network approach effective? In this final section we analyze some of the characteristics of the network organization of SARCS activities and how this influenced the conduct of the assessment and supporting science programs.

16.4.1 No control tower

Organizing the international scientific community is remarkably similar to herding cats. In the network model there is no control tower. The "Science Coordinator" for START in Southeast Asia had virtually zero direct control over human resources. The only way to mobilize was to use dialogues, whether workshops or electronic exchanges, to build on shared interests, trust, and concerns about sustainability. Collaborative grants and stipends for authors provided some additional incentive.

One of the most important positive features of the SARCS network organization has been its flexibility. Networks are coalitions of convenience. They allow a wide variety of levels of commitment and contributions. Most find that you get out what you put in. In this way you often get highly motivated people to contribute. The ability to draw in new members and create sub-networks as problems, political opportunities, and challenges arise is a strength of a network approach for both science and its evaluation.

The emphasis on cooperation, inclusion, and collaboration, however, has some undesirable side effects. Quality control, competition, and critical dissent at times wane when they are really needed to improve the quality of the assessment. In these cases, independent reviews and feedback often have to be obtained beyond the current networks or meetings with manufactured debates have to be set up.

This organizational model with no centralized budget, no strong lines of command, and distributed leadership of contributing projects and administrative functions, but with only very modest funds for coordination support is contrary to the domestic experiences of most researchers in Southeast Asia. However, the history of SARCS suggests that it is a viable model, and probably one that is becoming more so, with the diffusion of improved information technologies in the newly industrializing economies.

16.4.2 Diverse interests, languages, and politics

An international region with a complex mix of mature and developing economies provides extra challenges for an assessment. First, and most obvious, are differences in the current environmental challenges, which vary from providing basic water and sanitation facilities to problems of toxic waste and air pollution, or from congested traffic to wasteful consumption and packaging. Second, cultural and language differences make agreement on common or shared trans-boundary problems even more difficult to resolve. The ability to communicate effectively in English, the second language for all and the language for international cooperation and dissent, is undoubtedly a powerful tool in setting agendas, getting attention and projects that serve your needs. Third, political systems and cultures vary widely. Fourth, environmental assessments are subject to nuanced politics of scale from the choice of metrics through to issues of coverage and boundary setting (Lebel 2006b).

Over the past decade, in most countries in the region, even those with apparently active democratic systems, there have been times when direct criticism of government development policy was dangerous or an ineffective strategy to bring about awareness or change. Regional projects very easily slip into a mode of assuming that what is right for one is right for all. Context matters. The different levels of economic development result in huge disparities in human and technical resources available to engage in assessments. This makes it hard, without a conscious effort by the facilitators and leaders of an assessment process, for the less powerful, poorer voices to properly be heard on issues of priorities, problem definition and, especially, appropriate goals and response strategies.

Even with capacity in-house, international networks and their projects are "captured" by a very select group of non-governmental organizations (NGOs), government agencies, and universities. These are the organizations that have the pre-requisite capacities to interact effectively, maybe as a result of historical linkages, not just because of language and scientific or policy expertise, but also because of their capacity to "put-on" an international face for the eloquent foreign experts that bring the funding, technology, and promises of collaboration.

These selection biases create serious challenges of problem definition and representation when trying to implement an integrated regional assessment that cuts across national boundaries. SARCS has struggled with these challenges in its organizational history for both science program development and assessment activities. Early on it placed an emphasis on national representation in the committee. This proved to be not as effective as might be hoped because of a lack of continuity between meetings as different senior representatives were sent to meetings, few of whom had a real interest or commitment. Over time the committee began focusing more on areas of active research coordination, which worked for science, but then left problems of under-representation of the weakest nations where capacity building and assessment priorities were concerned. A two-tiered system with a more policy-oriented council and a more science-oriented steering committee has been introduced, but in almost 3 years the council has only met once.

As of August 2007 the SARCS Committee itself has not met in full session for more than 3 years without officially being disbanded, reflecting a move towards more Asia-wide organization of research. The division of START into South,

Southeast, and Temperate East Asian committees was political, reflecting competition between China and India, and finding a way for the substantial global change
research community in Taiwan to be engaged (via SARCS). Although these tensions
have not disappeared entirely scientific cooperation is moving ahead much more
broadly. The Monsoon Asia Integrated Regional Study (Fu and de Vries 2006),
provides an appropriate framework for such a re-scaling. The challenges facing the
science community, in short, parallel those in the diplomatic and macroeconomic
circles within and beyond ASEAN.

16.4.3 Support and influence from outside

The regional capacity building and integrated regional assessment efforts have benefited tremendously from linkages with the international global change programs.
In many cases, the time and effort of some of the best scientists in the world has
been voluntarily given to countries in the region, as a result of program linkages.
These international networks bring a tremendous knowledge base of state-of-the-art
science (Figure 16.1). Almost all the major core projects of IGBP and IHDP have
made an important contribution to the Southeast Asia region in the past decade.
They have done so largely by engaging the most relevant nodes in the SARCS networks with the result that individuals sitting on committees often belong to multiple
networks (Figure 16.2). There is both competition and cooperation in these relationships as organizations and individuals are pulled towards one research agenda
or another. The more successful organizations affiliated with SARCS have had to
be strategic with respect to the opportunities and time and other resource demands
placed on them in requests to participate in international science synthesis and
assessment activities.

International networks can be helpful to assessment processes in unexpected
ways because they build credibility and provide other ways for highly sensitive
information to be exposed and discussed. A good example of this was the ability
for the epistemic network to hold meetings and workshops involving Malaysian scientists during the 1997–98 trans-boundary haze episodes, whereas within Malaysia
academics were forbidden to discuss the issue with the media (Murdiyarso *et al.*
2004).

16.4.4 Labeling and ownership

One of the peculiar features of a distributed network organization like START is
that there is a lot of fuzziness in ownership and attribution for projects and findings.
For the most part the network tries to be inclusive, looking for credit where it can,
in part to justify its own existence, while at the same time being wary of going

too far beyond its stated priorities. On the other hand, researchers use the network as a label for credibility-raising when seeking funds, but later may play-down the linkages to take the credit for themselves. Although the SARCS committee debated, and at various times adopted, procedures for the endorsement of projects this was never implemented in a systematic way. As a consequence counting products and claiming credit is difficult. For "assessment products" this can cause problems of legitimacy as it is unclear who should take the credit (or blame).

16.4.5 Engaging the public policy process

By far the most glaring weakness of the IRA process so far has been the lack of attention given to creating more formal opportunities to engage the policy community and the wider public in a dialogue on assessment findings related to global environmental change. There were several reasons for this, some of which could not be easily anticipated at the beginning. First, and as already described, documents written for wider public consumption were not completed. Second, the START-APN assessment was primarily science-driven, in the sense of improving system understanding, exploring limits, and gauging risks, rather than focusing on specific policy issues such as regulating emissions or carbon trading under the Kyoto Protocol. The interest of policy was never likely to be very high for countries not yet needing to make decisions on such matters.

Third, and most important, national governments and regional organizations, like the Association for Southeast Asian Nations, have not, until 2006–7, begun to show more than mild interest in the issue of climate change and the global environment. The demand for assessment within the region has been very modest and primarily focused on meeting the communication requirements of international treaties rather than of being any relevance to national or regional development strategies.

To date most interaction has been on an individual and ad hoc basis through involvement of policy makers or bureaucrats in workshops or meetings, the attendance of scientists in government committees and regional policy discussions, or through scientists speaking to the mass media. The "effects" of these sporadic forays into public policy are hard to gauge and even harder to attribute to the activities of a particular network or one of its activities. This is doubly so because there has been little effort after the initial planning meetings to promote or highlight the SARCS Integrated Study as an integrated regional assessment. To date the findings of various synthesis and assessment-like products have been communicated in piecemeal fashion undermining some of the key initial intentions of fostering integration, but reflecting the realities of voluntary collaboration in large scientific networks.

16.5 Conclusions

Regional scientific networks have many advantages as organizational forms both for support of the science and conduct of the integrated regional assessments. By building regional collaboration on science they may help build foundations for other more politically sensitive discussions related to use and management of natural resources, such as trans-boundary water-sheds and air-sheds. But without clear links to regional institutions or engagement of decision makers and resource managers they will often struggle to reach an appropriate balance of legitimacy, credibility, and saliency (cf. Clark *et al.* 2002; Social Learning Group 2002) with the consequence that most of the knowledge produced will never get used.

Integrated regional assessments driven by informal epistemic networks are not a substitute for intergovernmental endorsed processes. However, the extra freedom to define terms of reference means that, by taking issues on that are not yet "popular" they could be highly complementary – a sort of early warning system for resilient societies. But to have that kind of role a high level of communication skills and some avenues to bring important issues into the public policy agenda are needed.

By necessity policy insights that come out of these exercises will have to be simple and general, perhaps in the form of principles or imperatives, for example, maintaining diversity, experimenting, cultivating, and protecting sources of innovation, rather than suggestions on specific taxes, land-use systems, and so on. Indeed the findings of an integrated regional assessment, apart from being able to provide guarded consensus on some trends, causes, and processes, will more usually be dealing with uncertainties and tradeoffs. For many issues, participatory mechanisms, where stakeholders can negotiate and alter their preferences or debate value changes, are essential for moving from an understanding of research-based knowledge to action. If well done an integrated regional assessment may start a dialogue, but it does not end it.

References

Angel, D. P. and M. T. Rock (eds.) 2000. *Asia's Clean Revolution: Industry, Growth and the Environment*. Sheffield: Greenleaf.

ASEAN 1997. *First State of the Environment Report*. Jakarta: ASEAN Secretariat.

Bello W., S. Cunningham, and Li Kheng Poh 1998. *A Siamese Tragedy: Development and Disintegration in Modern Thailand*. London: Zed Books.

Clark, W., R. Mitchell, D. Cash, and F. Alcock 2002. *Information as Influence: How Institutions Mediate the Impact of Scientific Assessments on Global Environmental Affairs. Kennedy School of Government Faculty Research Working Paper Series*, RWP02–044. Cambridge, MA: Harvard University.

Contreras, A. 2003. *The Kingdom and the Republic: Forest Governance and Political Transformation in Thailand and the Philippines*. Quezon City: Ateneo de Manila University Press.

Desanker P. V., P. G. H. Frost, C. O. Justice, and R. J. Scholes 1997. *The Miombo Network: Framework for a Terrestrial Transect Study of Land-use and Land-cover Change in the Miombo Ecosystems of Central Africa. IGBP Report 41.* Stockholm: International Geosphere–Biosphere Programme.

Dhakal, S. 2004. *Urban Energy Use and Greenhouse Gas Emissions in Asian Mega-cities: Policies for a Sustainable Future.* Urban Environmental Management Project. Kitakyushu: Institute for Global Environmental Strategies.

Dixon, C. 1999. *The Thai Economy: Uneven Development and Internationalisation.* New York: Routledge.

Dore, J. 2003. The governance of increasing Mekong regionalism. In K. Mingsarn and J. Dore (eds.), *Social Challenges of the Mekong Region.* Chiang Mai: Social Research Institute, Chiang Mai University, pp. 405–440.

Fu, C. and F. W. T. P. de Vries (eds.) 2006. *The Initial Science Plan of the Monsoon Asia Integrated Regional Study.* MAIRS-IPO, Beijing.

Fuchs, R., H. Virji, and C. Fleming 1998. *START Implementation Plan 1997–2002. IGBP Report 45.* Stockholm: International Geosphere–Biosphere Programme (IGBP).

Giri, C., P. Defourny, and S. Shrestha 2003. Land cover characterization and mapping of continental Southeast Asia using multi-resolution satellite sensor data. *International Journal of Remote Sensing,* 24(21), 4181–4196.

Global Carbon Project 2003. *Global Carbon Project. Science Framework and Implementation.* Canberra: Global Carbon Project. Earth System Science Partnership (IGBP, IHDP, WCRP, DIVERSITAS).

Hirsch, P. 2006. Water governance reform and catchment management in the Mekong region. *The Journal of Environment and Development,* 15, 184–201.

Hsiao H.-H., C.-H. Liu, and H.-M. Tsai (eds.) 2002. *Sustainable Development for Island Societies: Taiwan and the World.* Chung-li: SARCS Secretariat, National Central University, Taiwan.

Jomo, K. S. (ed.) 1998. *Tigers in Trouble: Financial Governance, Liberalization and Crises in East Asia.* London: Zed Books.

Knight, M. 1998. Developing countries and the globalization of financial markets. *World Development,* 26, 1185–1200.

Koch, G. W., R. J. Scholes, W. L. Steffen, P. M. Vitousek, and B. H. Walker 1995. *The IGBP Terrestrial Transects: Science Plan. IGBP Report 36.* Stockholm: International Geosphere–Biosphere Programme.

Lasco, R. D. 2002. Forest carbon budgets in Southeast Asia following harvesting and land cover change. *Science in China Series C,* 45 (Suppl.), 55–64.

LBA Science Planning Group 1996. *Large-scale Biosphere–Atmosphere experiment in Amazonia (LBA).* Science Plan.

Lebel, L. 2002a. Southeast Asia: economic globalization as a forcing function. In P. Tyson, R. Fuchs, C. Fu, *et al.* (eds.), *The Earth System: Global–Regional Linkages. Global Change: IGBP Series.* Heidelberg: Springer-Verlag.

Lebel, L. 2002b. Acid rain in Northeast Asia. In P. Noda (ed.), *Cross-sectoral Partnerships in Enhancing Human Security. Third Intellectual Dialogue on Building Asia's Tomorrow,* Bangkok, June 2000, ch. 3. Tokyo: Japan Centre for International Exchange.

Lebel, L. 2004a. Transitions to sustainability in production–consumption systems. *Journal of Industrial Ecology,* 9(1), 1–3.

Lebel, L. 2004b. Social change and CO_2 stabilization: moving away from carbon cultures. In C. Field and M. Raupach (eds.), *The Global Carbon Cycle: Integrating Humans, Climate, and the Natural World.* Washington, DC: Island Press, pp. 371–382.

Lebel, L. 2005a. Carbon and water management in urbanization. *Global Environmental Change*, **15**, 293–295.

Lebel, L. 2005b. Environmental change and transitions to sustainability in Pacific Asia. In S. Tay (ed.), *Pacific Asia 2022: Sketching Futures of a Region*. Tokyo: Japan Center for International Exchange, pp. 107–143.

Lebel, L. 2005c. Institutional dynamics and interplay: critical processes for forest governance and sustainability in the mountain regions of northern Thailand. In U. M. Huber, H. K. M. Bugmann, and M. A. Reasoner (eds.), *Global Change and Mountain Regions: an Overview of Current Knowledge*. Berlin: Springer-Verlag, pp. 531–540.

Lebel, L. 2006a. Multi-level scenarios for exploring alternative futures for upper tributary watersheds in mainland Southeast Asia. *Mountain Research and Development*, **26**, 263–273.

Lebel, L. 2006b. The politics of scale in environmental assessment. In W. V. Reid, F. Berkes, T. J. Wilbanks, and D. Capistrano (eds.), *Bridging Scales and Knowledge Systems: Concepts and Applications in Ecosystem Assessment*. Washington, DC: Island Press, pp. 37–57.

Lebel, L. 2007. Adapting to climate change. *Global Asia*, **2**, 15–21.

Lebel, L. and W. Steffen (eds.) 1998. *Global Environmental Change and Sustainable Development in Southeast Asia. Science Plan for a SARCS Integrated Study*. Bangkok: Southeast Asian Regional Committee for START: Chulalongkorn University.

Lebel, L. and C. W. Tang 2006. Building the capacity for assessment and deliberation. *IHDP Update*, **3**, 18–21.

Lebel, L., Nguyen Hoang Tri, Amnuay Saengnoree, Suparb Pasong, Urasa Buatama, and Le Kim Thoa 2002. Industrial transformation and shrimp aquaculture in Thailand and Vietnam: pathways to ecological, social and economic sustainability? *Ambio*, **31**(4), 311–323.

Lebel, L., A. Contreras, S. Pasong, and P. Garden 2004. Nobody knows best: alternative perspectives on forest management and governance in Southeast Asia: politics, law and economics. *International Environment Agreements*, **4**, 111–127.

Lebel, L., E. Nikitina, V. Kotov, and J. Manuta 2006. Assessing institutionalized capacities and practices to reduce the risks of flood disasters. In J. Birkmann (ed.), *Measuring Vulnerability to Natural Hazards: Towards Disaster Resilient Societies*. Tokyo: United Nations University Press, pp. 359–379.

Lebel, L., C. T. A. Chen, A. Sndvongs, and R. Daniel, (eds.) 2009. *Critical States: Environmental Challenges to Development in Monsoon Asia*, Selangor: SIRD/Gerakbudaya.

Lepers, E., E. F. Lambin, A. C. Janetos, *et al.* 2005. A synthesis of information on rapid land-cover change for the period 1981–2000. *Bioscience*, **55**, 115–124.

Manuta, J., S. Khrutmuang, D. Huaisai, and L. Lebel 2006. Institutionalized incapacities and practice in flood disaster management in Thailand. *Science and Culture*, **72**, 10–22.

Marcotullio, P. J. 2003. Globalisation, urban form and environmental conditions in Asia–Pacific cities. *Urban Studies*, **40**, 219–247.

Millennium Ecosystem Assessment 2005. *Ecosystems and Human Well-being: Synthesis*. Washington, DC: Island Press.

Murdiyarso, D. and L. Lebel 2006. Southeast Asian fire regimes and land development policy. In J. Canadell (ed.), *GCTE Synthesis*. Berlin: Springer-Verlag, pp. 261–271.

Murdiyarso, D. and L. Lebel 2007. Local to global perspectives on forest and land fires in Southeast Asia. *Mitigation and Adaptation Strategies for Global Change*, **12**, 2381–2386.

Murdiyarso, D., L. Lebel, A. N. Gintings, *et al.* 2004. Policy responses to complex environmental problems: insights from a science-policy activity on transboundary haze from vegetation fires in Southeast Asia. *Journal of Agriculture, Ecosystems, and Environment*, **104**, 47–56.

Nikitina, E. 2006. Success and failures in flood risk reduction programs across Asia: some lessons learned. *Science and Culture*, **72**, 72–83.

Rock, M. T. 2002. *Pollution Control in East Asia: Lessons from Newly Industrializing Economies*. Singapore: Institute of Southeast Asian Studies.

Rock, M. T. and D. P. Angel 2005. *Industrial Transformation in the Developing World*. Oxford: Oxford University Press.

Root, H. L. 1996. *Small Countries, Big Lessons: Governance and the Rise of East Asia*. Hong Kong: Oxford University Press.

Scholes R. J. and D. B. Parsons 1997. *The Kalahari Transect: Research on Global Change and Sustainable Development in Southern Africa. IGBP Report 42*. Stockholm: International Geosphere–Biosphere Programme.

Siamwalla, A. 2003. Globalisation and its governance in historical perspective. In K. Mingsarn and J. Dore (eds.), *Social Challenges of the Mekong region*. Chiang Mai: Social Research Institute, Chiang Mai University, pp. 13–44.

Sidle, R., M. Tani, and A. Ziegler 2006. Catchment processes in Southeast Asia: atmospheric, hydrologic, erosion, nutrient cycling and management effects. *Forest Ecology and Management*, **224**, 1–4.

Social Learning Group 2002. *Learning to Manage Global Environmental Risks*, vol. 1. *A Comparative History of Social Responses to Climate Change, Ozone Depletion and Acid rain*. Cambridge, MA: MIT Press.

Tay, S., J. Estanislao, and H. Soesastro (eds.) 2000. *A new ASEAN in a New Millennium*. Jakarta: Centre for Strategic and International Studies.

Thanapakpawin, P., J. Richey, D. E. Thomas, *et al.* 2006. Effects of land use change on the hydrologic regime of the Mae Chaem river basin, NW Thailand. *Journal of Hydrology*, **334**, 215–230.

Tyson, P., W. Steffen, P. A. Mitra, C. Fu, and L. Lebel 2001. The Earth system: regional–global linkages. *Regional Environmental Change*, **2**, 128–140.

17

Institutions for collaborative environmental research in the Americas: a case study of the Inter-American Institute for Global Change (IAI)

DIANA LIVERMAN

17.1 Introduction

The agreement to establish the Inter-American Institute for Global Change Research (IAI) was signed by eleven countries in Montevideo, Uruguay in May 1992. Now supported by 19 countries across the Americas, the IAI is dedicated to the better understanding of regional dimensions of global environmental change in the Americas through collaborative research, networks of scientists, education and training, and free and open exchange of relevant data. The overall goals are to augment scientific capacity, improve the understanding of global change and its socio-economic implications, and promote the dissemination of related information (including to policy makers in a useful and timely manner). The IAI's mission is to develop the capacity to understand the integrated impact of present and future global changes on regional and continental environments in the Americas and to promote collaborative research and informed action at all levels.

This chapter sets out to examine the Inter-American Institute as an example of a regional intergovernmental organization for scientific cooperation and to assess its achievements and challenges. It describes the history and structure of the IAI and then evaluates some of the indicators of success and problems that were encountered in the development and programs.[1] I begin by examining what the published literature has to say about the theory and rationale for regional and international scientific cooperation on the environment.

Although there is a rapid proliferation of the literature on international environmental policy, few scholars have explicitly focused on the role or characteristics of scientific collaboration and institutions as a basis for policy making on the environment. Perhaps the best known theoretical debate has emerged from the

---

[1] The analysis is based on the newsletters and reports of the IAI, on an evaluation conducted by the AAAS (2007) and on my experience as a member of the IAI Scientific Advisory Committee (SAC) for six years. The opinions expressed are of course my own and not those of IAI or its SAC.

international relations literature, where Peter Haas promoted the concept of "epistemic communities" of scientists that played key roles in the creation of environmental agreements and regimes for international ozone protection and regional cooperation to protect the Mediterranean Sea (Haas 1994). These epistemic communities are trans-national networks of scientists or non-governmental organizations that work to create a consensus and pressure governments to act on environmental issues. This cognitive approach to international environmental relations contrasts to neo-realist, liberal institutional, and structural perspectives.

The neo-realist explanation assumes that international agreements develop when a dominant (hegemonic) state such as the United States has the self-interest and resources to support environmental cooperation, perhaps because environmental degradation threatens national security or because cooperation will increase the relative balance of power with respect to other nations (Keohane 1986; Gilpin 1987). The liberal analysis assumes an anarchic system of interdependent nation states where international institutions or regimes are developed as a rational way to manage global problems (Keohane and Nye 1977; Krasner 1983). Structural or political economy explanations see both realist and liberal theories as overly state-centered and ignorant of the underlying force of economics and capital accumulation and cognitive approaches as too behavioral (Hurrell and Kingsbury 1992; Lipshutz and Conca 1993). In this case, international environmental agreements help maintain the conditions for the global production, distribution, and consumption of goods and services and need the support of major international corporations, multi-lateral development banks or powerful economic interests to succeed.

From the cognitive perspective the IAI might be seen as an effort to expand and construct an epistemic, regional community of global change scientists in the Americas in order to build support for environmental regimes such as the UN Framework Convention on Climate Change (UNFCCC). If as Haas has proposed (Haas 1994), we combine different theoretical perspectives, the IAI can also been seen as strongly influenced by the USA's hegemonic power in the hemisphere reflected in the US influence on some of the operating rules and funding mechanisms of the institution. The USA has self-interest in opening up the flow of data, especially because of links between USA and hemispheric climatic, ocean, and ecological systems, and in fostering consensual knowledge and alliances in negotiating international environmental regimes. As Keohane (1994: 26) notes, "what appears to be a process of science driven increases in consensual knowledge may have deeper sources in the classical political variables of interests, power, and institutional organization."

The literature on international environmental policy also provides a long list of reasons for international environmental cooperation (Haas *et al.* 1993; Porter and Welsh-Brown 1993; Caldwell 1996; Young 1997), among which scientific cooperation plays a small role. A 1992 Carnegie report (Carnegie Commission on Science Technology and Government 1992) on international environmental research suggests the following arguments for international scientific cooperation:

- efficiency and economies of scale to avoid duplication of effort and save on costs of monitoring and equipment;
- acceleration of the speed of learning through a diversity of experience and shared lessons from local innovations and applications;
- support for nations whose input and knowledge is needed by the US or world community but that cannot afford investments without assistance;
- to create informed consent for international environmental agreements and negotiations;
- to assist in strengthening national research capacities and thus the ability to influence local decision makers and environmental managers;
- providing training to build community of scientists contributing to international literature and expertise;
- analyzing regional problems that cannot be studied only from a local perspective or where cooperation brings interdisciplinary insights.

One argument not noted by the Carnegie report is the increasing probability, at least in Latin America with new democratic governments, that scientists who participate in collaborations will become decision makers in governments. There is certainly evidence of this in Latin America where global change and environmental scientists have taken senior decision making posts in governments (e.g., Julia Carabias and Carlos Gay in Mexico).

The literature on science in Latin America describes the way in which certain countries (Argentina, Brazil, Mexico, and to a lesser extent Chile, Colombia, Costa Rica, and Venezuela) have invested in science and technology for reasons of nationalism, regulation of imported technology, export development and the belief that a strong science establishment underlies economic progress (Segal 1987). But even these countries face formidable barriers to developing scientific capacity that include political and economic stability, a brain drain, and economic recession and debt since the 1980s. Other constraints that have been identified in many Latin American countries include a rigid higher education structure that has emphasized professional rather than research degrees, a lack of instructors with PhDs, poor public and policy understanding of science, weak connections between academia, government, and the private sector, and isolation and lack of contact within the region (Segal 1987).

17.2 History and structure of the IAI

The IAI arose from a number of initiatives including a recommendation from the International Geosphere–Biosphere Programme (IGBP) in 1988 to establish regional observatories and a related initiative to set up the SysTem for Analysis Research and Training (START) as a non-governmental network with the objective of stimulating cooperation between scientists, capacity building, and data exchange around the globe (Fuchs 1995). Other regional networks at that time included an Asia–Pacific Network (APN) and a Europe–Africa network (ENRICH) together with the intention of setting up a third major network in the Americas.

In 1990 the (first) Bush administration in the USA organized a White House Conference on Science and Economics related to Global Change and invited leaders from 70 countries to consider the idea of regional institutions for global change research. This was followed by a 1991 meeting in Puerto Rico to develop a framework and draft an agreement for a regional institute in the Americas, a second drafting meeting in Mar del Plata, Argentina, and by the Montevideo Agreement and Declaration that established the IAI and its implementation committee in May 1992. The initial Montevideo agreement was signed by representatives of the governments of Argentina, Bolivia, Brazil, Chile, Costa Rica, Dominican Republic, Mexico, Panama, Peru, the United States, and Uruguay, joined within a few months by Canada, Cuba, Ecuador, and Paraguay. Subsequently Colombia, Guatemala, Jamaica, and Venezuela signed the agreement to bring the membership to 19 countries. After some delays, all countries except for Bolivia, ratified the agreement which came into force after six ratifications in late 1993.

The decision to pursue an intergovernmental institution for the IAI was probably designed to strengthen the institution, and to support other intergovernmental agreements that emerged with the opening of political and economic links with democratization and market integration in the Americas. The disadvantages of this model included some initial tension with the non-governmental START and other regional networks, slow implementation of programs as a result of delays in ratification, the challenges of negotiating annual financial contributions, and ensuring that science rather than intergovernmental politics would drive the research agenda and allocations of funds.

The Montevideo agreement sets out objectives for the IAI that include:

- promoting regional cooperation for interdisciplinary research on aspects of global change related to the Earth, ocean, atmosphere, and the environment and to the social sciences with particular attention to impacts on ecosystems and biodiversity, and to mitigation and adaptation to global change;
- sponsoring scientific projects on the basis of their regional relevance and scientific merit;
- pursuing research on a regional scale that cannot be pursued by individual countries;
- improving scientific infrastructure and capability through training and data management;
- fostering standardization, collection, and exchange of scientific data;
- improving public awareness and providing scientific information for public policy;
- promote cooperation among research institutions within the region and with other regions.

Several of these objectives merit further discussion. For example, the decision to include the social sciences within the IAI recognized the growing interest in the human dimensions of global change and the recognition that many global change problems require interdisciplinary analysis, including the social sciences, to understand their origins and consequences and to propose policy solutions. The focus on regional-scale problems and relevance reflected both scientific and pragmatic justifications for research below the global scale but beyond national boundaries. The commitment to standardization and exchange of data sought to break down some of the longstanding traditions of guarding geophysical, ecological, satellite, and meteorological data in the name of national security, rising costs for some data, and to encourage some countries to contribute more to key observational networks, such as ocean and ozone monitoring, perhaps through using their military and other resources to gather the data. Weather forecasting, for example, especially the new efforts to base long-term forecasts on sea surface temperatures, relies on extensive observational networks.

Initially the IAI was administered from Washington through an office in the National Science Foundation with NSF Assistant Director for Geosciences, Bob Corell, as Interim Director and Mexican scientist Ruben Lara Lara as Executive Scientist. The organizational plan included a "Conference of Parties" (COP) composed of senior political representatives, mostly secretaries of environment and an "Executive Council" (EC) composed of country representatives, typically from national science foundations, global change programs, or government environment, meteorological, and geophysical institutes. The COP is the overall policy and oversight group for the IAI and all member countries are represented. The EC is elected by the COP and has nine members with two-year terms meeting twice a year to make recommendations for plans, programs, and budgets to the COP and provide guidance to the directorate. The Scientific Advisory Committee (SAC) is a group of 10 scientists elected by the COP serving in their personal capacities and not as national representatives. The SAC develops and makes recommendations to the COP and EC regarding the scientific agenda, long-range planning and results, and, in close cooperation with the directorate coordinates the peer review of proposals and other activities of the IAI.

The directorate, located in Sao Jose dos Campos, Brazil, has the IAI director and a small staff and is responsible for day-to-day operations of the IAI including the administration of the grants program, publicity, and general program development. The directorate location was selected after several countries submitted proposals and the Brazilian government made an offer of space and support on the INPE campus in Sao Jose. This generous contribution from Brazil to the IAI has nevertheless created a little difficulty because of the high cost of living in Brazil for IAI staff and travel costs from Sao Paulo to other countries in the region.

IAI is funded by several mechanisms including (1) core funding for the Directorate based on a sliding scale of contributions from member countries and (2) program funding to support research, data, and training from pledges from member nations. The IAI Directorate also writes proposals for specific research initiatives and training programs to agencies such as the World Bank GEF, the US National Science Foundation, the MacArthur Foundation or the Canadian International Development Research Centre (IDRC). There were significant problems in obtaining even the funding for the core office from member countries because of bureaucratic problems in financial transfers and economic crises of member governments. Many countries slipped behind in their contributions and few were able to contribute to program or project funding. However, nearly all member countries provided in-kind support to projects of the IAI network through academic institutions. The Argentinean government provided significant matching funds for approved projects under IAI Collaborative Research Network program (CRN) that included Argentinean investigators until the Argentinean economic crisis in 2001 made this support impossible.

17.3 The IAI science program

The initial science agenda for the IAI was developed in a series of workshops from 1992 to 1994 involving more than 800 scientists from across the Americas that identified a set of seven themes of relevance to the hemisphere including tropical ecosystems, climate and biodiversity, El Niño, ocean–land–atmosphere interactions, comparative coastal studies, comparative studies of terrestrial ecosystems, and high latitude processes. Human dimensions, training, communications, and modeling were identified as crosscutting issues (Brum 1994). This agenda was developed quite rapidly in the start up phase of the IAI and shows the stamp of the interests of scientists who were members of the initial science working group. The agenda was designed as an evolving mechanism to adapt to changing priorities and include emerging issues.

In order to simplify the IAI agenda so that it could be more clearly communicated to policy makers and the scientific community, and to make the importance of human dimensions research clearer, the science agenda was reviewed by the Science Advisory Committee in 1998. The SAC proposed four major themes:

- understanding climate variability;
- comparative studies of ecosystems, biodiversity, land use, and water resources;
- changes in the composition of the atmosphere;
- integrated assessments, human dimensions, and applications.

A second revision in 2003 further refined the agenda and the current (September 2007) four major themes are:

- understanding climate change and variability in the Americas;
- comparative studies of ecosystems, biodiversity, land use and cover, and water resources in the Americas;
- understanding global change modulations of the composition of the atmosphere, oceans, and fresh waters;
- understanding the human dimensions and policy implications of global change, climate variability, and land use.

The research phase of the IAI was inaugurated through a program of "start up" or planning grants up to US$50 000 funded and administered by the US National Science Foundation (NSF) to support travel and workshops to plan collaborative long-term research networks between in 1996–97. Thirty six grants were awarded totaling about US$2m. The topics ranged from tree ring analysis, biogeochemistry, remote sensing of land use and climate modeling of different regions in the Americas to coastal fisheries. This was complemented by the three rounds of grants for the Initial Science Program (ISP) for one time three year awards for research, training, data collection, and modeling of $30 000 to $120 000. A total of US$3.4m was awarded to 39 projects (some of which were renewals). In 1999 grants, totaling US$10 million, were awarded to 14 Collaborative Research Networks (CRN) including more than 170 scientists for initial periods of three to five years. The networks focused on biogeochemistry and land use, dendrochronology, deforestation and ranching, comparative ecosystems and biodiversity, ultra-violet radiation impacts on ecosystems, disaster vulnerability, global change impacts on mountain and grassland ecosystems, climate and health, Andean Amazon rivers, global change and coastal fisheries, and four major climate diagnostic and modeling studies in the trade wind convergence, Mercosur, western south Atlantic, and Caribbean/Central American region. From 2002 a new phase of funding provided almost US$1m in 38 small grants to collaborative groups and was followed in 2006 with a new set of 12 Collaborative Research Network (CRN II) grants. Following a recommendation of the IAI SAC during the CRN II evaluation, CRN II was complemented in 2007 with six projects under the Small Grants Program for the Human Dimensions (SGP-HD).

Education and training activities included a program funded by the World Bank Global Environment Facility through the World Meteorological Organization to support training and software for geographic information systems and weather data (almost $3 million over 3 years), and summer training courses. The IAI has also supported conferences and workshops, often in partnerships with other institutions and is increasingly visible in international programs such as the International Human Dimensions Programme (IHDP). The IAI has also contributed to the UNFCCC Subsidiary Body on Science and Technology Assessment (SBSTA) (as an observer), the Millennium Ecosystem Assessment and the Intergovernmental Panel on Climate Change (IPCC).

17.4 Evaluation

How well has the IAI performed according to its own objectives as well as the broader set of criteria for success of international intergovernmental organizations and scientific cooperation?

Several broad achievements were noted in annual reports (Inter-American Institute 1998) and in informal discussions about IAI. For example, the signing and ratification of the IAI agreement by 14 countries in the hemisphere is one measure of success, especially since it includes a commitment to the free and open exchange of data and to financial contributions to a core office. The overall network of scientists involved in IAI has also grown dramatically with more than 1000 scientists from the region attending IAI workshops and receiving IAI mailings (more than 3000 in 2007). This network includes physical, biological, and social scientists. For example, more than 200 proposals were received for the initial science program (round II) of which 23 were funded. The first collaborative research network competition received 70 pre-proposals and 33 full proposals of which 14 were funded involving 161 scientists and 87 institutions from 16 countries in the Americas (Wilcox 2000). For CRN II 99 pre-proposals were received and 37 full proposals with 12 projects funded. This submission rate is a good indicator of the growing awareness of IAI but also of the unmet needs for research funding in the region because the IAI was only able to fund a small proportion of the proposals submitted.

There is also an impressive number and range of researchers who are involved in IAI-funded activities. One concrete indication of this network and its productivity was the impressive showing at the 1999 American Meteorological Society (AMS) symposium on Global Change Studies in Texas where more than 90 papers (about 60%) had IAI grantees as authors, mostly discussing their IAI-funded research. This was preceded by a meeting at the 1997 AMS in Long Beach where 75 IAI researchers discussed their plans.

IAI's overall legitimacy also rose as a result of high climate variability during the 1990s, particularly the onset of an extremely severe El Niño in 1997–98 and several severe hurricanes, which increased government and public interest and concern about climate issues and the potential of forecasting. Some IAI scientists were involved in the production and dissemination of the seasonal climate forecasts that allowed some governments to prepare for El Niño. The IAI cosponsored five regional workshops with the US National Oceanic and Atmospheric Administration (NOAA) to discuss the climate forecasts and publicize them to stakeholders.

One other measure of success was the creation of new networks and the inclusion of new, younger scientists in international collaborations. Several researchers initially funded by IAI were invited to join proposals and projects funded by other agencies, to present their research at international meetings, and to submit their work to scientific journals. The IAI mailing lists were used to announce conferences, grant, and fellowship opportunities. IAI was able to work around some of the scientific hierarchies that mean that only senior scientists attend international meetings or receive research funds by insisting on education and training activities in proposals and using strict peer review to select proposals. There was also some progress in breaking down disciplinary boundaries by promoting interdisciplinary teams that include policy, applications, or social science issues.

IAI was also able to link up with other scientific programs including the World Meteorological Organization (WMO), the International Geosphere–Biosphere Programme (IGBP), the International Human Dimensions Programme (IHDP), and DIVERSITAS as well as other (regional) initiatives such as Asia–Pacific Network for Global Change (APN), the SysTem for Analysis, Research and Training (START), the Large-scale Biosphere–Atmosphere Experiment in Amazonia (LBA), Variability of the American Monsoon Systems (VAMOS), the Caribbean Basin AMIGO/Friend Project in Regional Hydrology (AMIGO), and the International Research Institute for Climate Prediction. The World Bank GEF grant permitted the training of scientists from 16 countries through fellowships, and the provision of equipment and software to most of these countries, focusing on the use of geographic information systems and meteorological data.

Many of these achievements are noted by the recent evaluation of the IAI by the AAAS (2007) that concludes that the IAI "has largely proven its worth" with high quality science and extensive capacity building. They point out the contributions of IAI climate scientists to the recent IPCC 2007 Fourth Assessment (IPCC 2007) as well as biodiversity researchers to the Millennium Ecosystem Assessment (MA 2005).

17.5 Problems and challenges

Despite these indicators of success, the IAI was also faced with some formidable financial, political, and scientific challenges. I have already noted the delays and problems in obtaining even the modest contributions for the administration of the directorate from member governments, and the dependence for program funds on the USA. The AAAS reports that of the 19 members of IAI only five countries are current in their contributions and 10 are more than four years in arrears (AAAS 2007). There were also substantial difficulties in distributing grants to successful proposals because of wide variations in financial institutions and transfers between governments and universities. Continuing economic crises and other priorities in almost all countries, including the USA where global change funding decreased over the 1990s, are a major constraint to the success of the IAI. The small directorate staff was overburdened with the tasks of proposal flow, review, grant administration, travel, and communications.

The IAI also had difficulties in raising funds from other sources, except for a grant from the World Bank-GEF for training. Most philanthropic foundations are concerned directly with issues of poverty or environmental conservation and direct much of their funding to environmental non-governmental organizations. In order to attract such funding the IAI needs to frame its agenda more clearly in terms of sustainable development and capacity building that makes science of much more direct relevance to local groups and resource managers. The IAI is also competing with other urgent international and regional policy concerns including economic stability, health (e.g. AIDS), drugs, and civil unrest.

The GEF project ran into some of the classic problems of technology transfer projects. Although many countries welcomed the computer equipment and software, it was not always placed in the most effective location, and many of the systems were not used as frequently or maintained as well as IAI would have hoped. The training in the use of the SPRING GIS software was very useful to some researchers, but others would have preferred to learn and obtain the more commercial systems such as ArcInfo or to have the training program driven by scientific questions rather than the technology.

Another challenge was the identification of researchers to bring into the IAI network from those countries with smaller scientific communities and less expertise in the international research arena. Although there are requirements to include two to four countries in most proposals the majority of the investigators came from the USA, Canada, Brazil, Mexico, and Argentina. This reflects long standing inequalities in the size and resources of science with the hemisphere, clearly reflected in the data reported by La Red Iberoamericana de Indicadores de Ciencia y Tecnología (RICYT), which shows these five countries with the largest number of scientists, largest investments in science, most MA and PhD graduates, and most publications cited in the international literature. But even in these countries the data is for science in general – including medicine and engineering – and the reality is that even in Argentina, Brazil, and Mexico earth and environmental research communities are relatively small and dominated by a few institutions such as Instituto Nacional de Pesquisas Espaciais (INPE) in Brazil or Universidad Nacional Autónoma de México (UNAM) in Mexico. Only a few scientists and institutions had the international contacts to become involved in the initial IAI workshops and this group, to some extent, dominated subsequent grant competitions. Lack of opportunities and experience in writing peer-reviewed proposals meant that some of the proposals received did not fare well in reviews, had inadequately specified goals, tasks, or budgets. Some attempts were made to remedy this by making proposal requirements clearer, organizing workshops to discuss proposal preparation, and requesting revision and clarifications of some proposals.

The proposal guidelines evolved, with advice from the Scientific Advisory Committee, to try to provide for a more equitable and transparent review process and to address differences in research cultures between countries. For example, earlier proposals often included high indirect cost (overhead) rates of 50% or more for US institutions as well as significant summer salary support for principal investigators. At the other extreme, many Latin American institutions included no overhead or salary support but requested funds for equipment. The salary differentials between USA–Canada and some regions of Latin America resulted in proposals where the bulk of the funds would end up in a US institution with small percentages for Latin American collaborators. A policy of no more than 10% overhead and limitations on salaries charged to grants was therefore established to encourage more balanced and lower cost proposals.

Other problems arose in ensuring a balance in regional and disciplinary representation in the review panels for major funding decisions because of the small size of the global change community in much of the region and the associated potential for conflict of interest as most reviewers and even SAC members were associated or had collaborated with institutions that were submitting proposals to the competition. Despite strict conflict of interest rules (leaving the room during discussions, not voting on proposals) there was some suspicion that certain research institutions and countries could have benefited from the composition of review committees. My experience suggests that criteria of scientific excellence and regional relevance drove the review process, and that in some cases great effort was made to identify proposals that might be worthy of funding from all IAI countries. Hundreds of scientists assisted in the review process. In some cases, no proposals that met even minimum standards of excellence were received from member countries. Nevertheless, it is certainly true that the major global change research institutions in the region tended to submit proposals of high quality and received funding. Over time, the balance of proposal submissions and success shifted from the USA and Canada, to Argentina, Brazil, and Mexico and the AAAS report (AAAS 2007) notes a shift over time from US- and Canadian-led projects (e.g. 21 of 26 initial grants) to Latin American-led (25 of the 38 small grants from 2002–2006). Out of the 12 projects under CRN II (2006–2011), eight are under Latin American leadership.

The proposal submission and review process was driven, to some extent, by IAI's dominant funding institution the US National Science Foundation. While this set a workable model, it did not always allow the flexibility that might be needed for a new multi-national collaborative effort. This has been largely improved with the implementation of the IAI Program Management Manual, which has NSF approval. Perhaps the greatest difficulty arose in connection with Cuba because of US government restrictions on funding Cuban research. Cuba is a full member of the IAI, and submitted good proposals, but NSF funds could not be used to fund Cuban participation, and because of the lack of support from other countries the IAI directorate had to stretch its own budget to help keep Cuba involved. Consequently (the very limited) non-US source funding had to be used to support any Cuban projects or project components.

Perhaps the most significant step to broaden participation in IAI projects to a larger group of countries was the decision to set aside some of the funds from the Collaborative Research Network (CRN) competition for the Program to Enhance Scientific Capacity in the Americas (PESCA). The PESCA program gave small grants to investigators (including several more junior scholars) from those countries with few IAI projects to join the existing CRN funded networks. More recently a Training Institute and Seed Grant (TISG) grants program has been initiated, but there is still no scientific program specifically targeted at early career scientists.

One barrier to full participation in the IAI was the decision to conduct most activities in English, including the meetings of key committees, review panels, and submission of proposals. Although the IAI newsletter is also published in Spanish, English is regarded as the international language of science, the dominance of English for communication in the IAI limited the contributions of some members of committees and disadvantaged some proposals.

Another major difficulty in expanding the network, program, and policy relevance of the IAI was the identification of social scientists working on global change issues in the Americas, and the satisfactory inclusion of social science perspectives in proposals and projects. Although many of the scientists involved in the establishment of IAI recognized the importance of a "human dimension" in global change issues, the initial science agenda did not highlight social science perspectives and very few social scientists attended the initial workshops to define the science agendas. Attempts to remedy this included a small workshop in Taxco, Mexico and a survey of social scientists in Latin America designed to inform them about the IAI and link them to the network. The revision of the science agenda to include an explicit social science theme was also designed to attract proposals with human-dimensions perspectives and an increased number of social scientists in the IAI SAC. However, very few social science or policy proposals were received in the IAI grants programs, and many of the science proposals that tried to incorporate applications or policy components did so inadequately.

There are several explanations for this difficulty in bringing a social-science component to the IAI. First of all, the IAI announcements initially did not reach many of the social-science newsletters, list-servers, and institutions or were not seen as likely sources of funding for social scientists. Most Latin American countries have been slow in developing a research agenda or awareness of human dimensions of global change issues and there is little awareness of global change issues in the social science academic communities. Of course, there are large numbers of social scientists working on environmental issues, but their focus tends to be on more local and urgent concerns – such as urban pollution and forest conservation – and they may not see ways to link their research to the IAI agenda. A second barrier is that in some countries there is little tradition and some suspicion of physical and social scientists working together. Reasons include the strong Marxist orientation of some social science work in Latin America, the legacy of authoritarian regimes and repression of social critics within academia, and the organization of universities and research institutes. When resources are scarce, interdisciplinary research, especially across the two cultures may seem very risky. On the other hand Latin America also hosts some outstanding interdisciplinary research institutes that have a tradition of environmental research that includes social issues. Moreover, US and Canadian physical scientists are not necessarily any more willing or aware of social science issues than Latin American colleagues and may view them as too political or "soft" for serious attention.

There are many reasons to continue the efforts to integrate social science and policy into the IAI including the need to show useful results for policy, develop science applications, and understand the social causes and consequences of global changes. The IAI directorate has used links with the International Human Dimensions Programme (IHDP) to build contacts and expertise in social science research by using IAI funds to support training workshops and conferences. In 2001, the IAI helped to organize the fourth international meeting of the human dimensions community (held in Rio de Janeiro, Brazil) as one way of building capacity and interest in social science within the IAI and has continued active collaboration with IHDP including joint training courses.

The recent SCOPE 68 book "How to improve the dialogue between science and society: The case of Global Environmental Change" marks a concerted effort by IAI to reach out to social-science and policy communities (Tiessen *et al.* 2007). It includes chapters on working with stakeholders and the media, and lessons learned from IAI efforts over a ten year period and is summarized in a policy brief (IAI 2007).

17.6 Conclusions

The IAI has been creating and reinforcing a new set of regional relationships and shared consensual knowledge about global change in the Americas that fits the model of an epistemic community and will facilitate future environmental policies and agreements. The institution has certainly had some success in identifying and supporting a diverse number of scientists to work together on important scientific problems and train new generations of scholars. It has also learned some important lessons and still faces formidable challenges. The lessons include the broader need to show policy relevance and demonstrate regional representation in committees and funding while maintaining criteria of scientific excellence in order to sustain legitimacy, as well as more practical considerations about processes of proposal review and communication. Some of the major challenges include the search for a more sustainable and diverse financial base, the enhancement of capacity in those countries with smaller scientific communities so that they can fully participate in research competitions and projects, and the more effective integration of social science and policy issues into projects.

Acknowledgements

I would like to acknowledge the contribution of Gerhard Breulmann, scientific officer of the IAI in helping me update this chapter.

References

American Association for the Advancement of Science International Office 2007. *Report of the External Review Committee Assessment of the Inter American Institute for Global Change Research*. Washington, DC: American Association for the Advancement of Science.

Brum, F. 1994. Development of the Inter-American Institute for global change research. *Ambio*, **23**(1), 98–100.

Caldwell, L. K. 1996. *International Environmental Policy: from the Twentieth to the Twenty-first Century*. Durham, NC: Duke University Press.

Carnegie Commission on Science Technology and Government 1992. International environmental research and assessment. Washington, DC: Carnegie Commission.

Fuchs, R. J. 1995. START – the road from Bellagio. *Global Environmental Change*, **5**(5), 397.

Gilpin, R. 1987. *The Political Economy of International Relations*. Princeton, NJ: Princeton University Press.

Haas, P. M. 1994. Regime patterns for environmental management. In P. M. Haas, H. Hveem, and K. R. O. A. Underdal (eds.), *Complex Cooperation: Institutions and Processes in International Resource Management*, vol. 2. Oslo: Scandanavian University Press.

Haas, P. M., R. O. Keohane, *et al.* 1993. *Institutions for the Earth: Sources of Effective International Environmental Protection*. Cambridge, MA: MIT Press.

Hurrell, A. and B. Kingsbury (eds.) 1992. *The International Politics of the Environment*. Oxford, Clarendon Press.

Inter American Institute (IAI) 2007. *How to Improve the Dialogue between Science and Society: the Case of Global Environmental Change*. www.iai.int/files/policy_brief/ Policy_Brief.pdf

Intergovernmental Panel on Climate Change (IPCC) 2007. *Working Group I Report "The Physical Science Basis."* http://ipcc-wg1.ucar.edu/wg1/wg1-report.html

Keohane, R. 1994. Against hierarchy. In P. M. Haas, H. Hveem, and K. R. O. A. Underdal (eds.), *Complex cooperation: institutions and processes in international resource management*, vol. 2. Oslo: Scandanavian University Press.

Keohane, R. O. (ed.) 1986. *NeoRealism and its Critics*. New York: Columbia University Press.

Keohane, R. O. and J. S. Nye 1977. *Power and Interdependence*. Boston, MA: Littlefield and Brown.

Krasner, S. D. (ed.) 1983. *International Regimes*. Ithaca, NY: Cornell University Press.

Lipshutz, R. D. and K. Conca (eds.) 1993. *The State and Social Power in Global Environmental Politics*. New York: Columbia University Press.

Millennium Ecosystem Assessment 2005. *Ecosystems & Human Well-Being*, vol. 2. *Scenarios*. Washington, DC: Island Press.

Porter, G. and J. Welsh-Brown 1993. *Global Environmental Politics*. Boulder, CO: Westview.

Segal, A. 1987. *Learning by Doing: Science and Technology in the Developing World*. Boulder, CA: Westview.

Tiessen H., G. Breulmann, M. Brklacich, and R. S. C. Menezes (eds.) 2007. *SCOPE 68: How to Improve the Dialogue Between Science and Society: the Case of Global Environmental Change*. Washington, DC: Island Press.

Wilcox, B. 2000. *Annual Progress Report to NSF – the Collaborative Research Network*. Sao Jose dos Campos, Inter American Institute (Unpublished).

Young, O. R. (ed.) 1997. *Global Governance: Drawing insights from Environmental Experience*. Boston, MA: MIT Press.

18

The Regional Integrated Sciences and Assessments (RISA) Program: crafting effective assessments for the long haul

ROGER S. PULWARTY, CAITLIN SIMPSON, AND CLAUDIA R. NIERENBERG

18.1 Introduction

Climate variability and change significantly influences the health, prosperity, and well-being of individuals, societies, and the environment. For the United States this has been demonstrated, most recently, by several high impact events such as the 1997–98 El Niño event, the hurricane seasons of 2004 and 2005, the ongoing drought since 1999 in the Southwest, falling Great Lake levels, and the worst drought in 100 years in the Southeast (2007). Over the past two decades there has been significant progress in understanding longer-term climate patterns that influence these events, such as the El Niño Southern Oscillation (ENSO), the Pacific Decadal Oscillation (PDO), the North Atlantic Oscillation (NAO), and the Arctic Oscillation (AO). Increasingly, attention is being paid to the cumulative impacts of regional climatic events driven by decadal-scale modulations of these phenomena.

Much recent research has shown that enabling effective responses to environmental variability and change requires knowledge assessments at both the global scale and at the appropriate scales of decision making i.e., the region and the locale (NRC 1999; Clark *et al.* 2001). As identified at the federal level and in academia, there is a need for credible, unbiased assessments of the status and trends of environmental patterns and processes (US Congress 1994). At the same time there are calls for more and better structured processes to identify, assess, and meet national, regional, private, and local climate-related needs, and to foster the timely adoption and effective use of commercially valuable information and technology throughout the US economy (US Congress 1998; US Congress 2007).

This paper outlines the development and evolution of a long-term US-based interdisciplinary program focusing on climate impacts assessments and regional and local decision support: the Regional Integrated Sciences and Assessments Program (hereafter RISA). The RISA program has existed for over ten years and

367

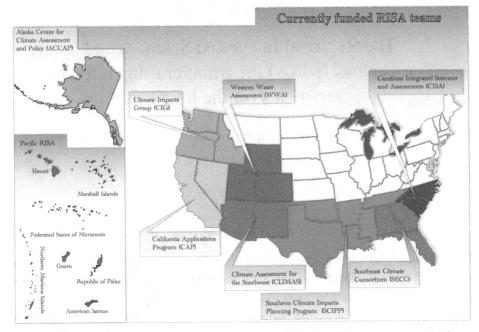

Figure 18.1. Present RISA activities: teams, critical problems, partners, and approaches. (http://www.climate.noaa.gov/cpo_pa/risa/)

tangible lessons may begin to be drawn from the experiences of the RISA teams and from program management. Many studies or even "assessments of assessments" aim at providing new frameworks, usually idealized, but offer little on how the assessment itself originated, is organized, and sustained. Put differently, knowing what to do is not the same as doing it. The RISAs have developed as decentralized scientific applications and policy experiments (Brunner 1996) providing traceable accounts of successful federal–state and local partnerships in interdisciplinary research, climate impacts assessment, and decision support.

18.2 The RISA program: history and maturation

At present, there are nine regional integrated sciences and assessments activities funded by NOAA. These activities are focused on the Pacific Northwest (CIG), the Southwest (CLIMAS), California and Nevada (CAP), Inter-Mountain West (WWA), Alaska (ACCAP), Hawaii and US-affiliated islands in the Pacific Ocean (Pacific RISA), the Carolinas (CISA), and the Southern (SECC, SCIPP) regions of the United States (see Fig. 18.1 and Box 18.1 for details). The final RISA configuration is envisioned to be an ongoing assessment system distributed across relatively large regions of the United States, consisting of integrated networks

Box 18.1

1. *The California Applications Program (CAP)*, led by researchers at Scripps Institution for Oceanography, studies the impacts of climate variability and change in California and the surrounding area. CAP evaluates weather and short-term climate forecasts and climate change projections, with particular attention to climate influences from the Pacific Ocean and western North America. An associated emphasis is to develop a better capacity for observing the climate over the complex landscape of the California region. CAP is working to improve climate information for decision makers in key sectors, including water, human health, and wildfire. http://meteora.ucsd.edu/~meyer/caphome.html

2. *The Carolinas Integrated Sciences and Assessments (CISA)* project aims to improve the range, quality, relevance, and accessibility of climate information for water resource management in North and South Carolina. CISA examines water resource issues at interannual, decadal, and longer scales to determine how decision makers use climate information to manage water and how current operational practices can benefit from new climate and water resource products. CISA investigates how best to present climate sciences that are relevant to water resource policy, and to foster understanding of climate variability, issues of forecast uncertainty, and risks associated with forecast failure. http://www.cas.sc. edu/geog/cisa/

3. *The Climate Impacts Group (CIG)*, located at the University of Washington, Seattle, examines the impacts of natural climate variability and global climate change in the US Pacific Northwest. CIG's goal is to increase the resilience of the region to climate fluctuations through research and interaction with stakeholders. Research emphasizes four key sectors of the Pacific Northwest environment: water resources, aquatic ecosystems, forests, and coastal systems. Focusing on the intersection of climate sciences and public policy, CIG works with planners and policy makers to apply climate information to regional decision-making processes. http://tao.atmos.washington.edu/PNWimpacts/

4. *The Climate Assessment for the Southwest (CLIMAS)* project fosters collaboration between university researchers, agency scientists, resource managers, educators, and decision makers throughout the region to understand climate and its impacts on human and natural systems in the US Southwest and adjacent USA–Mexico border area. CLIMAS investigates vulnerability to climate variability in both rural and urban areas, how to improve climate inputs for drought planning, and climate impacts on water resources, water policy, and wildland fire. CLIMAS studies how climate information is used by decision makers and works to evaluate and improve forecasts. http://www.ispe.arizona.edu/climas/

5. *The Pacific Islands RISA* supports emerging regional efforts to pursue an integrated program of climate risk management. With an emphasis on understanding and reducing Pacific Island vulnerability to climate-related extreme events such as drought, floods, and tropical cyclones, activities within this emerging RISA build

substantially on existing regional efforts in climate sciences and El Niño forecasting. Led by researchers at the East–West Center in Hawai'i, Pacific RISA works in close collaboration with stakeholders in water and natural resources, agriculture, tourism, and public safety and health. http://research. eastwestcenter.org/climate/risa/

6. *The Southeast Climate Consortium (SECC)* is a multi-institutional, multi-disciplinary team focusing on the vulnerability of agriculture, forestry, and water resources management to climate variability. SECC scientists are developing methods to translate regional climate forecasts into local forecasts, linking them with crop and hydrology simulation models in order to enhance understanding of decision makers so they can reduce risks associated with climate variability. The consortium is developing partnerships needed to build equitable outreach programs for farmers, forest managers, water resource managers, homeowners, and policy makers to enhance user familiarity with seasonal climate forecasts. http://www.coaps.fsu.edu/lib/Florida_Consortium/ http://fawn.ifas.ufl.edu/

7. *The Western Water Assessment (WWA)* provides information about climate variability and climate change to water resource decision makers with the goal of improving management of the Intermountain West's most critical resource, water. Through partnerships with key decision makers, WWA provides vulnerability assessments, climate forecasts, and paleoclimate studies designed to enhance short-term and long-term management decisions. WWA experts focus on the Colorado and Platte River Basins, researching policy options, streamflow forecasting, snowpack monitoring, drought planning, and reservoir management. http://cires.colorado.edu/wwa

8. *The Alaska Center for Climate Assessment and Policy (ACCAP)*, the newest RISA center, is being led by investigators at the University of Alaska. The primary functions of ACCAP will be (1) the synthesis of available data and information in order to quantify actual and potential impacts of changes in the seasonality of weather and climate on Alaskan people and ecosystems, and to determine corresponding needs for enhanced product delivery by agencies such as the National Weather Service; (2) research that will facilitate the product enhancement identified in (1); and (3) assessment of the vulnerability and adaptive capacity of various Alaskan sectors, together with a determination of the management and policy decisions that can reduce vulnerability and facilitate adaptation. The transportation sector will provide the initial prototype for this activity. http://www.uaf.edu/accap/

9. *The Southern Climate Impacts Planning Program (SCIPP)* was recently initiated in late-2008. It is centered at the University of Oklahoma and incorporates the states of Oklahoma, Texas, Louisiana, Mississippi, Arkansas, and Tennessee. Its emphases are on regional and cross-sectoral social and economic indicators or drought impacts and on decision support.

(discussed below) that enable local and regional capacity to address climate-related risks and opportunities.

18.2.1 Development of the RISA Program

In the late 1980s, the NOAA Office of Global Programs (OGP) was created to provide research support for the NOAA contribution to the cross federal agency US Global Change Research Program (USGCRP). The basic tenets of the 1991 "Our Changing Planet," which is the annual USGCRP report to Congress, were to:

(1) integrate science into the policy process;
(2) maintain partnerships among all participants;
(3) focus on interdisciplinary science and interactions.

While most agencies interpreted this directive to mean linking physical science models with economic and other models, the OGP leadership envisioned a more interactive process with decision makers. Additionally, the OGP support of climate variability research and seasonal forecast development resulted in an early emphasis on changes in the higher moments of the climate system in addition to changes in the mean state. For instance, both stochastic and deterministic elements of key processes were seen as fundamental to understanding ENSO variability and change (see also Peters *et al.* 2004). In this context the climatic timescale would come to be treated as a continuum rather than as completely discrete modes of variability in which change was a wholly separable component.

In the early 1990s, the NOAA Office of Global Programs began funding economics studies and then added human dimensions studies. The 1993 request for proposals written by one of this chapters' authors (CRN) described the goals of the human dimensions research as developing a greater understanding of human adaptation to past climate. A follow-up memo (1995, CRN) emphasized the need for regional assessments addressing climate change in the context of other environmental problems of significance. The Pacific Northwest Climate Impacts Group (CIG), the first RISA, was selected because of a focusing event (the 1994 closure of Columbia River salmon fisheries, resulting in the first ENSO-related disaster declaration in the USA) and the involvement of the Principal Investigator (PI) in the 1995 IPCC Assessment. The PI concluded that the IPCC process and products were not meeting regional needs at scales commensurate with decision making. This conclusion resonated with that reached earlier by disaster and development researchers. As important was the increasing recognition that definitions of community and region had been modified, beyond the physical unit of analysis, by trends toward democratic participation in planning processes and the need for mediating institutions, more recently called "boundary organizations" (Campbell

1969; Shackley and Wynne 1996; Linder 2005). Another major intervention, the US National Assessment of Climate Change Impacts on the United States (see Morgan *et al.* 2005 for an insightful review), was launched in 1998, three years after the NOAA Regional Assessment Program had been initiated. A major criticism of the National Assessment was that, while it was innovative in raising awareness of climate change-related risks, for many of the participants it was unable to sustain the follow-on interactions needed for effective learning and response. The now obvious conclusion is that complex environmental problems can seldom be dealt with by single discrete actions or policies but respond more effectively to sustained efforts.

Integrated knowledge about climate and climate impacts was being built on components of two major activities already being supported within NOAA Cooperative Institutes, with federal, state, and academic partners across the country and internationally. These activities were: (1) climate and environmental monitoring and research; and (2) economic and human dimensions research (now the Sectoral Applications Research Program) on vulnerability and on the usability of climate forecasts. Targeted sub-activities would form a third focus area on integrated risk assessment applications and decision support (i.e., the development and communication of relevant research results to meet specific needs). This third area would provide the basis for the RISA program. Integrated risk assessments were viewed as occurring over the many dimensions of a resource (e.g., surface, groundwater, humidity), as a component interacting with other systems (e.g., climate and ecosystems) and, with reference to broader interrelationships (e.g., negative and positive impacts of climate on social and economic development).

Advances in our appreciation of complex systems allowed for a reframing of the climate–society interface as a set of multi-dimensional problems in which studies of larger-scale climatic forcings and regionally-focused assessments of impacts and decision making contexts all needed to proceed simultaneously (Figure 18.2). In 2000, the name of the program was changed by the then Program Manager (RSP), in agreement with the project leads, from Regional Assessments to RISA to emphasize the fully interdisciplinary and contextual nature of conducting ongoing regionally-specific knowledge assessments in conjunction with decision support. Context here was taken to include situation, capacity, and also the ways in which a problem is constructed, perceived, and represented by different stakeholders.

As noted above, global- to sub-continental-scale climate initiatives are necessary to provide a foundation for knowledge use by society, but these initiatives are not, by themselves, sufficient to provide the potential users of climate information with the requisite capacity for use. The RISA program was initiated to make and secure the connection with the stakeholders, and to generate the regional science and capacity for learning from those connections.

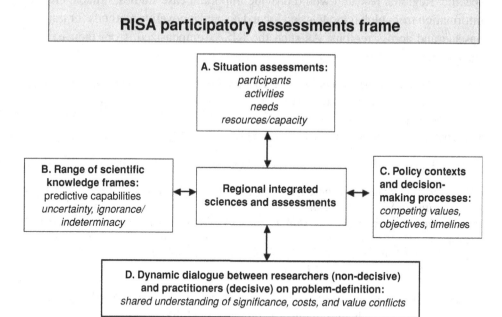

Figure 18.2. RISA participatory assessments framework.

The RISAs have come to function under several key empirically grounded observations, many of which had been articulated in disparate sources such as agricultural extension, disasters, and adaptive management (see White 1966; Holling 1978). These may be summarized as follows:

- adaptation involves a broad range of responses and practice of which climate sensitivity and information are parts;
- new knowledge, new problems, and opportunities continuously arise as events unfold;
- there is no uni-causal model of explanation of system behavior in a particular region; no set of dynamics holds identically true across all regions;
- predicted effects are highly uncertain, implying the need for better characterization of uncertainties, use of existing data on past and contemporary events, and stakeholder experience with those events;
- assessment of effects involves tradeoffs between the multiple interests of current and future generations;
- increased methodological complexity (see Toth and Hiznyik 1998) does not necessarily result in better assessments nor does better information always result in better decisions;
- there is a need to re-define relationships among federally-funded efforts with academic, non-governmental, and local partners (see below).

The key initial assumption was that public and political perceptions of the value of integrated science would be higher if research and products are regionally specific. Regional research would provide important case studies, reliable climate information from global models and local data, and innovative methods for transfer upscale and across regions. As such the RISAs emphasized, from their earliest stages, the:

- value of interaction with stakeholders when they and scientists are regionally co-located;
- climate research and information specific and scaled to regions (observations, forecasts, impacts, projections); and
- responses to agendas established by the USGCRP/CCSP, NRC, and IPCC.

In 2002, the US House of Representatives noted the following:

Other than a relatively small program [RISA] at NOAA, there is currently no structure or process within USGCRP to identify potential users, understand their needs, and connect them to the research agenda... RISA has been called a step in the right direction by some while others view it as a model that could guide larger efforts within USGCRP. Committee on Science US House of Representatives, New Directions for Climate Research and Technology Initiatives, April 17, 2002.

Similar observations have recently been made in National Academy reports on the US Climate Change Science Program (NRC 2007), an incarnation of the USGCRP (US Global Change Research Program).

Key to the RISA approach, and as observed by Brewer (1999), is the framing that problems should designate theory, not the reverse. The risk communication dialogue developed allowed articulation of contested values among resource users and researchers, and preferred outcomes in a particular setting. A fundamental and ongoing issue is to uncover the practical degree of flexibility within regions and communities to adjust decision making in climate-sensitive sectors based on the informed application of science-based information and past experience. This problem orientation allowed RISA the added impetus of having to outline alternative decision pathways and to legitimize discussions of the consequences of those decisions in public fora.

In the RISA context, "regions" exist at the nexus of the local to global continuum. Integrated scientific assessments constitute the sum of efforts to (1) characterize the state of knowledge of climate variations and changes at appropriate scales of interest, (2) identify knowledge gaps and linkages in selected climate–environment–society interactions, and (3) provide an informed basis for (a) responding to climate-related risks and for (b) establishing priorities in basic research investments to meet these needs. To achieve the goals of meeting evolving needs, assessments must be forward looking and anticipatory, and broad enough to evaluate the potential for scientific surprises.

Because of their initiation with a focus on climate variability and extremes, and now across variability and change, the RISAs are not tied to a specific scenario or set of scenarios but link assessments and impacts to emergent problems in the context of short and longer-term vulnerabilities. Their major innovation has been iterating climate impact and response assessments across timescales (ENSO, PDO, long-term trends and projections) and sectors (e.g. fisheries and hydropower) and as such presaged much of the recent climate change literature emphasizing information mainstreaming into existing practice and cross-scale response.

18.3 The structure of RISAs

Not all RISAs started the same way. The RISAs were deliberately allowed to develop different pathways to allow for experimentation and critical issue definition. Following Guston and Sarewitz (2002), we can articulate the RISA programs in terms of different scales, types of impacts and decisions, participation mechanisms, and organization. The user-driven dialogue in each case was designed and implemented by the individual teams and was given high and visible priority in the context of program goals by program management. As expected, those activities funded first have been making inroads to meeting the longer-term goals of the program, in large part because they have refined participation mechanisms over time. Those in pilot or preliminary stages focus on clarification of initially defined critical regional issues, integration of the team, developing cooperative stakeholder linkages, knowledge development intended to lead to user-inspired tools, and data assimilation. Many individual researchers had previously worked on integrated projects in their regions (sometimes supported by NOAA) before the formation of their particular RISA (see Shea *et al.* 2001). These researchers often came on board with science-stakeholder experience and moved more quickly through the early implementation phases.

Implementation of RISA projects has taken the form of several co-evolving tasks, including:

- team development: selection of physical and social science researchers; initial characterization of current state of knowledge of relevant climate, ecological, and hydrologic variability, and various units of analysis (watershed, urban etc.) depending on recent and other historical events;
- developing stakeholder linkages (build on earlier work) to deepen the identification of climate-related critical issues/problems within the region; further refine assessment goals and expectations; determine database and methods integration needs;
- assessing social, economic, and ecological impacts and vulnerability to climate on multiple timescales in selected cases vertically (linking assessment and management); to begin with, identify levels of critical decision-making needs within 1–3 important sectors and/or groups (e.g. water resources, energy, urban areas, agriculture, fisheries etc.) in the region;

- developing pilot projects and implementing prototypes for characterizing environmental information and enhance collaboration among researchers, decision makers, and the public;
- improving horizontal (i.e., across sectors) decision-support dialogues, openness, and developing awareness with respect to integrated climate impacts on regional and local system outputs;
- developing frameworks for structuring a process to articulate present knowledge and knowledge gaps; testing these in different fora, such as in responding to shorter-term events and extremes in the region and providing feedback into assessment design;
- refining mechanisms of interaction and learning among the research and resource management and planning communities;
- capacity building within user communities as needed to realize the benefits afforded by developments in climate research, products, and services.

Pilot efforts have undergone external reviews and evaluation before consideration of continuation and/or expansion to the next phase of assessments. Funds are also targeted in the program towards research on assessment design, including deliberate lesson drawing exercises, such as cross-RISA comparative studies of stakeholder engagement and transferability of approaches, etc. Vehicles for learning include the development of prototypes and more recently, exploratory implementation in other areas and regions. Prototyping allows for sensitivity to context and explicitly draws on lessons from the diffusion of information and incorporation of those innovations into the evolution of assessment design and management.

Each set of investigators within a region was asked to design a research agenda in partnership with stakeholders in their particular region. For example, in the Southeast, the "problem" was defined in terms of the vulnerability of important crops to climate variability and access to climate information for risk reduction and opportunistic planning; in the Pacific Northwest, the problem was variations and changes in ocean and hydrologic variability and land use in relation to thresholds in fisheries (Salmonids) and hydropower. The scope of a "region" would be refined through interaction with decision makers networked across a relatively broadly defined area facing climate-sensitive challenges. Whereas the signal problem that the USGCRP Program addressed was establishing and characterizing predictability of the climate system, the RISA program established a mechanism to legitimize the pursuit of climate-sensitive problems, and identifying stakeholders' needs in combination with scientific capacity to bridge the information gap between needs and problems. This emphasis grew out from several studies (funded by the NOAA economic and human dimensions and other programs) showing that many resource management decisions are made under time, financial, institutional, and other constraints that limit the utility of comprehensive modeling exercises. In addition, the modeling approaches focus primarily on efficiency from an economic perspective and may not be able to accommodate other management objectives such as equity (see Pulwarty and Melis 2001; Rayner *et al.* 2005).

After ten years of experience, the (general) temporal phases involved in program implementation may be characterized as:

Years 1–2: team integration and building to ensure interdisciplinary approaches; more defined regional characterization; development of core capacity (e.g., core offices with 1–2 full-time personnel for some of the larger teams); start-up pilot projects;

Years 2–3: Clarifying issue criticality, vulnerability, regional climate sensitivity assessments; beginning vertical integration;

Year 4 and beyond: Fully integrating partnership networks and lines of communication and research developed in preliminary studies; standards and practices for vertical integration; collaborative development of tools for application.

Depending on the regional issues addressed, RISA research components include: statistical and dynamical climate analyses; hydrologic, agronomic, fisheries, forestry, or other impact modeling; reservoir operations modeling; stakeholder and researcher interviews and historical studies, surveys, institutional mapping, and policy analysis, economic and decision analysis, and vulnerability and evaluation studies. RISA knowledge assessments have been complemented by assessments of climate services provided and of the management systems involved in mediating climatic risks within their various portfolios (fisheries, water resources, forestry etc.). Each RISA activity has fulfilled these roles to varying degrees, depending on the respective start time, the relevance and visibility of impacts in the region, and scale of operation. Efforts are being made (see below) to draw tradeoffs and lessons between comprehensive assessments across a range of issues (such as in the Pacific Northwest and the Southwest) and the vertically integrated study of one or two particular questions (such as the early single sector focus on climate and agriculture in the Southeast).

18.4 Scientific achievements

Each RISA project has succeeded in developing pathways in integrated climate science and impacts assessment, awareness building, and decision support. RISAs have matured to the extent that they are creating linkages and acting as coordinators among federal, state, and local agencies in different regions to identify, undertake, and evaluate integrated research on climate-sensitive issues. Successful RISAs create science–society research elements that monitor interdisciplinary integration around impacts and elicit changing knowledge and perceptions among both stakeholders and researchers.

A useful illustration of the approach articulated above is provided by early research undertaken in the Pacific Northwest (CIG) and the Southwest (CLIMAS) that are still bearing fruit. Early studies are chosen for illustration since their impact and robustness may be better established than recent efforts. By integrating information about oceanic, atmospheric, ecological, and hydrologic processes, (employing monitoring, forecasting, and observational systems funded by NOAA) the Pacific Northwest RISA team (CIG) has led to a clearer understanding of "natural" versus human-caused fluctuations in Pacific salmon numbers. This issue has been a source of great conflict in the Northwest. In addition, their work in this area has successfully contributed to prioritizing the PDO as an important area of focus for basic research (Box 18.2).

Box 18.2
The Pacific Decadal Oscillation (PDO): an early and ongoing priority

The phrase "Pacific Decadal Oscillation" was first coined by Mantua *et al.* (1997), within the Pacific Northwest RISA Group. In its positive phase, the PDO is a pattern of Pacific sea surface temperature (SST), with cold anomalies in the central northern Pacific and warm anomalies along the eastern edges of the basin (i.e., the west coast of North America). The PDO was in the negative phase from 1900 (when the first reliable SST records are available) to 1925 and from 1945–1977 and in the positive phase from 1925–1945 and from 1977. The RISA team established that the PNW climate signal is dominated by a combination of the ENSO phenomenon (cool/wet, warm/dry) on a seasonal/interannual timescale, and the PDO. The most pervasive climate-driven impacts are generated by the PDO, and the impacts are magnified whenever the PDO and ENSO are in phase with each other. Annual streamflow is the single most sensitive terrestrial signal of climate variability in the PNW, and almost all climate impacts are mediated through the regional hydrology. Depending *solely* on whether the PDO is in the cool or warm phase, small changes in temperature ($-0.11\,^{\circ}\mathrm{C}$ to $+0.17\,^{\circ}\mathrm{C}$) and precipitation ($-4\%$ to $+2\%$) generate large changes in: snowpack (-15% to $+17\%$), streamflow (-9% to $+6\%$), survivability of Washington coho salmon (-16% to $+19\%$), and frequency of forest fires (-49% to $+65\%$). Depending *only* on ENSO, the impacts are on snowpack (-14.7% to $+9\%$) and streamflow (-12% to $+8\%$). When PDO and ENSO conditions are *in phase*, the impacts are enhanced for snowpack (-29.7% to $+26\%$) and streamflow (-17% to $+14\%$). More than half of the variations in annual salmon catch in the USA are associated with the PDO. The general pattern is that Alaskan fisheries do worse during the negative phase of the PDO (e.g., 1945–1977) and better in the positive phase (e.g., 1925–1945), while fisheries in Washington, Oregon, and California do worse during the positive phase and better in the negative phase. The implications of these findings are that management of many western salmon stocks is more vulnerable to (and constrained by) climate variations than managers had realized. These efforts of the Pacific Northwest Climate Impacts Group have fed directly into research priorities for the US Climate Variability and Predictability (CLIVAR) program.

The RISAs have also helped identify problems unrecognized or acknowledged by mission agency programs and disciplinary university structures. An example is synergistically blending the perspectives of many sciences on urban growth in the US Southwest and past changes in water supply reliability and future demand (Box 18.3).

RISA activities depend on innovative partnerships among a spectrum of interests (federal, state, local, and private etc.) to enable organizational capacity within a region to develop and test experimental climate information services delivery on an ongoing basis. As such, the RISA program relies heavily on consolidating the results and data from ongoing NOAA and other agency research already funded in a region, into an integrative framework. Table 18.1 shows the existing interactions and critical issues being addressed across the various climatic timescales within the Western Water Assessment (Webb and Pulwarty 2006).

These and other efforts at engaging stakeholders have led to what has best been described by RISA leadership as "A Sea Change in Perceptions." They have resulted in a dramatic change in stakeholder perceptions of the value and relevance of information about climate variability and change. Miles (personal communication), illustrates the evolution of awareness in (Pacific Northwest) Climate Impacts Group (CIG) case, as follows:

1995: few managers saw role for climate information, recognized predictability of climate, or possessed a conceptual framework for applying climate information;

1997–98: El Niño and concomitant media attention stimulated widespread interest in information about climate variability and in CIG; most stakeholders unfamiliar with potential impacts of climate change and unprepared to use such information;

2001: senior-level water resources managers recognize climate change as a potentially significant threat to regional water resources; acknowledge climate change information as critical to future planning;

2001/2: 50-year drought brings intense media attention to the issue and CIG's work → public and private pressure on state agencies to include climate change impacts in long-term planning → significant involvement of CIG in multiple planning(??) efforts;

2003 to present: continued significant breakthroughs with stakeholder groups.

As is widely acknowledged, single interventions do not settle problems once and for all, nor are credibility and relevance instant products of a workshop (or two). Key to the CIG's evolution has been a focus on the timing and form of climatic information (including forecasts) developed, providing access to expertise to help incorporate the information and projections in decision-making processes. In many instances the latter has been shown to be as or more important to individual users than improved forecast reliability (Pulwarty and Melis 2001; Orlove *et al.* 2004; Rayner *et al.* 2005).

Box 18.3

Sensitivities of the Southwest's urban water sector to drought: Arizona case studies

The CLIMAS project analyzed the water budgets of five Arizona cities to determine the degree of severity of impacts from the deepest one- (1900), five- (1900–1904), and ten-year droughts (1946–1955) on record. Case study sites included the Phoenix Active Management Area and the Tucson Active Management Area (AMA).

AMAs are areas in Arizona where stringent groundwater management is mandated under the 1980 Arizona Groundwater Management Act. The CLIMAS study showed that in each of these areas, even under assumptions of continuation of "average" climate conditions, issues persist regarding the possibility of achieving safe-yield (i.e., renewable supply and demand are in balance) by the year 2025, as articulated in the Act. The water sectors in the Phoenix and Tucson AMAs are constrained by availability of both surface water, including Colorado River water delivered via the Central Arizona Project (CAP), and groundwater resources. Phoenix continues to be one of the fastest growing urban areas in the country. Here, serious water conservation efforts are notably lacking, even in the context of a relatively arid environment and continued high population and economic growth. The Tucson AMA encompasses the second-largest population concentration in the state. This AMA remains reliant on groundwater, although much of the area is making the transition to blending recharged CAP water with groundwater. Groundwater levels have fallen as much as 60 m in the AMA since 1940.

In the Phoenix AMA, the capacity to draw upon multiple sources of surface water, groundwater, water banked under the Arizona Water Banking Authority, and (potentially) large amounts of effluent, provides an important buffer to drought. However, there are significant localized differences within the AMA. Each of the 31 large and nearly 80 small, water providers has a unique portfolio of water supply sources and customer base, as well as a more or less complex web of arrangements regarding treatment and recharge facilities. Longer-term, relatively severe droughts have potential to cause considerable problems in some areas, particularly those where groundwater pumping is the sole source of supply.

Unlike the situation in the Phoenix AMA, changes in water management in the Tucson AMA promise a decrease in the rate of groundwater overdraft anticipated in the near future. However CAP water is expected to account for most of this progress toward achieving safe-yield. As is abundantly clear from both paleo and historical records, the Southwest and Colorado River streamflow are characterized by very high degrees of climatic variability over annual and decadal timescales. Even if agriculture were eliminated and aquifer overdraft cut by half, withdrawals would continue to exceed renewable supplies under the drought scenarios used in this study. These results became a major point of attention by the Arizona Department of Water Resources and the Governors Planning Commission in revising and reauthorizing the State Groundwater Management Act.

Table 18.1. *Western Water Assessment partners and products across the continuum of climate timescales.*

Event characteristics	Objectives: understand, explain, predict, assess, communicate, evaluate	Climate processes	With whom does WWA work?
Short-term extreme events	Develop experimental forecasts, monitoring, and application products. Experimental attribution assessments of regional extremes.	Sub-seasonal variability, Arctic outbreaks, monsoon, floods, heat waves, tornados, hurricanes.	Reclamation, Fish, and Wildlife Service, CBRFC, Office of Hydrology, CPC, HPC, regional councils, wildfire managers.
Drought: seasonal to multi-year	Develop drought forecasts, monitoring, paleoclimate reconstructions, and application products. Assess social, environmental, and economic impacts.	Flash droughts, snowpack evolution, soil moisture evolution, El Niño and La Niña, multi-decadal ocean variability.	Western Governors Association (WGA), NIDIS, NWS, RFCs, NCDC, RCCs, NDMC, USDA, NRCS, USGS, NASA, regional councils, state and municipal agencies.
Decadal climate variability	Develop experimental monitoring, attribution, and application products. Assessments of regional trends and risks to inform adaptation strategies.	Pacific decadal variability, Atlantic multi-decadal variability, short-term influences, regional trends.	Regional councils, wildfire managers, NCAR, regional watershed councils, municipal agencies (e.g., Denver).
Climate change	Develop experimental attribution assessments of hemispheric to regional trends. Assess social, environmental, and economic risks (e.g., Colorado Compact).	Observed, current and evolving trends, enhanced hydrologic cycle, high elevation change.	CCSP, Reclamation, EPA, USGS, IPCC, NCAR, NASA, regional watershed councils, municipal agencies.

In 2007, the CIG produced an "Adaptation Guidebook" (CSES/CIG, 2007) in the context of climate change impacts. The observed acceptability of such a book, focused on a particular region and co-produced with non-academics, could only have resulted after years of engagement in studies on extremes and variability, studies on the impacts of salient events, and on maintaining necessary partnerships and social interactions at stakeholder driven meetings and events.

The RISAs continue to play significant and steadily increasing roles regionally, nationally and to some extent internationally. This is evidenced by the substantial contributions of RISA members to the IPCC Fourth Assessment, in Congressional and other briefings on the status of climate risks and adaptation options, and the development of state and utility drought and watershed plans (see for example Cayan *et al.* 2003; Colby *et al.* 2005; Brekke *et al.* 2007). They also contribute to fundamental research on the environment–society interface (Orlove *et al.* 2004; Dow and Carbone 2007). Recently, several RISA researchers and managers were recognized formally by the State of California for their roles in facilitating climate services delivery.

18.5 What is being learned by the RISA teams?

RISAs have proven to be particularly innovative at organizing the dialogue between scientists and practitioners (see Ingram *et al.* 2006 for an example from agriculture) and identifying critical entry points for information through the various calendars of decision making (Pulwarty and Melis 2001). RISAs have experimented with public fora, regular and sustained meetings, proactively seeking opportunities to participate in technical or professional meetings, one-on-one technical assistance, working with research partners who sit in resource management agencies, disseminating material through web sites, local and national media targeted publications, among other techniques.

Insistence by program managers that the research team members would be primarily resident in their region of study, so that they were also seen as stakeholders (albeit non-decisive ones), contributed to their understanding of context and their acceptability. Participatory, integrative social science that fully includes stakeholders requires commitment of resources and client-agency interest towards the development of reciprocal partnerships with stakeholder communities (Box 18.4). This provides the foundational elements from which usable research and successful awareness-building projects subsequently emerge. The process often involves or requires transcending existing bureaucratic boundaries, such as those between federal agencies, a notably difficult task.

A key issue within the RISAs was to establish a dialogue of risk communication that would be richer than the traditional model of providing information and data without considerations of context or interpretation or the model of a consultancy-based two way approach of attempting to provide only what is requested. As illustrated by the RISA experience the information supply and demand model applied by some (see, for instance, McNie *et al.* 2007) is a limited construct in complex situations where the demand is not always well-defined and learning processes and fora are needed for conducting collaborative framing and implementation exercises.

> ## Box 18.4
> ## The RISA teams have uncovered or confirmed many important insights about research–stakeholder partnerships
>
> - Development/maintenance of stakeholder partnerships can only take place with researchers at the local to regional level.
> - Partnerships cannot focus on climate variability alone: efforts must be interdisciplinary and focused on the integration of the multiple stresses relevant to the stakeholder and the region.
> - Stakeholder partnerships, once established, must be sustained; failure to do so will jeopardize the partnership and reduce hard-earned credibility.
> - Stakeholders cannot be considered solely as individuals or only within the context of single economic sectors; regional assessments must be able to accommodate integration across individuals and sectors.
> - Stakeholders need demonstration that their needs, ideas, and concerns are central to problems investigated in regional climate assessment and science.
> - Stakeholders need the guarantee that the quality of climate knowledge, particularly at their regional- to local-scales of interest, will be ever-improving; regional stakeholder-driven science represents a major gap in climate funding.

As the individual RISAs mature, they have adopted a project-by-project mode in order to refine their work vertically and at specific entry points of interest to stakeholders. More mature RISAs move toward horizontal integration across sectors and/or sets of stakeholder issues. However, defining measurable goals and targets has not been adequately stressed within the projects (see below). It remains a difficult avenue to pursue given limited resources, ongoing events, and rising stakeholder demand for new information. RISAs have, however, achieved one of the program's major goals, which was to demonstrate in practice the potential utility of climate information in very specific contexts. Through this vehicle they have empirically demonstrated the value of a sustained regional focus in revealing environmental uncertainties most critical to decision making. As anticipated (or more accurately "hoped"), many of the assessment projects themselves are becoming more successful at garnering support from regional constituents for cooperative research and applications. A more rigorous understanding of their successes and failures will require a concentrated effort on evaluation (both internal and external) than has yet been undertaken.

18.5.1 Potential pitfalls in the RISA approach: a risk assessment

In this section the authors hope to offer some brief insights into the difficulties and possible limitations of taking the above approach. It is important to note that the authors, as active and former administrators of the RISA program, recognize the pitfalls of concluding on the efficacy of interventions in which they are engaged, and which, to some extent, they advocate.

Individual RISA projects were and are purposely initiated with relatively modest funding in order to focus on the proof of concept within any given region, to encourage the establishment of key relationships with a small number of stakeholders, and to derive clear and hopefully replicable lessons for practice. One innovation was the emergence of certain efficiencies once a core number of centers were established. In other words, expertise in fire risk, or climate–hydrology interactions, or water banking analyses could be tapped into rather than having to develop locally or from fundamentals in every instance. The RISAs have also become fairly skilled at attracting federal and in some cases state funds outside of NOAA. From its inception, the program envisioned NOAA support as providing integration and seed resources upon which the teams would leverage other resources. While this may represent success on their part, it may imply a certain failure in the federal context to support these kinds of assessments commensurate with their acknowledged value in a truly cross-agency framework. Each federal agency has its core missions, and a regional approach (as opposed to sectors or topics) is harder to reconcile i.e., maintain autonomy and accountability, within those missions.

The RISA program managers worked from the onset to forge partnerships with the emerging research teams and deliberately create an environment of experimentation and learning. While some RISA team members are based at government research facilities and non-profit organizations, the research team members are primarily based at universities. Barriers to interdisciplinary research and problem-oriented approaches within university settings have been well documented (Brewer 1999; Guston and Sarewitz 2002). Yet, strong interdisciplinary programs are built from the foundations of strong disciplinary programs. Many of the difficulties lie in the "intransigencies" of the research setting (Campbell 1969). Political realities also mitigate against thorough or critical evaluation of risks and reforms (Brunner and Ascher 1992). NOAA program management understood that the initial projects could not have gotten off the ground through traditional means of scientific advisory bodies and an open competition around questions defined a priori. Program managers needed to outline an acceptable process (to academic partners and federal offices) for experimenting with both interdisciplinary integration and stakeholder engagement.

A major risk in RISA initiation in the early days was that capacity might not yet exist in a particular locale, or that there was not a community of decision makers interested in climate information. As noted above, that landscape is changing rapidly. Another risk was (is) that it is difficult for many university-based investigators to invest in the long start-up time of a project like RISA. In addition, for other than the most advanced in their careers, many academics may not find it sufficiently professionally rewarding to embed themselves in the applied research and stakeholder processes, necessary for informing adaptation practice. Many in academic settings, of necessity, find themselves leveraging support from several sources to meet the needs of particular projects, with attendant divided attention among projects.

A major issue surrounding evaluation has been the inability to fully articulate the end-to-end utilization of information on a particular problem. This has usually been the result of the proprietary nature of information and its use. More effective and formal mechanisms for acknowledging and documenting information importance, use, and outcomes are needed. While there has been increasing focus on the processes by which knowledge has been produced, less time has been spent examining the capacity of audiences to critically assess knowledge claims made by others for their reliability and relevance to those communities. Finding outside evaluators from the resource management and other relevant communities has proven even more difficult, but this is slowly changing.

Although the experimental approach was key to the success of the program it also resulted in a lack of program specificity about project goals early in the process. RISA programs have not yet been wholly successful at developing effective feedback and lesson drawing mechanisms into the larger monitoring and research programs, i.e., the strategic components, of the federal enterprise. Maintaining coordination within interagency groups is widely acknowledged to be important but can conflict with mission agency priorities and has not been as fruitful as program management anticipated. It is clear that the climate information services in support of adaptation must sustain an ongoing and well-coordinated suite of regional, sectoral, national, and global-scale assessment activities to meet statutory, programmatic, and scientific requirements.

18.5.2 Team leadership

A notable gap in most studies of assessments has been in elucidating the critical area of within-team leadership (see NRC 2007). Many scientific researchers may underestimate or fail to comprehend the need for management and may actually conclude that it is beneath their level of attention. It is however, where things (read "integration") fall apart. The integration of multiple disciplines (beyond academic exercises) and multiple perspectives remains challenging to generate and sustain in practice. Much of this has to do with how individuals are recognized, rewarded, and reinforced within academic settings. While this appears to be evolving, the changes are driven by particularly resourceful and innovative individuals (the RISA team leaders) and not by a "sea change" in academia. That being said, several RISA leaders may now be able to offer insights to the broader assessments community on overcoming this long-standing problem. All RISA-type activities have been shown to require the integrative leadership of a talented individual (in RISA and in the program agency). Most importantly, a leader defines a broad scope of work to allow the salient problems to emerge and also to maintain a level of independence from the client-agency and even from stakeholders to avoid pressures to produce desired answers. The leader ensures the continuity of relations with clients and his/her team to develop trust and a deeper understanding of the problem being faced.

As in other adaptive assessments (Walters 2007), the RISA team leaders have been individuals who: (i) have a broad overview of the decision making and implementation process, along with intimate knowledge of administrative details involved in each step; and (ii) are persistent in the face of lack of early interest (or after a missed forecast). Most importantly, they are willing to devote a significant portion of their careers to the implementation process, and to create attractive career paths for team members. This is also true for the program managers involved in the client-agency, a usually underappreciated fact.

Based on the above discussion, academic settings, by themselves, may prove to be sub-optimal for conducting and sustaining assessments in the context of decision support for numerous stakeholders. They are credible but not necessarily legitimate (as in meeting legal requirements) purveyors of risk-based information in the public domain. Universities do, however, allow expanded commitments to the education and training of scientists and stakeholders. RISA members have come to include State Climatologists, Regional Climate Center members, and extension specialists, among others. Where the university-based researchers have, over time, partnered with operational agencies, including the research branches of these agencies or with extension networks, tangible contributions to specific management goals have been documented. However, it is clear that much more needs to be done to incorporate the operational arms of federal, state, and tribal entities (and the private sector) to transition research and applications technologies into day-to-day operations, early warnings, and long-term capacity building. RISAs have provided informed pathways, but it is clearly up to the governance infrastructure (local, tribal, state, and federal) to support and advance the RISAs goals while being informed by their ongoing discoveries and mistakes.

18.6 RISAs and climate services: what are the lessons?

The orientation towards providing an informed basis for "services" is one of the major distinctions between RISA efforts and the experience of previous climate impact assessments beginning with the SCOPE (1986) "Report on Climate Impacts

Assessments" (Gilbert White, personal communication, 2002). RISAs are experiments in the design and implementation of climate and environmental services. They are not the service itself. They do however provide useful insights for climate services implementation (see below). Implementation of service activities relies upon the specific programs and activities derived from the mission responsibilities and unique assets and experience of the climate and global change programs member agencies. At present (spring 2009) there are several Congressional Bills advocating the formation of a National Climate Services or at least a better infrastructure for delivering accountable decision support. The first National Climate Program Act was introduced in 1978 (US Congress 1978).

The National Research Council defines "Climate Services" as *"The timely production and delivery of useful climate data, information and knowledge to decision makers"* (NRC 2001). To achieve such a service the NRC further recommends that relevant agencies develop "regional enterprises" designed to expand the nature and scope of climate services, a much larger construct than "decentralized policy experiments." The RISAs have, importantly, expanded the "climate services" concept to include a network of activities that maintain well-structured paths from observations, modeling, and research to the development of relevant place-based knowledge and usable information.

Miles *et al.* (2006) effectively expands on the RISA themes and experience to outline the functions of a National Climate Service, as follows:

- integrate global, national, and regional observations infrastructure to produce information and assessments of use to stakeholders and researchers;
- develop models for decision support; perform basic and applied research on climate dynamics and impacts relevant to stakeholder interests;
- create and maintain an operational delivery system and facilitate transition of new climate applications products to NCS member agencies;
- develop and maintain a dialogue among research teams, member agencies, and stakeholders for developing information relevant for planning and decision making;
- identify climate-related vulnerabilities and build national capacity to increase resilience;
- represent regional and national climate issues and concerns in regional and national policy arenas and facilitate regional–national communications on NCS needs and performance;
- outreach to stakeholder groups.

Creating acceptability of a new services design requires moving beyond a flow chart of institutional components, especially if those institutions were not involved in the design. The RISA experiments illustrate that at a minimum such a framework should:

(a) produce practical and acceptable design principles and a coordination framework for regional climate services;

(b) be credible and acceptable to private and public partners and to NOAA leadership by being both academically and institutionally sound;

(c) if possible achieve consensus on evaluation requirements and strategies to maximize the applicability of results and to foster program improvement.

RISA contributions have been to attempt to bridge, directly, the inadequate fit between what the research community knows about the physical and social dimensions of uncertain environmental hazards and what society chooses to do with that knowledge. An even larger challenge has been to consider how different systems of knowledge about the physical environment and competing systems of action can be brought together in pursuit of resilience and other diverse and competing goals that humans wish to pursue (Jasanoff 1996; Mitchell 2006). A more systematic RISA contribution towards understanding how such a service for a complex system would be governed and implemented involves:

- developing a mixed portfolio of products research, communication approaches, and applications credible to scientific and operational communities;
- assessing impediments to the flow of knowledge among existing components (RISAs, Regional Climate Centers, NWS Field Offices, State Climatologists, and extension arms of other agencies etc.);
- assessing policies and practices that can give rise to failures of the component parts working as a system;
- assessing opportunities for and constraints to learning and institutional innovation;
- developing capacity for local actors to design their own institutions and partnerships "public entrepreneurship";
- identifying transactions costs involved in implementing service components including international and national assessments activities.

The RISAs may over time contribute, beyond their list of risk assessment and management "projects," to a broader dialogue on constitutive issues surrounding the development of information services in the context of adaptation to global change. Understanding how "learning" is documented and becomes incorporated into practice is not a straightforward task. To set this goal in the larger context requires an understanding of the conditions governing the continuity or transformation of social systems and structures (Giddens 1986). As noted some time ago it is easier to evaluate abrupt and decisive interventions since a gradual response may be indistinguishable from the background secular change (Campbell 1969). Behind the specific questions of the transparency of risk, are broader questions about the public sphere (Jasanoff 1996; Mitchell 2006): What public goods will be provided by governments (and how will they be funded), what public goods will be provided by private organizations in civil society, what will and will not be provided to market actors? How do we move beyond framing outcomes in terms of winners and losers to securing partnerships in knowledge production and use?

Though the RISAs constitute only a small subset of a much larger climate research enterprise, and an even smaller subset of the human resources needed to produce and convey usable climate information, they demonstrate the profound importance of investment in the spaces between how we experience climate, what we know about it, and the varieties of responses.

18.7 Conclusion: a continuing voyage of discovery

An oft-heard assumption is that increases in knowledge about environment–society interactions will result in improvements in the quality of public and private decisions (a decidedly idealized view). Much recent work has shown that this expectation is most difficult to meet when decision stakes are high, uncertainty is great, technologies are new, experience is limited, and there are unequal distributions of burdens and benefits. Enabling successful information interventions at any point in time requires a critical mass of accessible, credible, and legitimate information and the capacity to apply knowledge and evaluate consequences of its use. The goals and successes of the RISA program are and have been in informing the development of place-based decision support and services by expanding the range of practical choices available to those affected by climate-related risks and environmental stresses, exacerbated through human activities or otherwise.

The RISAs continue to draw lessons from a variety of sources and events, such as from the 1997–1998 and other ENSO events, from Endangered Species Act declarations, the National Assessment, wildfire events, the IPCC process, and from multi-year droughts. Their experience shows that effecting change in risk management is most readily accomplished when at least three conditions are met: (1) a focusing event (climatic, legal, or social) occurs and creates widespread public awareness and calls for action; (2) leadership and the public become engaged; and (3) *a basis for integrating monitoring, research, and management is already established and supported.*

The RISAs have proven themselves to be more than just a client-based consultancy seeking to answer received questions. RISA projects do not necessarily advocate one set of policy options over another but seek to evaluate and make transparent the implications of different choices under varying and changing climate conditions (see the Appendix). Assessment of critical climate-sensitive issues, in this setting, is the iterative process of integrating interdisciplinary knowledge and experience about risks and vulnerabilities in a region commensurate with the design and support of effective responses. Such activities require innovative partnerships among a spectrum of interests (federal, state, local, and private, etc.) to enable

organizational capacity within a region for developing accurate balanced syntheses (i.e., identifying risk characteristics, uncertainties, critical knowledge and data
gaps, social and environmental vulnerability) and services.

The RISA experiment, because of its evolutionary approach to learning and
implementation, offers unique opportunities to continually assess and construct
post-audits of evolving events to inform longer-term risk reduction strategies. It
is our contention that evolutionary or learning-based approaches to "assessment"
as designed and developed by RISA-type programs are more effective at entering
into national, regional, and local plans of action for responding to complex environmental problems than traditional, periodic integrated knowledge assessments.

Acknowledgements

This chapter owes a significant debt to J. Michael Hall, the former Director of the
NOAA Office of Global Programs, now retired. The RISA program and support
for innovations described above would not have existed without his vision and
leadership. We would also like to acknowledge the roles of James Buizer, Harvey
Hill, Juniper Neill, Josh Foster, and Hannah Campbell in advancing the RISA
program over the past decade. Finally, the success of the RISA program relies on
the dedication, expertise, and openness of the RISA teams. The opinions expressed
in this chapter are those of the authors and not those of the National Oceanic and
Atmospheric Administration.

References

Brekke, L., B. Harding, T. Piechota, *et al.* 2007. *Review of Science and Methods for
    Incorporating Climate Change Information into Reclamation's Colorado River
    Basin Planning Studies. Colorado Basin Final Environmental Impacts Statement.*
    Washington, DC: US Department of Interior Bureau of Reclamation.
Brewer, G. 1999. The challenges of interdisciplinarity. *Policy Sciences*, **32**, 327–337.
Brunner, R. 1996. Policy and global change research: a modest proposal. *Climatic
    Change*, **32**, 121–147.
Brunner, R. and W. Ascher 1992. Science and social responsibility. *Policy Sciences*, **25**,
    295–331.
Campbell, D. 1969. Reforms as experiments. *American Psychologist*, **24**, 409–429.
Cayan, D., M. Dettinger, K. Redmond, *et al.* 2003. The transboundary setting of
    California's water and hydropower systems – linkages between the Sierra Nevada,
    Columbia River, and Colorado River hydroclimates. In H. F. Diaz and B. Woodhouse
    (eds.), *Climate and Water – Transboundary Challenges in the Americas*, ch. 11,
    Advances in Global Change Research, vol. 16. Dordrecht: Kluwer pp. 257–262.
Clark, W., J. Jager, J. v. Eijndhoven, and N. M. Dickson (eds.) 2001. *Learning to Manage
    Global Environmental Risks*, vol. 1. *A Comparative History of Social Responses to
    Climate Change, Ozone Depletion and Acid Rain*. Cambridge, MA: MIT Press.

Colby, B., J. Thorson, and S. Britton 2005. *Negotiations Over Tribal Water Rights. Fulfilling Promises in the Arid West.* University of Arizona Press.

CSES/CIG 2007. *Preparing for Climate Change: a Guidebook for Local, Regional, and State Governments.* http://cses.washington.edu/cig/fpt/guidebook.shtml

Dow, K. and G. Carbone 2007. Climate science and decision making. *GeographyCompass* **1**(3), 302–324. doi:10.1111/j.1749–8198.2007.00036.x

Giddens, A. 1986. *The Constitution of Society: Outline of the Theory of Structuration.* University of California Press.

Guston, D. and D. Sarewitz 2002. Real-time technology assessment. *Technology in Society*, **24**(1–2), 93–109.

Holling, C. (ed.) 1978. *Adaptive Environmental Assessment and Management.* New York: Wiley.

Ingram, K., J. Jones, J. O'Brien, J. Paz, and D. Zierden 2006. AgClimate: a climate forecast information system for agricultural risk management in the southeastern USA. *Computers and Electronics in Agriculture*, **53**, 13–27.

Jasanoff, S. 1996. Beyond epistemology: relativism and engagement in the politics of science. *Social Studies of Science*, **26**, 393–418.

Linder, S. 2005. The adoption of adaptation measures. In K. Ebi, J. Smith, and I. Burton (eds.), *Integration of Public Health with Adaptation to Climate Change.* London: Taylor and Francis, pp. 242–257.

Mantua, N. J., S. R. Hare, Y. Zhang, J. M. Wallace, and R. C. Francis 1997. A Pacific interdecadal climate oscillation with impacts on salmon production. *Bulletin of the American Meteorological Society*, **78**, 1069–1079.

McNie, E., R. Pielke Jr., and D. Sarewitz 2007. *Climate science policy: lessons from the RISAs. Workshop Report.* Decisionmaking Under Uncertainty Project National Science Foundation.

Miles, E., A. Snover, L. Whitely Binder, *et al.* 2006. An approach to designing a national climate service. *Proceedings of the National Academy of Sciences*, **103**, 19 613–19 615.

Mitchell, J. 2006. The primacy of partnership: scoping a new national disaster recovery policy. *Annals of the American Academy of Political and Social Science*, **604**, 228–255.

Morgan, G., R. Cantor, W. Clark, *et al.* 2005. Learning from the US National Assessment of climate change impacts. *Environmental Science and Technology*, **39**, 9023–9032.

NRC 1999. *Our Common Journey: a Transition Toward Sustainability.* Washington, DC: National Academy Press.

NRC 2001. *The Science of Regional and Global Change. Putting Knowledge to Work.* Washington, DC: National Academy Press.

NRC 2007. *Evaluating Progress of the US Climate Change Science Program: Methods and Preliminary Results.* Washington, DC: National Academy Press.

Orlove, B., K. Broad, and A. Petty 2004. Factors that influence the use of climate forecasts: evidence from the 1997–98 El Niño Event in Peru. *Bulletin of the American Meteorological Society*, **85**, 1735–1743.

Peters, D., R. Pielke Sr., B. Bestelmeyer, *et al.* 2004. Cross-scale interactions, nonlinearities and forecasting catastrophic events. *Proceedings of the National Academies of Sciences*, **101**, 15 130–15 135.

Pulwarty, R. and T. Melis 2001. Climate extremes and adaptive management on the Colorado River. *Journal of Environmental Management*, **63**, 307–324.

Rayner, S., H. Ingram, and D. Lach 2005. Weather forecasts are for wimps: why water resource managers do not use climate forecasts. *Climatic Change*, **69**, 197–227.

SCOPE 1986. *Climate Impact Assessment*. Scientific Committee on Problems of the Environment ICSU Secretariat.

Shackley, S. and B. Wynne 1996. Representing uncertainty in global climate change science and policy: boundary-ordering devices and authority. *Science, Technology and Human Values*, **21**, 275–302.

Shea, E., G. Dolcemascolo, C. Anderson, *et al. Preparing for a Changing Climate: the Consequences of Climate Variability and Change for Pacific Islands*. Honolulu: East–West Center.

Toth, F. and E. Hiznyik 1998. Integrated environmental assessment methods: evolution and applications. *Environmental Modeling and Assessment*, **3**, 193–210.

US Congress 1978. *National Climate Program Act*. Public Law 95–367. Amended in 2000 through Public Law 106–580.

US Congress 1994. *Federal Environmental Research: Promises and Problems*. Committee on Science, Space and Technology. House of Representatives 103rd Congress No. 126 May 1994. Washington, DC: Government Printing Office.

US Congress 1998. *Unlocking our Future: Towards a New National Science Policy*. House of Representatives 105th Congress House Committee on Science 105B. Washington, DC: Government Printing Office.

US Congress 2007. "Climate Service" Bills introduced in the US Congress Fall 2007: S 2307 *Global Change Research Improvement Act*; HR906 *The Global Climate Change Research, Data and Management Act*; S 2355 *Climate Change Adaptation Act of 2007*.

Walters, C. 2007. Is adaptive management helping to solve fisheries problems? *Ambio*, **36**, 304–307.

Webb, R. and R. Pulwarty 2006. *Presentation to Congressional Staffers*, July 2006. Boulder, CO: NOAA Earth Systems Research Laboratory/Physical Sciences Division.

White, G. F. 1966. Optimal flood damage management: retrospect and prospect. In A. Kneese and S. Smith (eds.), *Water Research. Resources for the Future*. Baltimore, MD: The Johns Hopkins Press, pp. 251–269.

Appendix

Acronyms

ACCAP	Alaska Center for Climate Assessment and Policy
AO	Arctic Oscillation
CAP	California Applications Program
CBRFC	Colorado Basin River Forecast Center
CCSP	Climate Change Science Program
CIG	Climate Impacts Group
CISA	Carolinas Integrated Sciences and Assessments
CLIMAS	Climate Assessment for the Southwest
CPC	Climate Prediction Center
ENSO	El Niño Southern Oscillation
EPA	Environmental Protection Agency

HPC	NOAA Hydrologic Prediction Centers
IPCC	Intergovernmental Panel on Climate Change
NASA	National Aeronautics and Space Administration
NCAR	National Center for Atmospheric Research
NDMC	National Drought Mitigation Center
NGO	Non-governmental Organization
NIDIS	National Integrated Drought Information System
NRCS	USDA Natural Resources Conservation Service
NOAA	National Oceanic and Atmospheric Administration
NAO	North Atlantic Oscillation
NWS	National Weather Service
PDO	Pacific Decadal Oscillation
RCC	Regional Climate Center
RISA	Regional Integrated Sciences and Assessments
SECC	Southeast Climate Consortium
BoR	US Bureau of Reclamation (Department of the Interior)
USDA	US Department of Agriculture
USGCRP	US Global Change Research Program
USGS	US Geological Survey
WGA	Western Governors Association
WWA	Western Water Assessment

19

Integrated regional assessment: reflections on the state of the art

JILL JÄGER AND C. GREGORY KNIGHT

19.1 Introduction

For more than a decade there has been a growing interest in the regional implications of global environmental change. This is manifested in the multitude of
studies carried out under the auspices of intergovernmental processes, international programs, national initiatives, and individual studies. The chapters in this
book illustrate the approaches that are being taken and refer to many past and
present processes referred to collectively as "integrated regional assessment."

The purpose of this concluding chapter is to reflect on the material presented
in this book with reference to the questions posed in the introductory chapter. In particular, however, we focus on the implications for future work in this
area.

Integrated regional assessment has been defined in the first two chapters of
this volume. It is an interdisciplinary, iterative process that involves scientific
researchers, policy makers, and societal stakeholders. Its aim is to promote a better
understanding of – and more informed decisions on – how regions contribute to
and respond to global environmental change. Thus integrated regional assessment
considers regions as holistic, integrated entities and thereby provides scientists and
stakeholders with a framework that helps them comprehend the totality of regional
variations in the causes and consequences of global environmental change. Understanding the regional implications of global environmental change is important,
because it is at regional levels that global environmental change mitigation must
be practiced and that human impacts will be felt.

19.2 Methodological challenges

As shown by Malone (Chapter 3, this volume), *qualitative and quantitative issues*
are particularly significant in integrated regional assessment. No research is purely

one or the other, of course, so the degree of interdependence or overlap is particularly interesting. Strategies for integrating the two general approaches often produce uneasy compromises. However, integrated regional assessment provides opportunities for strong collaborations in addressing specific problems in specific places. Quantitative research rests on judgmental, qualitative assumptions about how the world works, what suitable categories are for data, what constitute good data, and the validity of scientific procedures. For future projections, the role of qualitative assumptions is even more marked. Qualitative research, on the other hand, if it is to make sense of the world at all, must weigh and measure, at least in a comparative way, must judge what is important and what variables are critical in human development and change, and what constitute exhaustive and exclusive categories. Whether or not numbers are used, these are essentially quantitative tasks. Furthermore, the ability to scope and assess issues quantitatively provides the qualitative researcher with bases for framing problems.

Integrated regional assessment clearly needs both approaches. On the one hand there are the qualitative dimensions, such as building collaborations; including policy makers and other stakeholders in the research process; debating the starting assumptions; and being willing to re-examine categories, assumptions, and data of the research program as it goes along, all of which are illustrated throughout this book. On the other hand, quantitative approaches include defining numerical data sets that will be of use in scoping and defining the research question; undertaking the hard work of gathering, checking, evaluating, and validating data; performing adequate statistical analyses or developing relevant models; and presenting results of quantitative analyses clearly and comprehensively.

The examples of integrated regional assessment in this volume suggest that ways to use both quantitative and qualitative research exist in the collaborations, networks, and interdisciplinary programs themselves. There are still challenges, however, in particular because there remain large differences between different scientific disciplines on the credibility of qualitative approaches.

Many chapters in this book identify *scalar dynamics* as an important issue for integrated regional assessments. Unfortunately, there is little guidance for operating and analyzing scalar dynamics in integrated regional assessments in general, abstract terms. An empirical literature on this topic is developing, the theoretical foundations of which are varied. The empirical examples described by Polsky and Munroe (Chapter 4, this volume) demonstrate how "spatial" violations of the common statistical assumption of independence of observations can be interpreted as evidence for substantively important (yet poorly understood) information about the process under study.

No single quantitative approach will ever likely be able to capture all the dynamics and dimensions of coupled human–environment processes. More research and analysis is warranted to evaluate the intersection of multiple, competing sources of scale effects and scalar dynamics.

Uncertainty is at the core of integrated regional assessment; management of uncertainty is a key challenge. As van Asselt (Chapter 5, this volume) shows, uncertainty is inevitable in the assessment of complex problems that lie across, or at the intersection of, many disciplines. There is also inherent uncertainty, for example in cases for which necessary historical records or monitoring systems are lacking, with questions referring to human behavior, with questions pertaining to the future, and questions that involve value judgments.

There is no recipe or best practice for uncertainty management in integrated regional assessment. However, concepts and methods are available that allow serious attention to be paid to uncertainty in regional assessment endeavors. Two examples cited by van Asselt illustrate how management of uncertainty could look in practice, both in integrated assessments performed by experts and in more participatory regional assessments. These examples also demonstrate that further work is needed, for example in the area of communication about uncertainty.

In recent years there has been increased attention to the *vulnerability* of human–environment systems, using approaches of great relevance to the focus of this book on integrated regional assessment. As Leary and Beresford (Chapter 6, this volume) show, a rich literature of vulnerability case studies has been accumulating and adding to the body of knowledge about vulnerability to environmental change. An overarching and robust result from this literature is that vulnerability is not a simple function of exposure but is caused and shaped by multiple interacting environmental and human processes. The state and dynamics of these processes vary from place to place and from person to person, generating conditions of vulnerability that differ in both character and degree. Consequently, people exposed to similar environmental stresses are not impacted to the same extent.

However, the state of knowledge about vulnerability to environmental change and methods for assessment of vulnerability are incomplete. Assessments have been based on varied concepts and frameworks and have applied many methods. This variety and experimentation has been a necessary stage in the development of vulnerability assessment as a system of inquiry into the differential consequences of environmental change. But comparing and synthesizing results across case studies, and finding robust explanations of the causes of vulnerability, have been hampered by the diversity of approaches.

There is need for a robust conceptual framework and development of analytic methods and tools to make further advances in understanding that can inform vulnerability-reducing actions. Useful frameworks for vulnerability assessment have been developed in recent years. Future assessments of vulnerability need to address a number of critical gaps in our knowledge and methods. Methods are needed to realistically and credibly downscale environmental changes to better match the spatial scales relevant to decision making. Cross-scale interactions among different components of coupled human–environment systems need to be explored to better understand how they shape vulnerabilities, capacities to respond, and the availability and feasibility of response options. Evaluations of the cumulative effects of multiple stresses operating over time and interacting with each other are a pressing need. An important motivation for vulnerability assessment is to inform more effective responses for lessening the potential harm from environmental change. This has led to assessment approaches that are user-oriented and that engage stakeholders in the assessment process. These important developments will be a central feature of future vulnerability work. To facilitate advances in this direction, research is needed to better understand decision processes of different classes of actors for managing environmental risks, the roles of institutions, the knowledge needed to make good decisions, the transformation of information into relevant and usable knowledge, and the design of assessment processes that can contribute to deliberation and decision making.

Strongly related to the research needs for vulnerability are the research needs in relation to *adaptation* (Dickinson, Bizikova, and Burton, Chapter 7, this volume). Again, the linkages between knowledge and action at the local and regional scales are essential elements of research. Policy makers at the local level are in the difficult situation of trying to reconcile a wide diversity of local development visions with tradeoffs over limited resources, at a time when more actions, both in mitigation and adaptation, will be needed in order to tackle future climate change impacts, as well as to protect us from climate-related surprises. Therefore, viewing adaptation and mitigation as separate fields of action and policy without direct linkages may work against the implementation of opportunities that are perhaps not the most significant contribution to emission reduction, or avoided climate damage, but which can still offer tangible local benefits. Largely lacking, however, is an integrated policy analysis of the various options that might be considered for the integration of adaptation. The need for research and practical experiments on how to integrate adaptation will be a significant part of the integrated regional assessment agenda in coming years.

In recent years there has been increasing recognition of the need to integrate *stakeholders* into integrated regional assessment processes. The judgments of stakeholders provide valuable input to the process and at the same time participation can improve stakeholders' acceptance of the need for actions and their support of recommended actions. Experience is accumulating on how stakeholders can be effectively engaged in assessment processes, and Fisher and Kasemir (Chapter 8, this volume) provide two examples of such experience. There is and will be no standard procedure for stakeholder engagement. The procedure depends on the issue at stake, the overall context of the assessment and the resources available for the assessment process. Research on innovative approaches to stakeholder engagement will continue to be important.

Meeting all of the methodological challenges addressed above and in the first part of this book requires a *framework* for examining the causes and consequences of global climate change at the regional level. Knight and colleagues (Chapter 9, this volume) describe the framework that was developed at the Center for Integrated Regional Assessment (CIRA) to guide a team of researchers addressing regional contributions to global change and regional impacts and response choices to those changes as they might occur. The framework was also intended to illustrate to the stakeholder and policy-maker communities the logic through which integrated regional assessment was being carried out, without necessarily being a "wiring diagram" for the actual processes of change.

It is expected that the CIRA framework will be revised as experience with integrated regional assessment grows. In some settings, a relatively simple, informal integrated regional assessment process will be enough to make the issues clear and identify desired actions. In other situations (for example, for longer time spans and larger regions), it will be helpful to refine the modeling approach to handle the multiple complex interactions and goals. The framework could also support a modeling approach to climate change impacts at regional scales, taking the experience in numerical modeling in cascading end-to-end assessment described in the second part of this book into a modeling system complete with feedback elements and the ability to experiment with alternative futures.

19.3 Experience with integrated regional assessments

The second part of this book provides examples of integrated regional assessments that use models. In Chapter 10, Jäger examines the international context of integrated regional assessment, setting the stage for the subsequent chapters. She points out the development of new modeling approaches to IRA, as well as expanding concern for human–environment challenges to sustainability beyond global climate change. The importance of capacity building is also emphasized; many of the activities described in the following chapters have capacity building as a focal component.

Kainuma and colleagues (Chapter 11, this volume) describe the Asia–Pacific Integrated Model (AIM), which was developed by an Asian collaborative project team composed of the National Institute for Environmental Studies (NIES) and Kyoto University in Japan, and a number of research institutes in China, India, Korea, Thailand, Indonesia, and Malaysia. It estimates the emission and absorption of greenhouse gases in the Asia–Pacific region and the impact that they have on the natural environment, society, and economy. AIM comprises three models: the AIM/emission model for predicting greenhouse gas emissions, the AIM/climate model for estimating global and regional climate change, and the AIM/impact model for estimating the impacts of climate change.

The AIM model has been applied to the assessments of emissions scenarios, climate change scenarios, climate impact scenarios, mitigation costs, and new policy integration in order to respond to the research needs of international organizations, national governments, and non-governmental organizations. These assessments include IPCC emissions scenarios, mitigation scenarios for global climate stabilization, land-use related mitigation scenarios, implications and economic impacts of the Kyoto Protocol, the effect of the United Nations Framework Convention on Climate Change (UNFCCC) Clean Development Mechanism in competition with emission trading, the potential of GHG reductions in Asian developing countries, and policy design for Asia–Pacific collaboration. Kainuma *et al.* conclude that further studies are needed to evaluate policy options to mitigate global climate change within the context of increasing economic activities and reducing local environmental problems.

Shukla and colleagues (Chapter 12, this volume) have also used the AIM model linked to other models to examine how greenhouse gas mitigation goal achievement could be considerably enhanced through energy cooperation. This analysis highlights the emerging understanding that national energy and climate change futures would benefit from a regional sustainability perspective. These have to coordinate and balance bottom-up driven national systems such as democratic governance to take people's aspirations and expanding needs into consideration, with top-down driven processes such as regional cooperation, energy security, and federal structures to ensure regional balance in development and equitable availability of fruits of development such as social and physical infrastructures.

A further study in Asia (Yin, Chapter 13, this volume) takes a different approach to regional assessment. The Integrated Land Assessment Framework (ILAF) adopts a systems analysis approach assisted by simulation modeling and other impact assessment methods, analytic hierarchy process, and goal programming. The ILAF was applied in the Yangtze Delta case study that focused on the identification and specification of regional sustainability goals/indicators and their relationship with climate change impacts and vulnerabilities. This study provides procedures for integrating regional sustainability goals/indicators development and a range of economic sectors, in order to systematically investigate the impacts of climate change scenarios on regional sustainability. Yin concludes that the model developed in this study needs further testing, since testing and validation are critical to ensure that potential users have confidence in model results.

The testing of models and acceptance by users are well illustrated by Warrick (Chapter 14, this volume), who describes the integrated modeling system *SimCLIM*. The roots of SimCLIM stretch back to the early 1990s. SimCLIM was developed from a more targeted, location-specific model called CLIMPACTS, designed for assessing impacts and adaptation to climate variability and change specifically in New Zealand. CLIMPACTS, in turn, provided a foundation for other special-ized applications, both within New Zealand and internationally. CLIMPACTS was essentially a "researcher's" tool that was used mainly by those who developed it. The system was "hard-wired" – that is, the program could only be expanded and maintained by the model developers themselves. The primary focus was on biophysical impacts, with minimal explicit attention to the human dimensions of climate change, the risks of variability and extreme events and the adaptation options for reducing the risks. These "drawbacks" of CLIMPACTS led to the development of the SimCLIM system. SimCLIM has a much more open structure, which can be customized and maintained by the user. Other software models can now be linked to the system. The scenario generating capacity has been expanded to include a unique sea-level scenario generator. There is an enhanced capacity for local-scale modeling of impacts and adaptation, including the incorporation of environmental, social, and economic data. Functions to assess extreme events and risks have been improved. Initial developments of economic impact and evalua-tion tools have been introduced. Overall, with such modifications SimCLIM has become accessible to end-user groups, such as regional councils, environmental consultants, and universities, and has recently been made available to them.

As these examples show, during the last two decades model-based global envi-ronmental assessments have become established as tools to evaluate the potential consequences of different environmental and societal trends. Such assessments inte-grate different disciplinary perspectives in understanding and seeking a solution to a problem. There are still many uncertainties in developing environmental assess-ment models and plausible scenarios of global change. There is a need to link the coarse top-down approaches with regionally specific, bottom-up approaches that realistically capture the diversity of possible responses at all discrete scale levels of the organizational dimension. To achieve this, local, regional, and global assessors have to collaborate closely.

As Leemans (Chapter 15, this volume) demonstrates, new approaches to link different dimensions of global change have to be developed. Although there likely remains a significant role for traditional system-analysis approach models, other approaches have to be explored. Despite the fact that some models already include some stochastic effects, such as interannual variability in climate forcing, these traditional models generally simulate only smooth transitions. Recent insights show that many societal and environmental dynamics show rapid transitions, phase shifts, and multiple stable stages caused by such gradual changes. New innovative approaches are therefore called for. Candidates include actor-oriented modeling, the syndrome approach, and advanced cellular automata that link human behavior with the development of spatial patterns.

19.4 Regional programs

The third part of the book documents some examples of regional collaboration. Lebel (Chapter 16, this volume) evaluates some of the limitations and benefits of networks for conducting integrated regional studies. The analysis is based on the experiences of the Southeast Asian Regional Committee for START[1] (SARCS) in developing and promoting research and assessment of global environmental change in the Southeast Asia region. Lebel concludes that regional scientific networks have many advantages as organizational forms both for support of the science and conduct of the integrated regional assessments. By building regional collaboration on science they may help build foundations for other more politically sensitive discussions related to use and management of natural resources. But without clear links to regional institutions, or engagement of decision makers and resource managers, collaborations will often struggle to reach an appropriate balance of legitimacy, credibility, and saliency with the consequence that most of the knowledge produced will never get used. Thus, integrated regional assessments driven by informal epistemic networks are not a substitute for intergovernmental endorsed processes. Lebel's analysis suggests that for many issues, participatory mechanisms, where stakeholders can negotiate and alter their preferences or debate value changes, are essential for moving from an understanding of research-based knowledge to action. This will be picked up again below.

A further example of regional collaboration within the global change research community is provided by Liverman (Chapter 17, this volume), who examines the Inter-American Institute (IAI), a regional intergovernmental organization for scientific cooperation. The IAI has been creating and reinforcing a new set of regional relationships and shared consensual knowledge about global change in the Americas. Liverman's analysis suggests that the institution has had some success in identifying and supporting a diverse number of scientists to work together on important scientific problems and train new generations of scholars. It has also learned some important lessons and still faces formidable challenges. Some of the major challenges include the search for a more sustainable and diverse financial base, the enhancement of capacity in those countries with smaller scientific

---

[1] START stands for the global environmental change program, "The Global Change SysTem for Analysis, Research and Training." See www.start.org

communities so that they can fully participate in research competitions and projects, and the more effective integration of social science and policy issues into projects.

Finally, Pulwarty, Simpson, and Nierenberg (Chapter 18, this volume) provide an example of the development and evolution of a long-term interdisciplinary program focusing on climate impacts assessments and regional and local decision support, the Regional Integrated Sciences and Assessments Program (RISA). The goal and success of the RISA program are and have been in informing the development of place-based decision support and services by expanding the range of practical information available to those affected by climate-related risks and environmental stresses, exacerbated through human activities and decisions. The RISA experiment, because of its evolutionary approach to learning and implementation, offers unique opportunities to continually assess and construct post-audits of evolving events to inform longer-term risk reduction strategies. As Pulwarty *et al.* show, evolutionary or learning-based approaches to "assessment" as designed and developed by RISA-type programs are more effective at entering into national, regional, and local plans of action than traditional, discrete integrated knowledge assessments.

19.5 New frontiers in integrated regional assessment

19.5.1 Improving integrated regional assessment of climate change

Understanding global environmental change and developing options to respond to it require analysis at the regional level and require an integration of social science and natural science approaches, an integration of expert and stakeholder knowledge, and consideration of the overall global context in which regional changes unfold. This presents a number of methodological challenges that have been discussed in the first part of this book. While enormous progress has been made in addressing these challenges, much remains to be done in improving both the methods and tools used in integrated regional assessment and in linking knowledge to action.

With regard to the latter point of ensuring that the large body of existing knowledge actually leads to policy development and implementation, research in recent years has shown that institutional aspects of assessment are critical to better understanding and structuring the connection between science and policy in the environment arena. It is now clear that some assessments have more influence than others due to the *process* by which they are developed rather than just their *products* (see Box 1.3, this volume). Attention therefore has to be given to the design of assessment processes to ensure that there is effective knowledge communication, that mutual understanding of the issues is developed and that flexible, iterative, and reflexive processes contribute to effective management of environmental risks.

19.5.2 Beyond climate change to sustainability

Most of the examples in this book have focused on the issue of climate change, since this is an area in which integrated regional assessment has been developed over the past two decades, during which the issue of climate change has received much attention in the scientific and policy arenas. However, global change is much more than climate change. As vulnerability assessments have recently emphasized, human–environment systems are subjected to a wide range of interacting environmental and socio-economic stresses. Thus, a move from considering single issues such as climate change to assessments of sustainability is a logical although difficult step.

How can we take advantage of our experience in integrated regional assessment of climate change to develop approaches for "integrated sustainability assessment?" One example of progress in this direction is provided by the recently completed MATISSE project, which has developed both methods and tools for Integrated Sustainability Assessment (ISA).[2] ISA is intended as a pro-active, strategic, and potentially transformative process to give an explicit sustainability orientation to policy making and other undertakings that are expressly intended to address persistent complex problems of unsustainable development and take up opportunities for achieving more sustainable development. It has been defined as follows:

ISA is a cyclical, participatory process of scoping, envisioning, experimenting, and learning through which a shared interpretation of sustainability for a specific context is developed and applied in an integrated manner in order to explore solutions to persistent problems of unsustainable development.[3]

The essential design requirements for ISA stem directly from its intended role as a transformative process for exploring and supporting reframing and reorientation. ISA represents a new mode of knowledge production that offers a forum for:

- defining "socially- and ecologically-robust" targets and thresholds;
- integrating these as elements of operational, context-specific sustainability interpretations; and
- exploring alternative pathways of transition.

ISA brings together an integrated systems analysis and a participatory process involving a selection of relevant stakeholders and actors. The integration of stakeholders selected to represent different perspectives and interests is a basic requirement of ISA in order to develop a rich and robust interpretation of sustainability for a specific context. Thus, ISA as defined above builds on experience in integrated

---

[2] The MATISSE (Methods and Tools for Integrated Sustainability Assessment) project was funded by the European Commission. For further information, see www.matisse-project.net

[3] See www.matisse-project.net

regional assessment and its use at the regional level advances some of the method-ological developments described in this book, using some of the tools described in the second part.

19.5.3 The importance of capacity building

All of the advances in integrated regional assessment of climate change and devel-opments in Integrated Sustainability Assessment point, however, to the past impor-tance and continuing significance of capacity building. Several chapters in this book point to the importance of capacity building carried out under the auspices of the global change research programs, in particular START, and related organizations. There is a clear value in building the capacity of various actors to be involved in producing assessments and to understand the findings of assessments.

Assessment processes have gained influence with wider audiences by establish-ing a long-term goal and process to enhance the capacity of a range of scientists to participate substantively in assessments. Capacity building among potential users is equally important. Providing enough human and financial resources for capacity building for assessment processes remain a major challenge for the coming years.

19.6 A concluding note

It is our hope that this volume contributes to the progress and influence of integrated assessment of environmental challenges and change. In this first decade of the twenty-first century we see evidence of major changes in the way individuals and society view planet Earth as an holistic system that hosts human society as an integral part of nature. By careful assessment of the current human role in environmental change and of the role of humans in future developments over which they may have strong influence, we may contribute to achieving a sustainable relationship between people and Earth.

Acknowledgements

The editors acknowledge with thanks the continuing encouragement of this project by START, and in particular by Roland Fuchs and Hassan Virji from that orga-nization. Many entities from the local to regional and global scale have provided financial assistance and institutional support that made possible the work reported here by chapter authors. To all of them we express our gratitude.

Index

Note: common terms throughout this book (such as assessment, climate, environment, region, space, and systems) are not indexed.

Printed in the United States
By Bookmasters